INTEGRATED MANAGEMENT OF WATER RESOURCES WITH REFERENCE TO BIODIVERSITY AND LIVELIHOOD

INTEGRATED MANAGEMENT OF WATER RESOURCES WITH REFERENCE TO BIODIVERSITY AND LIVELIHOOD

EDITORS
Dr. S.N. Dwivedi
Prof. Praveen Tamot
Dr. Asfa M. Yasin

BIOTECH
2012
BIOTECH BOOKS

ISBN 978-81-7622-236-5

[Note: The research paper and views expressed by the authors in their papers are their own, editors do not owe any responsibility]

Published by

BIOTECH BOOKS

4762-63/23, Ansari Road, Darya Ganj,
New Delhi-110002 (India)
Phone: 23262132
E-mail: biotechbooks@yahoo.co.in

Laser Typesetting by : Deep Graphics, Delhi

PRINTED IN INDIA

PREFACE

Water as most precious natural resource is vital for sustaining life and livelihood of people. Different uses of water arc inter-dependent thus the key issue is integrated water resources management which is essential in the present scenario. On our earth, of water is spread as sea 97% and freshwater 3%, of the total freshwater 87% is not accessible to us, remaining 3% accessible to us, out of which 0.4% is for human consumption.

The fact and figures show that in developing countries there is acute water crisis both quality and quantity wise. Water shortage in over 40 countries of the world is faced by 2 billion people. Half the population of developing world is affected by water pollution problem leading to suffering from diseases. The projections show that increase in population from 6 billion to 9 billion, will be the main thrust for water resources management for the next 50 years. In India, improper disposal of domestic and industrial waste is the main cause for pollution of lakes, wet lands and rivers. Therefore optimal sustainable use of water, abatement of pollution, conservation of biodiversity in aquatic ecosystem including village ponds, hill stream, rivers, etc. are of paramount importance. This requires nation-wise efforts of scientists, scholars, social organization, fisheries and aquaculture project managers, administrators and policy makers for making honest efforts in planning and operationalisation of integrated water resources management. It is with this aim papers from eminent scholars, scientists, researchers, teachers and students were invited and compiled in the present book. We hope that these will be of immense support in the water resource management in the country.

In this book research papers and theme papers highlights various aspects of fisheries management, culture practices, biodiversity, ecology, limnological studies, fisheries based livelihood practices, socio-economic aspects related to integrated management water bodies, geographical information systems and seasonal impact on water quality of ground water.

It is hoped that this incredible collection of work done by eminent researchers, scholars, given in the book will be of great significance and interest of fisheries, fish entrepreneurs, students, administrators, policy makers in planning and formulations of strategies for Integrated Management of Water Resources considering biodiversity and livelihood related issues. We extend our grateful appreciation to all the authors for this significant contribution.

S.N. Dwivedi
Praveen Tamot
Asfa M. Yasin

CONTENTS

1

INDUCED SPAWNING OF CARPS AND HATCHERY MANAGEMENT

P. K. Varshney, S. N. Dwivedi and W. S. Lakra*

Abstract: The synergistic hormonal and environmental approach of induced spawning in carps is very successful. Sexually mature fish may be injected with appropriate hormone and then placed in an environmentally controlled hatchery system to spawn naturally. Circular and Modern carp hatchery CIFE D-81 & D85 models are excellent spawning and hatching systems with continuous water flow through gravity to create the natural environmental conditions conducive for breeding and hatching. These hatchery systems are boon especially to small and medium aquafarmers and revolutionized the trade of fish seed production in the state of Andhra Pradesh, Madhya Pradesh, Tamil Nadu, Punjab, Haryana and Uttar Pradesh.

Keywords: Synergistic, Modern Carp Hatchery, Aquafarmers

Introduction

Progress with respect to induced spawning of carps in captivity has been made. In India it has been initiated by Chaudhuri and Alikhuni (1957) successfully through the hypophysation of pituitary. Today most of the states are self dependent in seed production of carps which was one of the major constraints in the development of aquaculture. Subsequently for the control of the environmental parameters low cost, portable hatchery system especially for small fish farmers were introduced by Dwivedi (1985-86) during early 80s. This was a breakthrough in the field of induced spawning and larval rearing and revolutionized the trade of fish seed production in the country.

Most of the important cultivable species of fish spawn naturally in their natural habitats. Thus making it necessary for a fish farmer to collect the stocking material directly from nature till

National Bureau of Fish Genetic Resources, Aquaculture Research & Training Unit, Chinhat, Lucknow

* Former Additional Secretary, DOD, Government of India & Director General, MPCST, E1/ 106 Arera Colony, Bhopal

recently. Although the breeding requirements of some of these species are understood, much remains to be known before they can be bred in confined environments.

There are two approaches to induce the fish for spawning. (1) Hormonal and (2) Environmental. The two approaches may be combined to achieve the best results. After treatment, the fish in each approach either spawn naturally or stripped for artificial fertilization. Whether the ovulated eggs are fertilized artificially or naturally, the fertilized eggs are incubated and the hatched larvae raised to the spawn stage in the hatchery.

Hormonal Approach

It is based on an understanding of the neuro-endocrine control of oocyte maturation, ovulation and spawning behavior (Goetz, 1983; Liley & Stacey, 1983 and Stacey, 1983) in the female and of spermiation and seminal hydration (Billard *et al.*, 1982; Scott and Sumpter, 1983) in male.

Environmental Approach

This is based on the premise that for each species of fish, there may be a specific or a set of specific, environmental cues which trigger final oocyte maturation and ovulation to spawning behavior and spawning (Lam, 1983 and Stacey, 1984). In most species, particularly the tropical species, the specific spawning cues have not been identified, only suggestions have been made based on the general observations without experimental observations. Despite this paucity of specific information, the environmental approach to spawning induction has been practiced successfully in several species (Dwivedi & Reddy, 1985).

Several commercially important freshwater fish species of carps, barb, gouramies and catfish are spawned by environmental manipulation. In India Indian, Major Carps (*Catla catla, Labeo rohita* and *Crrihinus mrigala*) are induced regularly to spawn in bundhs where flooding is simulated (Sinha *et al.*, 1974). In gold fish gravid fish can be induced to spawn by raising the water temperature from 12°C to 20°C and placing vegetation in the tank (Stacey *et al.*, 1979 b).

Combined Approach

Early stages of fish are much more sensitive to aquatic environmental factors such as temperature, pH, dissolved oxygen, suspended solids, light and water current. As a result about one per cent of the eggs produced survive only in nature. Since then, need had been felt to develop a system where environmental parameters can be controlled. The system where these environmental parameters are controlled for the development of fish eggs to spawn stage is known as hatchery. The carps do not breed in captivity due to unfavorable environmental conditions and needs inducement for spawning. In India induced breeding through hypophysation of pituitary gland was initiated by Chaudhuri and Alikhuni, 1957. Apart from this various other synthetic preparations Ovaprim

(Nandeesha *et al.*, 1990 a & b and 1991) HCG alone and in combination with PG (Dwivedi *et al.*, 1984-86 and Lam, 1984) practiced successfully especially in carps.

Hatchery Systems and their Management

Although the environmental approach to induced spawning is practical and usually gives good results as far as egg quality is concerned, the level of control involved can not match that of the hormonal approach. The two approaches can be combined to obtain good effect (Fig. 1). Sexually mature fish may be injected with the appropriate hormone and then placed in an environmentally conducive tank to spawn naturally. The combination of hormone and environmental factors is very successful in spawning of carps. In hatchery simulated environmental conditions are created for breeding and hatching. In the present note a brief account of commercially important hatcheries is given. Based on the circulation of water current two hatchery systems are known.

1. Horizontal hatchery system.
2. Vertical hatchery system.

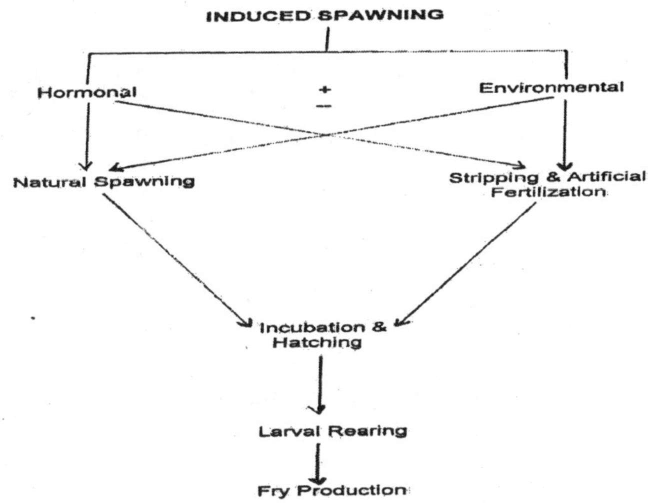

Fig 1: Hormonal and Environmental stimulation for Induced spawning in fish

1. Horizontal Hatchery System

In this hatchery system horizontal flow of water is made to keep the eggs in suspension. Therefore, it is known as horizontal hatchery system. In this type of hatchery system continuous flow of water by gravity is made to breed the carps and hatch the eggs. Examples of this system are Chinese hatchery, Circular hatchery etc.

Circular Hatchery

It is a low cost commercial hatchery with horizontal flow of water especially for small and medium fish farmers (Fig. 2). It's portable model is also available. The system comprises of the following components.

(i) Breeding and Hatching pool.

(ii) Spawn collection tank.

(i) Breeding and Hatching pool

It is a circular cement tank of various diameters. Usually it is 7 feet in (2m) diameter and 1 – 1.5 m deep. The bottom of the pool slopes to the centre (4-6 inches), which leads to the egg/spawn collection chamber through a pipe. The base of the tank is provided with diagonally fitted inlet pipes for circular water flow that creates an artificial riverine condition required for fish breeding. All these pipes opens in the junction box. This junction box is connected by the overhead tank through a inlet pipe. The inner chamber is made of steel frame of 4 feet diameter fitted with a net 40 × 40 or 40 × 20 mesh. At the bottom it is fixed in the groove. In the centre of this unit there is overflow pipe of 3 inch diameter. It has a coupling & joint in the centre to adjust the water level in the breeding and hatching pool. At the bottom this overflow pipe passes through the junction box and outlet opens into the spawn collection chamber. At the bottom also it has couplings.

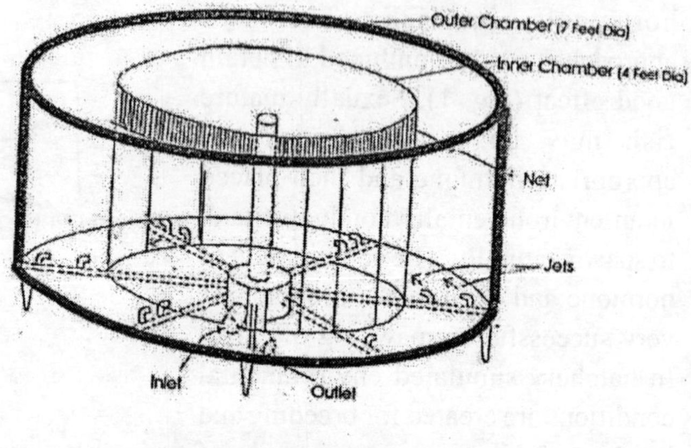

Fig 2 : CIRCULAR HATCHERY

After breeding the spent brooders are removed and fertilized eggs remains in suspension in the pool. To increase DO content in the tank a perforated galvanized iron pipe can be fitted above about half the length of the breeding pool.

(ii) Spawn Collection Tank

It is a rectangular cement tank. The spawn collection tank is constructed below the surface level of breeding and hatching pool. It helps in collection of spawn through gravity. The hapa are fixed in it for the collection of spawn.

Function

Brooders are kept in outer chamber after filling the water in the hatchery up to the desired depth. Mild continuous flow of water is maintained in the hatchery during spawning period. After injection in due course of time breeding takes place. The water level is increased after the swelling of the eggs to provide sufficient space. The eggs remain suspended in the water column due to the continuous horizontal water flow and current created by the water entering the hatchery through inlet via nozzles. In about 16 to 18 hours of duration depending on temperature hatching takes place. Due to continuous flow of water egg shells break into pieces and pass through the net and washed away through the overflow pipe and only hatchlings remain in the outer chamber. Hatchlings are kept in hatchery for about 72 hours in running water. Till 72 hours yolk sac serves the food to hatchlings. On absorption of yolk sack the spawn are ready to be stocked in well prepared nursery.

For collection of the spawn from the hatchery inner chamber is lifted out and overflow central pipe is also unscrewed and removed. As a result spawn get spread through the breeding and hatching pool. Then the spawn is collected in a hapa, fixed in the rectangular spawn collection chamber through outlet pipe.

The continuous flow of water during breeding and hatchery period helps in maintaining oxygen level in the system and removes the metabolites generated. The mild current produced by continuous flow of water in hatchery helps in breeding.

Chinese Circular Hatchery

Chinese hatchery consists of three main components.

(i) Spawning Pool

(ii) Incubation Pond

(iii) Spawn Receiving Pond

These type of hatcheries are also referred as eco-hatcheries. This system is designed for spawning and incubation. An 8.00 m diameter spawning pond with 3.60 m diameter incubation pond gives one crore of spawn with about 95% hatching survival in this system.

(i) Spawning Pool

It is a circular cement pond having inside dimension of 8.00 m diameter. The inside depth at periphery is 1.20 m which slopes down to the center at 1.50 m. The water supply to the pond is from overhead tank by means of 7.50 cm diameter pipe. The water supply line is laid along the outside the tank, using 20 mm diameter pipe with nozzle mouth all in one direction. They are fixed to the vertical wall and nozzle mouth is flush with cement plaster face and near the bottom along the periphery of the pond. In the center an outlet pipe of 10.00 cm diameter is fitted, through which on opening the valve, fertilized eggs along with water are transferred into incubation pond for hatching.

(ii) Incubation Pond

There are two circular incubation ponds each of 3.60 m internal diameter with two chambers in each. The outer dimension of the pond is 4.00 m having outer cement wall. Another circular wall with fixed M. S. screen is provided at 0.74 m clear distance from the outer wall. The inner chamber is provided with 10.00 cm diameter vertical outlet with holes at different heights for taking out excess water for dewatering the incubation pond. The spawn along with water flows from these ponds to spawn collection tank.

(iii) Spawn Receiving Pond

After completion of incubation the spawn is received in a pond of various dimensions. This is a rectangular cement tank. The inside dimension are 4.00 m × 2.50 m × 1.20 m. This is located at a lower elevation than the incubation pond, so as to drain out the water from it by gravity. M. S. hooks are fixed in two opposite walls of the pond for fixing net for collection of spawn.

Operation

Mature brooders are loaded in the spawning pond for about 4 - 8 hr for conditioning. At a time 40 kg females and 60 kg males are kept. Water jets are started for circular motion of the water. After about 8 – 10 hr of inducing hormone spawning takes place. One crore of eggs can be incubated at a time in one operation. The eggs are collected from the bottom and these eggs are transferred to the incubation pools through pipes by opening the valves.

Arrangements are made to circulate the water in the incubation pools. In four days time the spawn gets ready and is to be collected through spawn collection tank and transferred or stocked in well prepared nursery pond for further rearing till fry stage.

2. Vertical Hatchery System

The Central Institute of Fisheries Education (CIFE) has designed a vertical hatchery system, where the flow of the water is in vertical direction. Hence, it is called as Modern carp hatchery model CIFE, D-81 and its advance model as CIFE D-85.

Modern carp hatchery CIFE D-81

The hatchery system consists of the following components.

(i) Hatchery buckets.

(ii) Breeding pools.

(i) Hatchery Buckets (CIFE, D-81 Model)

It is the main component of the system. These are low-density polyethylene (LDP) jars. The upper part of the jar is 440 mm diameter. The height of the jar is 625 mm and the water capacity is 40 L. Each jar has a inlet pipe and control value at its bottom and a overflow outlet at upper end. There is inner basket of

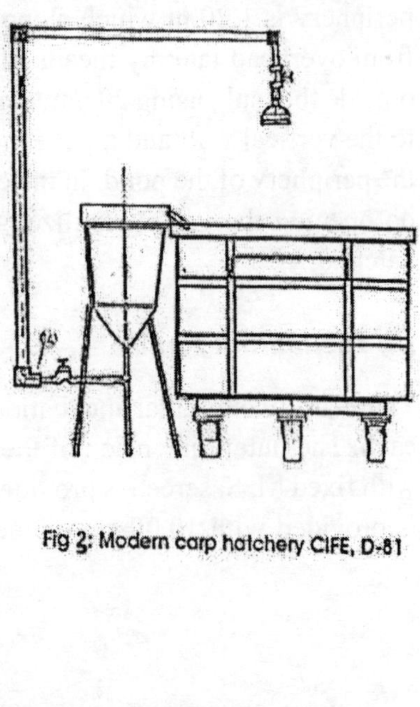

Fig 2: Modern carp hatchery CIFE, D-81

steel frame with mosquito netting for separating the egg shells. Direct flow of water is maintained in these jars at the rate of 2-4 L / m. Egg loading capacity of each jar is about 0.2 million (2 lacks). In one unit of CIFE, D-81 model generally six hatchery buckets and two plastic pools are used. However, number of buckets can be increased or decreased as per requirement (Fig. 3).

CIFE, D-85 Model

CIFE, D-85 bucket model is designed for large-scale production. The principle and design are the same as in D-81 model. The only difference is the size of hatchery buckets. The hatchery buckets has the egg loading capacity of 0.8-1.0 million (8-10 lakhs) eggs and about 200 L water capacity.

(ii) Breeding Pool

Breeding is done in a plastic pool in this hatchery. It's dimension is 6 feet diameter and 3 feet height with an over flow outlet on the upper side of the pool. A circular hapa is fixed on a metallic frame for breeding and the collection of eggs. A shower is provided over the pool to create the rainy condition as well as to maintain the dissolved oxygen level.

Function

Fish are induced bred by hormonal injection in breeding pool. The eggs are collected from hapa after they swell and loaded in hatchery bucket. In hatchery bucket there is continuous vertical flow of water to keep the eggs in suspension. Hatching takes place is 16-18 hours depending on temperature. After hatching, hatchling passes through the mosquito netting and collected through a lastic pipe channel again in hapa fitted in plastic pool. The egg shells left over in buckets are removed and water supply is discontinued in buckets after the hatching completed.

There is continuous waters supply to the hatchlings in plastic pool through shower to maintain oxygen level and removal of metabolites. In 72 hours yolk sac is absorbed & there is no other food supply to the spawn in plastic pool. Now the spawn is ready to be stocked in freshly prepared nursery.

Overhead Tank

It is a common feature in all the hatcheries. It helps in proper storage of tube well water and it's continues supply through gravity. Size of the overhead tank depends on the size of the hatchery. The tube well water is the most suitable for breeding and hatching operations. It has optimum temperature range (26-28°C) required for spawning as well as it is free from silt and pollution.

The two hatcheries i.e. circular hatchery and modern carp hatchery CIFE D-81 to 85 have been specially designed & useful for commercial seed production by small and medium aqua-farmers. The portable models of these hatcheries are also available. These hatcheries are easy to operate. While the Chinese circular hatchery is a big model for large scale commercial see production.

The CIFE- D series models have been widely adopted in the states of Andhra Pradesh, Madhy Pradesh, Tamilnadu, Punjab, Haryana and Uttar Pradesh. It has revolutionized the trade of fis seed production during 1980s through demonstration of the technology of induced spawnin specially among the small and medium aqua-farmers. The live demonstration of the induce spawning and hatchery operation in the state level science exhibitions organized by Uttar Prades Council of Science and Technology at Gorakhpur and Allahabad bagged First and Second prize during 1997-98 and 1998-99.

References

- Billard, R., A. Fostier, C. well and B. Breton, 1982. The endocrine control of spermatogenesis i teleost fish, *Can. J. Fish. Aquat. Sci* 39 65-79
- Chaudhuri, H. and K. H. Alikunhi, 1957. Observations on the spawning of Indian carps by hormon injection, *Curr. Sci,* 26 (12) 381-2.
- Dwivedi, S. N., 1985-86. Hi-Tech Aquaculture-Intensive Rearing Systems, CIFE Bulletin No.8, 1-1
- Dwivedi, S. N., C. S. Chaturvedi and P. K. Varshney, 1984-85. Human Chorionic Gonadotropin substitute for pituitary for induced fish breeding, *J. Indian Fish. Assoc.* Vol. 14 & 15, 75-79.
- Dwivedi, S. N., C. S. Chaturvedi and P. K. Varshney, 1986. Breeding of silver carp *Hypophthalmicth molitrix* in combination of Human Chorionic Gonadotropin and constant doses of Pituitary hormon *Agri and Biol. Res.*, Vol 2, 8-11.
- Dwivedi, S. N., and A. K. Reddy, 1985. Fish breeding under controlled environment, Carp Hatche CIFE D-81, Presented at symposium on aquaculture of carp and related species, PARIS (EVRY), 2- Sept, 1985.
- Goetz, F. W., 1983. Hormonal control of oocyte final maturation and ovulation in fishes. In Fis Physiology (W. S. Hoar, D. J. Randall and E. M. Donaldson, Eds), Academic Press, New York, Vo 9B, 117-161.
- Lam, T. J., 1983. Environmental influences on gonadal activity in fish. In Fish Physiology (W. S Hoar, D. J. Randall and E. M. Donaldson, Eds), Academic Press, New York, Vol. 9B, 65-116.
- Liley, N. R. and N. E. Stacy, 1983. Hormones, pheromones and reproductive behaviour in fish, In Fis Physiology (W. S. Hoar, D. J. Randall and E. M. Donaldson, Eds), Academic Press, New York, Vo 9B, 1-49.
- Nandeesha, M. C., K. G. Rao, R. Jayanna, N. C. Parker, T. J. Varghese, P. Keshavanath and H. P. C Shetty, 1990 (a). Induced spawning of Indian major carps through single application of Ovaprim. Proceedings of the Second Asian Fisheries Forum, In R. Hirano and I. Hanyu, (Eds.) pp. 581- 58 Asian Fisheries Society, Manila, Philippines.
- Nandeesha, M. C, S. K. Das, D. E. Nathaniel and T. J. Varghese, 1990 (b). Project report on breedin of carps with ovaprim in India, *Asian Fisheries Society*, India. Special Publication No. 4, pp. 1-41.
- Nandeesha, M., Ramacharya and T. J. Varghese, 1991. Further observations on breeding of carps wi Ovaprim, Special Publication No. 6, *Asian Fisheries Society, Indian Branch* 1-41.

- Scott, A. P. and J. P. Sumpter, 1983. The control of trout reproduction basic and applied research on hormones. In Control Process in Fish Physiology (J. C. Rankin, T. J. Pitcher and R. T. Duggan, Eds), Croom Helm London & Canberra 200-220.
- Sinha, V. R. P., V. G. Jhingran and S. V. Ganpati, 1974. A view on spawning of the Indian Major Carps, *Arch. Hydrobiol*; 73, 518-536.
- Stacey, N. E., 1983. Hormones and reproductive behaviour in teleosts, In Control Process in Fish Physiology (J. C. Rankin, T. J. Pitcher and R. T. Duggan, Eds), Croom Helm London & Canberra 117-129.
- Stacey, N. E., 1984. Control of the timing of ovulation by exogenous and endogenous factors, In Fish Reproduction Strategies and Tactics (G. W. Potts and R. J. Wootton, Eds.), Academic Press London 207-222.
- Stacey, N. E., A. F. Cook and R. E. Peter, 1979. Spontaneous and gonadotropin induced ovulation in the goldfish, *Carassius auratus* L, effects of external factors, *J. Fish. Biol,* 15, 349-361.

2

ENGINEERING INTERVENTION FOR GROWTH OF AQUACULTURE SECTOR

*Bimal Chandra Mal**

Abstract During the last century, fishery sector made a significant development, mainly, with the advent of knowledge in fishery science. However, the progress is still not upto the mark and not comparable to that of developed countries. Even India lags much behind China in terms of yield and total production. In the fifties scientists and planners realized the importance of engineering intervention for the development of the agriculture sector of the country. Similarly, in the eighties the need of engineering intervention for the development of the fishery sector of the country was felt. Engineering knowledge can help in the site selection, planning and design of an aquacultural farm including its components such as sluice gates, dykes, canals and other structures. The hatchery and its components also require proper engineering design. In addition, engineering intervention can help in different culture techniques and management of fish and prawn farms and hatcheries, design and operation of fishing crafts and gear for harvesting, preservation, processing, transport and marketing of fish and fishery products. Apart from this knowledge of fishery engineering can be useful in aquaculture system and management practices, aeration systems, filtration systems, water management, breeding and hatchery systems, remote sensing and GIS applications in aquaculture, fish and feed processing and aquatic ecosystem and environment management. Engineering intervention needed for sustainability of fish production of the country is briefly presented in this chapter.

Key words: Engineering, sustainable, fish production, fish harvesting, fish processing.

Introduction

Aquaculture is currently playing, and will continue to play a big part in boosting fish production and in meeting rising demand for fishery products. The role of aquaculture is increasingly important in fish production for human nutrition and poverty alleviation in many rural areas. Aquaculture, in

* Professor, Department of Agricultural and Food Engineering, IIT Kharagpur, Email bmal@agfe.iitkgp.ernet.in

common with all other food production practices, is facing challenges for sustainable development. “Sustainability” is a concept that has been gaining popularity since the 1980s. The most commonly cited definition of sustainability is that “Development which meets the needs of the present without compromising the ability of future generations to meet their own needs”. This means development without deteriorating resources and environment. Most aqua-farmers are continuously pursuing ways and means of improving their production practices, to make them more efficient and cost-effective. There is significant potential for continued expansion and growth of aquaculture and culture-based fisheries through enhancement of inland fish production through integrated aquaculture-agriculture farming systems and integrated utilization of small and medium size water bodies. Additional opportunities/strategies for further development and increased food production include intensification of production, specialization of production, rehabilitation of existing production facilities, improved fisheries enhancement methods, diversification of production, combining on-farm and off-farm activities etc. Awareness of potential environmental problems has increased significantly. Efforts are under way to further improve human capacity, resource use and environmental management in aquaculture. The promotion of sustainable aquaculture development requires that “enabling environments”, in particular those aimed at ensuring continuing human resource development and capacity building are created and maintained. In that situation the development of aquaculture needs specialized human resources. The Aquacultural system is a bio-engineering system and the role of Aquacultural Engineers becomes vital in sustainable development of the system. As sustainable development of aquaculture is possible only by proper development of infrastructure it needs proper engineering input.

Aquacultural Engineering is a specialization of Agricultural and Food Engineering which deals with the design and development of Aquacultural Systems for marine and Freshwater including harvest and post harvest facilities. Aquacultural engineers are specialized in the following areas a) planning, design, construction and maintenance of fish and prawn farms and hatcheries, b) different culture techniques and management of fish and prawn farms and hatcheries, c) design and operation of fishing crafts and gear for harvesting, d) preservation, processing, transport and marketing of fish and fishery products. Apart from this they are required to have research experience in one or more areas from the followings 1) aquaculture system and management practices, 2) aeration systems, 3) filtration systems, 4) water management 5) breeding and hatchery systems, 6) fishing gear and crafts, 7) remote sensing and GIS applications in aquaculture, 8) fish and feed processing and 9) aquatic ecosystem and environment.

Global Aquaculture Scenario

Aquaculture has been the world’s fastest growing food production system for the past decade (growth rate of about 10% per year). Aquaculture production increased from 12 million tonnes in 1986 to 34 million tonnes in 1996, valued at US$ 47 billion. In 1996, over a quarter of the world’s food fish supply was derived from aquaculture. Global projections for future supplies from aquaculture production include estimates of 47 million tonnes by the year 2010. Aquaculture production continues to expand, however primarily in developing countries. Asian aquaculture farmers contribute about 90% of the world's aquaculture production, and more than 82% of total aquaculture yield is being produced in low-income food-deficit countries. The growth rate of the aquaculture sector in low-income food-deficit countries between 1984 and 1996 was six times faster than that for non-low-income food-deficit countries. In 1997, despite the 0.09 percent decrease in the output of the fishing industry, the aquaculture sector still posted a 6.7 percent increase in production. With globally stagnating yields from capture fisheries and increasing demand for fish and fishery products, expectations for aquaculture to increase its contribution to the world's production of aquatic food are very high. Throughout its long history, aquaculture has contributed significantly to food supply and rural development, particularly in many communities of resource poor farmers in Asia. Moreover, with more than 80% of the world's population expected to be living in developing countries by 2010, there is much hope that aquaculture will continue to strengthen its role in contributing to food security and poverty alleviation in many developing countries. The vast majority of finfish produced by aquaculture is produced in extensive and semi-intensive freshwater culture systems, with Chinese and Indian carps contributing more than 42% of the global total aquaculture production. Coastal aquaculture is dominated by production of aquatic plants (seaweeds) and molluscs. The production share of crustaceans and finfish is rather low (i.e. less than 15%), however their relative contributions to the value of total coastal aquaculture production is significant (i.e. 50%). Over the last three decades, the global aquatic systems have been subjected to massive pressures from fishing and other types of fishery resources exploitation with indicative fall of the marine capture production and an evident growth in aquaculture (FAO, 2001). Worldwide per capita fish consumption nearly doubled from about 8 kg in the early 1950s to about 15.8 kg in 1999. Fish exports from developing countries have surpassed traditional export of crops such as sugar, beverages, and meat. According to FAO, net foreign exchange earnings from fisheries by developing countries rose from $5.1 billion in 1985 to $16.4 billion in 1996 (FAO,1999). At present, the fisheries sector in most developing countries continue to exhibit steady growth in production, consumption, and trade. Technological advances in aquaculture, changes in legal and institutional regimes and market demands have contributed to the changing structure of supply and demand patterns for fisheries products in both developing and developed countries.

Aquaculture Resources of India

India has a total coastline of 8129 km (including Andaman and Nicobar Islands and Lakshadweep Islands). The potential brackish water area available for aquaculture is approximately 1.2 million hectare. Out of which an area of approximately 1, 57,400 hectares (14% of the total potential area) is under brackish water aquaculture. The country has Inland water resources in terms of lakes and reservoirs of 2.09 million hectares and rivers and canals of 0.16 million kilometers. Apart from this fresh water earthen ponds of 2.25 Mha and abandoned paddy field suitable for aquaculture of 2.3 Mha are available. These aquaculture resources are either underutilized or need development for sustainable aquaculture.

Scientific Design of Farms

In the recent past Aquacultural Industry faced a lot of problems related to public nuisance, pollution and diseases due to improper site selection and design along with biological shortcomings. This has reduced the pace of development of fast growing aquaculture industry in India. This type of problem can be tackled by proper site selection, scientific design of farms and proper management. An Aquacultural Engineer may play an important role in design and construction of such farms in compliance with the environment regulations. A well designed fish farm will have a great impact on the quantity of production, cost of production, judicial use of water, energy requirement, maintenance requirement, ease of operation and overall pollution control. For sustainable development of aquaculture, it is essential to develop the farms in accordance with the environmental regulations. A scientific design of fish farm depends on proper site selection which involves the proper study of the soil, water source, quality of water, agro climatic condition, flood level, transport facilities, manpower, pollution level, competition from other industries, local tradition, political situations and other supporting facilities. If the site is selected with due consideration of the above factors, the aquacultural project is expected to run hassle free. Next important step is the scientific design of aquacultural farms. This involves design of peripheral dykes, partition dykes, sluice gates, supply channels, drainage channels and sedimentation & biofilter tanks (optional). Proper designs of these basic units are very important as they have a great impact on the life span of the project as well as economical operation of the farm. Based on the topographical condition of a site, a farm should be designed to reduce the cost of project as well as its economical operation. The selection of appropriate pumps is very important as it has direct impacts on cost of production and energy saving. The cost of pipe lines and channels can be reduced by scientific design. Use of aerators is also becoming popular among the fish farmers as it improves the water quality and provides scope for intensification of culture. Selection of suitable aeration devices, estimation of requirements of number of aerators, their power and placement in the ponds are very important. A well designed farm will have minimum water loss and good water quality can be maintained in culture tanks to avoid disease problem.

Scientific Design of Hatcheries

Breeding of different species are done in the hatchery by creating natural breeding conditions. A healthy and disease free seed grows fast in culture system to yield better production. Thus quality of seed is a very important parameter for sustainable development of aquaculture. In India inbreeding is surfacing as a major problem for the hatchery industry. Therefore, brood stock management and germplasms maintenance have become very important for sustainable development. The design and construction of hatchery systems require sound technical skills. Very good skill is required for water quality management. Efficient and economical water filtration and disinfection system are needed for this. A good design helps to produce healthy seed economically with ease of operation. The hatchery consists of water quality section, breeding section and larval rearing section. There should be good coordination between the sections. The aquacultural engineers can play significant role in design and construction of hatcheries. At IIT Kharagpur some works in collaboration with CIFA and CIFE have been carried out on scientific design and operation of carp and prawn hatcheries (Agrwal *et al.* 1990, Polappa *et al.* 1995, Chandrakant *et al.* 2008, Mohanty *et al.* 2009.)

Design of Aerators and Aeration System

Dissolved oxygen apart from feed is probably the most important single variable regulating the production of fish in intensive aquaculture system. The culture system with low dissolved oxygen does not cause mortality directly to the fish life, but leads to the deterioration of water quality, which imparts stresses in fish body and in turn causes mortality. The supply of oxygen to the water body is essential to avoid the lethal action. The variations of rates of respiration, photosynthesis and diffusion process during different periods of the day change the dissolved oxygen level of pond water continuously. Ponds generally reach the minimum dissolved oxygen level during the mid-night or early morning. However, on cloudy days the dissolved oxygen concentration may reach the alarming level even during the daytime. Therefore, aeration is necessary to maintain adequate dissolved oxygen level in water bodies under culture.

The intensive fish culture system often needs the introduction of artificial aeration systems (Boyd and Ahmad, 1987). These systems vary from emergency aeration, operated only when the oxygen level drops to dangerous values through ordinary night-time aeration, to continually operating aeration systems in highly stocked ponds. The obvious role of aeration is to supply oxygen to fish. In addition, aeration of the water may affect a variety of other biological systems in the pond. In the biological treatment of wastewater, aeration is an important process employed to raise the dissolved oxygen (DO) level to allow aerobic bacteria to reduce biochemical oxygen demand of the effluent and thus to improve the water quality. The oxygen supplied must be at a rate sufficient to at least balance the rate of removal of the active biomass. Aerators are the devices used to supply oxygen to meet such demands (Moulick *et al.*, 2002).

Many different types of aerators such as paddle wheel, diffuser, propeller-aspirator-pumps, cascade etc. are used in aquaculture. But as yet, most of the aerators are imported as no standard design is available and no manufacturer in India has taken up manufacture of aerators on a commercial scale. The performance testing of these aeration systems is also very important in selecting the design features to provide cost-effective yet efficient as aquaculture pond aerators. The most widely used surface aerator is the paddle wheel aerator (Ahmad and Boyd 1988). It mainly consists of a number of paddle blades rotated in a vertical plane to spray large volumes of water into the atmosphere for oxygen transfer. The high velocity spray induces a turbulent flow on the surface creating a white water effect due to the large volume of entrapped air bubbles. This flow is also the driving force for convective mixing in the wastewater, which leads to uniform mixing of DO throughout the water. Extensive studies have been conducted on the design and performance of floating electric paddle wheel aerator (Ahmad and Boyd, 1988; Boyd *et al.*, 1988; Boyd and Watten, 1989; Moore and Boyd, 1992; Moulick *et al.*, 2001, 2002, 2005, 2009). The design protocol developed by Moulick *et al.* (2002, 2005 and 2009) at IIT Kharagpur could very well be used for optimum design of a paddle wheel aerator based on the requirement of field oxygen transfer rate. At low hydraulic loading packed column aerator functions efficiently and such an aerator was developed at IIT Kharagpur (Mohapatra *et al.* 1989).

For small ponds, propeller-aspirator-pumps (1-3 hp) are found to be most suitable. The propeller-aspirator-pump aerator consists of a rotating, hollow shaft attached to a motor shaft. The submerged end of the rotating hollow shaft is fitted with an impeller which accelerates the water to a velocity high enough to cause a drop in pressure over the diffusing surface. Hence, atmospheric air is drawn into the hollow shaft. This air passes through a diffuser and enters the water as fine bubbles which are thoroughly mixed into the pond water by the turbulence created by the propeller. Due to its low weight, the propeller aerator can be installed easily, immediately and anywhere. Boyd and Martinson (1984) worked on the design aspects of the propeller aerators and found that maximum aeration efficiency occurs at positional angle of 30. It has been found out that the aeration efficiency of propeller aerators increases up to 30 ppt of salinity (Ruttanosrigit *et al.*, 1991; Vinatea and Carvalho, 2007). Kumar *et al.* (2008, 2009) developed optimum geometric and dynamic conditions for a propeller-aspirator-pump aerator (Nan Rong Hai Co. Ltd., Taiwan).

If site constraints and hydraulic conditions permit gravity flow, the least costly method to raise dissolved oxygen (DO) levels is with the use of gravity aeration (Techobanoglous *et al.*, 2003). Gravity aeration can be performed using a simple weir, an inclined corrugated sheet or a stepped cascade. Stepped cascades have been used for a long time for the purpose of energy dissipation, aeration or removal of volatile organic components (VOC) particularly in case of dam spillway. The stepped cascade basically consists of a series of steps over which the water is allowed to flow as a thin film. During the fall of the water, bubbles rise up as air gets dragged in. Gas exchange occurs between the air in these bubbles and the water. Oxygen diffuses from the air into the water and helps to increase the DO content of the water. Stepped cascades could be used to reduce the dissolved nitrogen content. In the treatment of drinking water, cascade aeration is used for reoxygenation and removal of volatile organic components (VOC) such as methane and chlorine, dissolved iron and manganese, carbon dioxide, hydrogen sulphide, as well as the colour and tastes caused by volatile oils. At IIT Kharagpur, Moulick *et al.* (2009) developed a design protocol for a prototype rectangular stepped cascade system of typical height 3.0 m. The optimum number of steps, slope of the entire stepped cascade and hydraulic loading rate were found to be 14, 0.351 and 0.009 m^2/s respectively producing the maximum value of overall aeration efficiency of 0.90.

Development of Suitable Cage and Pen Culture

In India there is ample scope to utilize rivers, irrigation canals, lakes, reservoirs, marine & coastal water, backwater, wetlands (such as bheels, bherries etc.) for cage and pen culture. The natural average production of these water bodies is very low. The productivity of these resources can be enhanced by using the above mentioned enclosure systems of culture. It has the advantages of intensive utilization of space, safety from predators; ease of harvest and culture of selected species. The design of cage and pen is to a large extent based on specific site condition and nature of water bodies. Locally available cheaper construction material may be utilized for development of pen and cage culture. The aquacultural engineers can help in the design and construction of site specific cage and pen culture system using local materials.

Integrated Agri-aquaculture System

The concept of Integrated Aquaculture is gaining popularity nowadays as it is the system in which aquaculture is integrated with horticulture, agriculture, duckery, piggery, poultry, dairy etc., ie. a combination of two or more enterprises. In addition to the production of fish for consumption or sale, integrated aquaculture has a variety of benefits for the farmers. The byproduct of one becomes the input for other. For example, poultry waste becomes fertilizer for fish pond. Mud from the bottom of fish ponds is also an organic mineral-rich fertilizer for agriculture and horticulture. The irrigation water of agriculture and horticulture is cycled through fish pond to produce fish and drained water which is rich in nitrogen is utilized for irrigation. The system is gaining popularity among the Indian farmers as they wish to have multiple products and good profit from their farms. This also provides security in production and good income under diverse climatic conditions. At IIT Kharagpur, some works on integrated agri-aquaculture has been successfully carried out (Mal, 2009). An Aquacultural Engineer is equipped with the knowledge to design such projects with competence.

Wastewater Treatment and its Reuse

The recirculating aquaculture is the system to produce more fish with limited water, feed and space and at the same time have proper control over production. In this system intensive aquaculture is practised by recycling and renovating the wastewater without affecting natural aquatic environment. This system is becoming commercially popular in situations of water scarcity and strict environment regulations. Recirculating aquaculture is economically viable and can be adopted for fish culture and ornamental fisheries. This system requires efficient design of culture tanks, mechanical filtration, biological filtration, aeration, and disinfection systems. Aquacultural engineers can contribute to the design and construction of the system as well as in research and further development works. Work is in progress at IIT Kharagpur for the development of the full system. Earlier, at IIT Kharagpur a study was conducted in bench-scale culture tanks using water mixed with domestic sewage in different ratios under simulated solar radiation (Upadhyay *et al.* 1988). Results showed that the dissolved oxygen concentration and pH of the system were affected considerably and the dynamics of their changes initially followed the pattern of impulse disturbance. The changes in dissolved oxygen concentration and pH depended on the quality of the sewage and the extent of dilution.

Application of Remote Sensing and GIS in Aquaculture

The potential area suitable for aquaculture, the status of fish & shrimp culture, and their ecological and socio-economic impacts can be precisely known only through the application of RS and GIS as the area is widespread and can not be surveyed manually. At IIT Kharagpur some work (Rajitha *et al.* 2007) on the use of RS and GIS application for the above purpose has been taken up. A selected number of case studies were taken up on the application of RS and GIS for sustainable management of shrimp culture. The need for accurate prediction of water quality parameters within the context of sustainable shrimp culture demands the application of advanced methods like artificial neural networks (ANN) combined with RS and GIS. These are essential in the spirit of sustainable development and management, particularly in developing countries, which are often more vulnerable to environmental degradation. The GOS tree (Goal-Objective-Strategy tree) based approach using fuzzy-Delphi process can be applied as a tool for strategic planning for the Indian shrimp industry that has displayed a cyclic performance in the past and has shown signs of rejuvenation in recent times (Prusty *et al.* 2009). It helps in setting long term goals, specifying short term objectives and formulating strategies to achieve the objectives.

Fish feed

The quality of feed is a vital parameter for the growth of aquacultural species and water quality management. A good quality feed gives better feed conversion ratio and produces minimum adverse effect on water quality. The cost of aquaculture production mainly depends on the cost of feed as more than 60% of the production cost goes to feed only. For this purpose, high energy feed formulation using locally available cheaper materials are important. Apart from this the water stability and digestibility of the feed should be good as it has direct impact on water quality management. Extruder cooking and palletizing play important role in the quality of feed. IIT Kharagpur has developed an efficient extruder for manufacture of fish feed (Rout and Bandyopadhyay, 1999).

Fish Feeder

Automatic and demand feeders are very important equipment for dispensing metered quantity of feed in the ponds at different time intervals. Use of feeders reduces the wastage of feed, reduces manpower and helps in timely feeding. This gives good feed conversion ratio as well as helps to maintain water quality. IIT Kharagpur has developed an automatic feed dispenser using local materials and it works without any power source (Mal, 1996). CIFA, Bhubaneswar has developed a demand feeder. Aquacultural engineers can develop improved design and construct these and other types of feeders.

Fishing Crafts and Gear

Approximately fifty percent of total fish production of the country comes from capture fisheries. The modernization of fishing gear and crafts has resulted in efficient utilization of natural resources. Developments of marine engines and hydraulics, hull materials, deck machinery and electronics equipments have improved the efficiency and accelerated the pace of mechanization of fishing vessels in the country. At present, approximately 87 percent of the total landings are obtained from mechanized fleets. Use of synthetic fibers in fishing gear has brought about major changes resulting in increased efficiency of major and minor fishing gears. Infrastructure such as landing and berthing facilities has played a pivotal role in the development of capture fisheries. Now responsible fisheries have come up to regulate the sectors. This will be helpful for sustainable production and growth of natural resources. Aquacultural/Fishery engineers can contribute in the design and construction of fishing gears and gear accessories, development of energy efficient fishing gears and crafts, fishing techniques, modification of fishing gears and crafts, locating fishing zones etc. At IIT Kharagpur, some works on the design of efficient fishing crafts and gear have been carried out and more are in progress (Kathavarayan *et al.* 2002, 2004 & 2006; Narsapurkar *et al.* 1988; Mukherjee, 1998 and Sahoo *et al.* 1988).

Processing

Post harvest handling of fish is very important for increasing shelf life and retaining quality and nutritional values of fish and fish products. It requires development of energy efficient processing techniques and equipments. Nowadays concept of ready to eat food is gaining momentum over the ready to cook and others. Thus product development is gaining importance. The processing sector has played a vital role in increasing the exports of marine products from the country. Modem processing units with state of the art facilities such as individual quick freezing and canning plants have been established and taken a form of organized industry. These industries require suitable technical manpower to run and maintain the sophisticated equipments and machineries. Aquacultural Engineers can contribute to the design and construction of fish handling system, grading and packing facilities, installation and operation of processing equipment, fish product development, research and development of processing equipment, packaging materials etc. Works on value addition of aquacultural products through the application of post harvest technology is in progress at IIT Kharagpur (Rao, 2007).

References

- Agarwal, B. P., Mal, B. C. and Mitra, A. 1990. Improvement of Water Management System of a Carp Hatchery System – A Case Study. Proc. ISAE. ID/90-5 54-62.
- Ahmad, T. and Boyd, C.E., 1988. Design and performance of paddle wheel aerators. Aquacultural Engineering, 7(1) 39 – 62.
- Boyd, C.E. and Martinson, D.J., 1984. Evaluation of propeller-aspirator-pump aerators. Aquaculture, 36(30) 283 – 292.
- Boyd, C.E. and Watten, B.J., 1989. Aeration systems in aquaculture. CRC Critical Reviews in Aquatic Sciences, (1) 425 – 472.
- Boyd, C.E., Ahmad, T. and La-fa, A., 1988. Evaluation of plastic pipe paddle wheel aerators. Aquacultural Engineering, 7(1) 63 – 72.
- Chandrakant, M.H., Reddy, A.K. and Mal, B. C. 2008. Performance of Used Synthetic Sea Water on

Growth and Survival of Larvae of Macrobrachium Rosenbergii. All India Seminar on Recent Advances in Processing and Marketing of Fishery and Horticulture Products. The Institution of Engineers (I), Kolkata.

- Kathavarayan, E., Dora, K. C. and Mukherjee, C. K. 2002. Effect of Square Mesh Panels on Catches of Stationary Bag Net in Hooghly Estuary. Fishery Technology, 39(1).
- Kathavarayan, E., Mukherjee, C. K., Jha, M. K. and Sha O.P. 2004. Assessment of Bottom Trawl Discards of Medium Size Trawlers. Proceedings of MBR 2004 National Seminar on New Frontiers in Marine Bioscience Research, NIOT, Chennai. Pp 221-227.
- Kathavarayan E., Mukherjee C. K, and N, Neethiselvan. 2006. Evaluation of the Effect on Seasonal Trawl Fishing Ban on Fish Catches along the Chennai Coast. Environment and Ecology. 245(1) 208-212.
- Kumar, A., Moulick, S. and Mal, B. C. 2008. Aeration Performance of a Propeller Aspirator Pump Aerator. All India Seminar on Recent Advances in Processing and Marketing of Fishery and Horticulture Products. The Institution of Engineers (I), Kolkata.
- Kumar, A., Moulick, S. and Mal, B. C. 2009. Performance evaluation of a Propeller-Aspirator-Pump Aerator. Aquacultural Engineering (Accepted). Doi 10.1016/j.aquaeng.2009.12.001.
- Mal, B. C. 1996. Performance of Hawaii Type Automated Fish Feed Dispenser. Aquacultural Engineering (Elsevier). 15(4) 81-90.
- Mal, B. C. 2009. Techno-economic Feasibility of Integrated Agri-aquaculture—A Case Study. 96th Indian Science Congress, Shillong.
- Mohanty, B.B., Mal, B. C., Sharma, K.K. and Mohapatra, B. C. 2009. Water Requirements of a Portable Carp Hatchery—for Rohu Spawning and Eggs hatching. Fishing Chimes 29(5) 47-49.
- Mohapatra, R. K., Rout, R.K. and Bandyopadhyay, S. 1989. Performance of packed column aerator at low hydraulic loading. Aquacultural Engineering, 8 (6) 393-405.
- Moore, J.M. and Boyd, C.E., 1992. Design of small paddle wheel aerators. Aquacultural Engineering, 11(1) 55 – 69.
- Moulick, S., Mal, B. C. and Bandyopadhyay, S. 2001. Testing Procedures of Surface Aerators in Standard and Field Conditions. Journal of Inst. of Engineers (I) 825-8.
- Moulick, S., Mal, B. C. and Bandyopadhyay, S. 2001. Scale–up Criteria of Paddle Wheel Aerators. Proc. 3rd All India People's Congress, Forum of Scientists, Engineers and Technologists (FOSET) 16-25.
- Moulick, S., Mal, B. C. and Bandyopadhyay, S. 2002. Prediction of Aeration Performance of Paddle Wheel Aerators. Aquacultural Engineering (Elsevier). 25(4) 217-237.
- Moulick, S., Bandyopadhyay, S. and Mal, B. C. 2005. Design Characteristics of Single Hub Paddle Wheel Aerators. ASCE, Journal of Environmental Engineering. 131(8) 1147-1154.
- Moulick, S and Mal, B. C. 2009. Performance Evaluation of Double Hub Paddle Wheel Aerator. Journal of Environmental Engng, ASCE 135(7)562-566.
- Moulick, S., Tambada, N. V., Singh, B. K. and Mal, B. C. 2009. Aeration Characteristics of a Rectangular Stepped Cascade System. Water Science and Technology (UK)-accepted.
- Mukherjee C K, 1998. Development of model for drag coefficient of netting. Presented in International Symposium on Fishing Vessel & Fishing Gear Technique, Italy.

- Narsapurkar, R. R., Mukherjee, C. K. and Mitra A. 1988. Design of Longlines for Indian Coastal Waters. Presented in World Symposium on Fishing Gear and Fishing Vessel Design, Marine Institute, New Foundland, Canada.
- Polappa, N., Mal, B. C. and Mitra, A. 1995. Performance of a Modified Design of Circular Carp Hatchery. The Harvestor. 241-3.
- Prusty, S. K., Mohapatra, P. K. J. and Mukherjee, C. K. 2009. GOS Tree (Goal-Objective-Strategy Tree) Approach to Strategic planning using a Fuzzy-Delphi Processan Application to the Indian Shrimp Industry. Technological Forecasting & Social Change. (article in press).
- Rajitha, K., Mukherjee, C. K. and Chandran, R.V. 2007. Applications of Remote Sensing and GIS for Sustainable Management of Shrimp Culture in India. Aquacultural Engineering 36 1-7.
- Rao, P.S. 2007. Value Addition of Aquatic Products. International Conf. 'Aquaculture Today and Tomorrow'. New Delhi 38-57.
- Rout, R.K. and Bandyopadhyay, S. 1999. A comparative study of shrimp feed pellets processed through cooking extruder and meat mincer. Aquacultural Engineering, 19(2) 71-79.
- Ruttanogosrigit, W., Musig, Y., Boyd, C.E. and Sukchareon, L., 1991. Effect of salinity on oxygen transfer by propeller-aspirator-pump and paddle wheel aerators used in shrimp farming. Aquacultural Engineering, 10(2) 121 – 131.
- Sahoo, N., Mukherjee, C. K., and Mitra, A. 1988. Development of a Bamboo Boat for Small Scale Fisheries. Presented in World Symposium on Fishing Gear and Fishing Vessel Design, Marine Institute, New Foundland, Canada.
- Sarkar, S., Kamilya, D. and Mal, B.C. 2007. Effect of Geometric and Process Variables on the Performance of Inclined Plate Settlers in Treating Aquacultural Waste. Water Research (Elsevier). 41 (5) 993-1000.
- Upadhyay, A. S., Rai, V. K. and Bandyopadhyay, S. 1988. Preliminary study on dynamic characteristics of sewage-mixed water body for aquaculture. Aquacultural Engineering, 7 (3) 211-219.
- Vinatea, L. and Carvalho, J.W., 2007. Influence of water salinity on the SOTR of paddlewheel and propeller-aspirator-pump aerators, its relation to the number of aerators per hectare and electricity costs. Aquacultural Engineering, 31(1) 73 - 78.
- This article is not included in your organization's subscription. However, you may be able to access this article under your organization's agreement with Elsevier.
- www.davidsuzuki.org/Oceans/ Aquaculture/
- http//www.cefas.co.uk/media/317http//www.fao.org/focus/e/fisheries/sustaq.htm http//ftp.fao.org/FI/DOCUMENT/ministerial/1999/dar.htm http//www.fao.org/focus/e/fisheries/codecond.htm
- http//www.fao.org/focus/e/fisheries/icam.htm
- http//86/aqua1.pdf
- http//www.ca.uky.edu/wkrec/SUSTAQUA21ST.pdf

3

CAN HILSA FISHERS ADOPT TO NEW FISHING REGULATION POLICY OF WEST BENGAL, INDIA GOVERNMENT FOR SUSTAINABLE CATCH!

*Madhumita Mukherjee & Chinmoy Chakraborty**

Abstract: The Indian shad hilsa, *Tenualosa ilisha* is most abundant in the Hooghly estuarine system of West Bengal forming one of the lucrative commercial fisheries of a single species. Disappearance of hilsa fish in the up-rivers has precipitated an acute crisis due to damming, siltation, progressive increase in salinity and formation of sand bars. Hilsa fishers are one of the poorest groups with an ever deteriorating quality of life. Their livelihood is threatened by declining fish catches. Under these circumstances, the Department of Fisheries (DOF), Govt. of West Bengal had initiated a project to find the way of sustainable exploitation of hilsa after studying their socio-economic status. The main objectives were to identify the landless fishers, support the development of SHGs and drawing lessons from the past experiences. The study was based on the hundred percent door-to-door surveys. Hilsa fishery contributes 14-16% of marine catches and in inland sector it contributes 1.7-3.5%. DOF has already taken measures towards Awareness Generation to ban and sale of young juvenile hilsa. It was found that interest in "adding value" to raw fishery products has grown tremendously. Contract Farming can be very effective tool to link them in food processing industry and super stores.

Keywords: Livelihood, Hilsa fisher, Hooghly estuaries

Introduction

The state of West Bengal is the fourth most populas state in the Eastern Region in India. 82 million people stay in the State. It is 2.7% of India's area but about 7.8% of the country's population, this state ranks first in terms of density of 904 per sq. km. According to the Planning Commission, the

* Affiliation Aquatic Resource Health Management Center, Street *Pailan, P.O. Pailan Hat, P.S. Bishnupur, Kolkata-700104*; E-mail madmita_mukh@yahoo.co.in

proportion of population below the poverty line in 1999-2000 in West Bengal was 31.85%. The percentage of Schedule Caste (S.C.) and Schedule Tribe (S.T.) population and Minorities together account for more than half the population and they are the poorest groups in rural Bengal. This state contributes highest share (19.61%) in national fish production (7.13 million t) with 29% in inland and over 6% in marine sector (Ayyappan, 2009). Appropriate technological interventions as well as extension services provided by the Department of Fisheries, Aquaculture, Aquatic Resources and Fishing Harbour (DOF), Govt. of West Bengal (GOWB) in time and space has led to this success. Even after that, Aquaculture is further poised to reach new heights in terms of enhancing production, productivity and profitability. The Indian shad hilsa, *Tenualosa ilisha* is most abundant in the Hooghly estuarine system of West Bengal forming one of the lucrative commercial fisheries of a single species. In the light of this it becomes the prime duty to capture hilsa fish in a judicious way following sustainable norms for increased production and gainful livelihoods. Disappearance of hilsa fish in the up-rivers has precipitated an acute crisis due to damming, siltation, progressive increase in salinity and formation of sand bars. Hilsa fishers are one of the poorest groups with an ever deteriorating quality of life. Their livelihood is threatened by declining fish catches. Unfortunately, due to over exploitation of the nature, sometime by them and most often by the outsiders, environment is not favoring them to maintain their sustainable livelihood. The time has come to find alternative livelihood for them to make their social life sustainable. Under these circumstances, DOF had initiated a project to find the way of sustainable exploitation of hilsa after studying their socio-economic status. The main objectives were to identify the landless fishers, support the development of SHGs and drawing lessons from the past experiences. The fisher community in the state of West Bengal is part of the lower strata of the society. The major contributing factors for this situation are illiteracy, poverty and lack of knowledge of latest technology. The situation has become the worse due to lack of institutional support in terms of both infrastructure and finance on one hand and exploitation by middlemen, traders and contractors on the other (Chakraborty *et al.* 2004a). Although, information on traditional system of brackish water aquaculture ('bhery') in coastal area of West Bengal are available (Ghosh & Naskar, 1991; Ghosh *et al.*, 1997 and Sinha, 1998), but the socio-economic aspects of fishers engaged in hilsa catch system are rarely investigated in this area. In this perspective, the present endeavor was devoted to retrospective appraisal of socioeconomic conditions of hilsa fishers in the state of West Bengal.

Materials and Methods

The study was based on the hundred percent door-to-door surveys conducted in the 390 houses of 39 landing centers in three different zones of West Bengal for understanding their livelihood and standard of living. The questionnaires were prepared based on a preliminary survey, pre-tested and finalized for data collection. Both formal survey and non-formal survey with Participatory Rural Appraisal tools like interactive discussion, semi-structured interviewing, sharing of information and ideas etc. were carried out following Anon. (1992), Townsley (1993) and Schonhut & Kievelitz (1994). Data on various aspects were also collected from 'Gram Panchayat' and related Departments.

The information obtained from fishers pertained to general information on caste structure, size of ownership, organizational participation, demographic structure, education, technical qualification, experience, social and occupational status, land and assets, fishing requisites, infrastructure availability, issues of credit, technology adoption etc. The gathered information was tabulated and socio-economic status has been analysed using simple tabular analysis in MS-Excel to cater to the need of the objectives.

Result and Discussion

The economic activities in the coastal area of West Bengal are mainly shared by Agriculture followed by Forestry and Fisheries though last but not least. This particular coastal part of this state has a vast amount of valuable human resource. The people of this area struggle against the nature to maintain their livelihood. The community belongs to this region having expert agriculturists, skill fishers, skill honey collectors, laborious and hardy womenfolk, since time they are surviving by effective management of the nature and environment they are surrounded with. Fragmentation of land holding is also a main reason for not adoption of newly released technologies. The time has come to find alternative livelihood for them to make their social life sustainable. The survey on socioeconomic conditions of hilsa fishers of West Bengal has provided considerable information on this community.

i. General information

Self-education (in years) of fishers was at 6.64 indicated that the fisher communities as a whole was marginally bellow the secondary level. The average experience (Table 1) in hilsa catch activities was 25 years with arrange from 3 to 60 years in this study. Considering the mean experience, it appears that hilsa catch activity is an age old practice in West Bengal. Diversion to alternate livelihood and change of group of people from year to year was also noticed. The study revealed that only 17.5% of total fishers have received technical training in this aspect. Training programme on the technical aspects of hilsa catch was attended only by the group of traditional fishers. Fishers should have the necessary knowledge, skills and aptitude for conducting sustainable hilsa fishing. Education on the managerial aspects of aquaculture and fisheries is essential in the larger interest and the future of the industry (Chakraborty & Banerjee 1995). The caste structure (Table 2) revealed that the S.C. and S.T. population were found 58.77% & 6.28% respectively. S.C. and S.T. together account for more than half of the population. The general category people were estimated at 39%. It may be due to more and more commercialization of fish catch and entry of general category people in this enterprise from outside area. The study is not in agreement with the previous observation of Bhaumik *et al.* (1993). Chakraborty *et al.* (2003) reported that the erstwhile South and North 24 Parganas are recognised as backward areas inhabited by the people living below the estimated poverty line. Fishing and Fish culture were perhaps one of their main sources of livelihood because the soil being saline, agriculture was less productive. Improved system of sustainable

Table 1 Training, Experience and Education of Hilsa Fishers

Sl No	Characteristics	Unit	Average	Min	Max
1	Age of Hilsa Fishers	(Yrs.)	42.75	15	77
2	Level of education in school	(Yrs.)	6.64	0	12
3	Highest level of education of the family member	(Yrs.)	8.35	0	12
4	Technical training (self)	(%)	17.5	-	-
5	From fishermen community	(%)	100	-	-
6	Experience in Hilsa Fishing	(Yrs.)	25.08	3	60

Table 2 Caste structure of Hilsa Fishers (In %)

Sl No.	Caste	Overall
1	S. C.	58.77
2	S. T.	6.28
3	O. B. C.	2.15
4	General	32.80

aquaculture method or process of value addition as alternative livelihood, if introduced, can only provide them with economic upgradation and bring them above the poverty line.

The survey on organizational participation by fishers (Table 3) revealed that 62.75% of the fishers are member of any local Fishermen's organization or association, which look after their day to day problems. Group fishing was highly successful and tested among small-scale fishers in other states of India and abroad. The model served as an effective extension intervention to educate fishers on sustainability while helping them to improve their fishing systems. The fishers, including women, could be equipped with the technology by strengthening the fishers-extensionist-researcher-political-administrative linkages (Srinath *et al.*, 2000). This model may be beneficial for this state. The fishers associations, if democratically run, can provide a valuable function in influencing the formulation of fishing policy and legislation and bringing educational and training benefits to members, as well as facilitating collective initiatives on common issues, such as observation of the "Ban Period", protection, management and usage of the resources in a sustainable manner and stop over fishing during peak spawning migration.

Table 3 Organizational participation of Hilsa Fishers (In %)

Sl No.	Organization	Overall
1	Panchayet	2.91
2	Fishermen's organization/Association	62.75
3	Non member	34.34

ii. Family background

The demographic pattern (Table 4) revealed that the percentage of male ranged from 53% to 75% in different districts with an average at 56% in the hilsa fishing community of the state. The percentage of female ranged from 25% to 47% in different districts with an average at 44%. Among the study, the overall sex ratio and adult minor ratios were 10.80 and 10.60 respectively. The percentage of minors in family indicated awareness of family welfare measures adopted by the fishers. Family labour input in the fishing activity on part time and fulltime basis was also observed. Considering the number of adults, the families have high employment potential. Most of the families (69%) were joint and the remainders (31%) were in nuclear family (Table 5). Out of the total population, 89% of the fishers in this study were found to be native to the area. In the remaining cases (11%), the fishers had come from outside to enterprise in hilsa catch. The survey showed that about 89% of the fishers did not want to send children for hilsa catch primarily due to risk, tension and uncertainty associated in this enterprise. However, those wanted (11%) to send children for hilsa catch, few (9%) of them had a plan for further expansion and those did not want to send children for hilsa catch have opted for trade (11%), service (58%), agriculture labour (29%) or other miscellaneous activities (2%). Children's education care was mostly (100%) taken by their wives. None of the women in fisher's family were engaged in any sort of activities of hilsa catch. Women in their family are mostly engaged in different daily labour jobs (90%) and only 10% have regular jobs in the area. Their share in family income ranges from 5%-50%. Gender discrimination is common throughout the region. Women commonly experience a multitude of disadvantages including 1) having to work harder to secure a livelihood; 2) having less control over income and assets; 3) reduced opportunities for human development; 4) being subjected to violence and intimidation; 5) subordinate social position; 6) poorly represented in policy and decision making. For women engaged in fry collection, the problems are compounded by extreme poverty, poor housing, health and sanitation, low literacy rates, limited access to support services, resource opportunities and social exclusion. Women that engage in fry collection usually do so because they have no other viable alternative. The process of uncontrolled population growth has reduced the per capita cultivable land and created overcrowding and high rate of disguised unemployment. The consequent difficulties in ensuring livelihood have also led to a high rate of out migration.

Table 4 Demographic pattern of families of Hilsa Fishers

Sl No.	Sex	Age group	Overall
1	Male (%)	Adult (%)	32.83
		Minor (%)	22.80
	Total		55.63
2	Female (%)	Adult (%)	27.49
		Minor (%)	16.88
	Total		44.37
3	Sex ratio Male Female		10.80
4	Adult minor ratio		10.60

Table 5 Family background of Hilsa Fishers and role of women

Sl No.	Characteristics	Dimensions	Overall
1	Family norms (%)	Nuclear	31.43
		Joint	68.57
2	Residential status (%)	Native	88.57
		Outsider	11.43
3	Would they like to send children for fishing? (%)	Yes	11.76
		No	88.24
4	if yes, Expand/ Diversify (%)	Yes	8.57
		No	91.43
5	if no, other options (%)	Trade	10.71
		Service	58.24
		Agriculture labour	28.57
		Others	2.48
6	Children's education care taken by (%)	Wife	100
		Others	0

7	Role of women (%)	Students	21.40
		Household	70.47
		Employed	7.44
		Business	0.70
8	Sector/area of Job of women (%)	Daily labour	89.74
		Regular job	10.26
9	No. of working days/ year	Days	365
10	Duration of work (in hrs)	Min	6
		Max	14
11	Share in family income (%)	Min	5
		Max	50
12	Mode of commuting to the work place	Vehicle	0
		Walking	100.00

iii. Occupation

Among the fishers surveyed, the major (79.43%) source of income was hilsa catch (Table 6). Few of them were also engaged in trading business of aquaculture inputs. Earner (nos.) per family was 1.56. The other source of income was agriculture in all the areas where off-farm activities contributed marginally higher. Skilled manpower is highly essential for the development of sustainable culture. Rao *et al.* (2000) opined that the ICAR, State Agricultural Universities, DOF and the private sector should concentrate on human resource development. Sen *et al.* (1997) observed that more than 90% of the rural population in the fringes of the Sundarban is living below the poverty line and they are fully dependent on the natural resources of mangrove forest.

Table 6 Occupational structure of Hilsa Fishers

Sl No.	Characteristics		
1	Hilsa catch main Occupation (%)	Yes	79.43
		No	20.57
2	Average earner per family	(nos.)	1.56
3	Average income per month	INR	1300

CIBA (1997) conducted studies based in the sea based farms of Nellore district of Andhra Pradesh indicated that there was an increase of 2% to 15% in employment after the establishment of the aquaculture farms in the vicinity and 6% to 22% increase in income of the farm labourers. The study of Raman & Gopalkrishna (2002) in coastal Andhra Pradesh revealed that before aqua farming came into existence, the local labourers had to go to far-flung areas for earning their livelihood for a paltry Rs. 10-25 day^{-1}. But with the onset of aqua farming the local people were able to get jobs with daily wages of Rs. 100 day^{-1} and local women get employment in nearby processing plants. Hence group of fishers from this sector may be opted for aquaculture or value addition etc., as an alternative livelihood in this state.

iv. Land holding, assets and infrastructure availability

Land holding study showed that fishers owned with an average of 10.21 decimal land used for agriculture and seasonal cropping (Table 7). The lands were available with the fishers for agriculture and allied activities. Assets holdings of fishers revealed that 67% of fishers had no electricity connection (Table 8) in their houses. Drinking water source was bore well (82%) and only 35% of total fisher population in the study had own toilets. Source of fuel was firewood in most cases (78%) and very few (6%) had LPG connection and rests (16%) were using kerosene for this purpose especially. Most of the fishers (100%) were using bus as their main public transport in their locality. In this study, only 10% of total fishers had own vehicle which is bicycle. The other assets included radio (26%) and television (27%). All the fisher population had access to primary health center in their locality. This study clearly showed that hilsa fishing activities have very little contribution now a day in developing their livelihood and standard of living. The majority of the fishers (93.78%) belonged to the same area or to the vicinity of their enterprise making it easy for them to commute to their Landing Centre. In few cases it was found that distance of landing centres were very far from the auction market. The present study revealed that 77% fishers had own houses and rests (23%) were tenanted. Rao (2000) reported that the infrastructure facilities like roads, electricity and transport need to be developed to ensure quick packaging and transport.

Table 7 Utilization of land particulars

Characteristics	Overall
Owned land used for agriculture /crop (Decimal)	10.21

Table 8 Access to assets and public utilities of Hilsa Fishers (In %)

Sl. No.	Assets and Public utilities	Dimensions	Overall
1	Domestic Light	Yes	32.68
		No	67.32

2	Drinking water	Well	0.00
		Pipeline	17.65
		Bore well	82.35
		River	0.00
		Others	0.00
3	Toilets	Own	35.29
		Shared	3.27
		Public	61.44
4	Fuel	Kerosene	16.34
		Firewood	77.78
		LPG	5.88
		Electricity	0.00
5	Public transport	Bus	100.00
		Train	0.00
		Taxi	0.00
6	Own vehicle	Cycle	10.46
		Two wheeler	0.00
		Car	0.00
		Jeep	0.00
		No	89.54
7	Radio	Own	26.14
		Share	73.86
8	Television	Own	27.45
		Share	72.55
		Club	0
9	Primary health center	< 5 km	100.00
		>5 -10 km	0.00
10	Primary school	< 5 km	100.00
		>5 -10 km	0.00
11	High school	< 5 km	100.00
		>5 -10 km	0.00

v. Credit source

The fisher community in India is part of the lower strata of the society. They are generally poor and are in a disadvantageous position in the society. Because of their economic backwardness, illiteracy and insignificant political attention and social insulation, they remain indebted to private moneylenders, who advance money to meet their professional and other expanses (Gadhia *et al.*, 1999). Sources of credit (Table 9) availed by fishers revealed that 33% of fishers used to borrow money from commission agents with high interest rate and 42% availed credit from other sources *viz.* private money lenders etc. They were bound to sell their catch to that particular private commission agents or wholesaler. It was due to the fact that wholesalers had paid a sizeable amount as advance, which was utilised by them to purchase inputs of hilsa catch. The remainder (22%) was taking loan from relatives or commercial banks (3%). Among the studied fishers, source of credit did not include any co-operative. The price in some cases, received by them was less compared to prevailing market price. The conditions of fishers were poor due to lack of institutional support in terms of finance on one hand and exploitation by middlemen on the other. Hence, there should be provision for easy credit facilities (Katiha & Chandra, 1997; Saha *et al.*, 1997 and Karmakar *et al.*, 1999) and intensive extension activities for the upliftment of socio-economic condition of the people engaged in this particular sector.

Table 9 Sources of credit availed by Hilsa Fishers (In %)

Sl. No.	Credit Sources	Overall
1	Commission agents	33.33
2	Commercial Banks	2.61
3	Co-operatives	0.00
4	Relatives and friends	21.58
5	Others	42.48

vi. Technology adoption

The survey on fishing characteristics has provided considerable information on this commercially successful venture. The details of crafts and gears used by the hilsa fishers of West Bengal and their ownership are depicted in Table 10 and 11. The mesh size ranges from 15-140 mm. The study revealed that the practice of catching juvenile hilsa is still in vogue. DOF has already taken measures towards Awareness Generation to ban and sale of young juvenile hilsa. In inland Sector, the mesh size of net under 12 mm is strictly prohibited and the chat jal, behundi jal etc. are also strictly prohibited to use for catching hilsa and other fish from river throughout the year under Inland Fishery Act, 1984 and 1985 of section 46, part1 and 2. According to the above law, the mesh size of net under 25 mm is strictly prohibited for fishing during the breeding season of hilsa and other fish i.e. from 15th June to 30th September in river. In marine sector, ban period for catching hilsa has been settled by the DOF are from 15th April to 15th June in marine sector. It is also prohibited to use the gill net below with the mesh size of 85 mm and hilsa of under 500 gm of body weight are prohibited to catch throughout the year. However, the catch data both in marine and inland sector reveled that the catch has been increased since 2008-09 due to awareness generation in ban implementation by DOF.

Nearly 49% of fishers involved in fishing possess neither craft nor gear. Only 28% of the profit from catch goes to major group of crew members (depending from 10-17). A meager amount of Rs. 1100 -1500 per month per head was the income of crew members. Fishery activities during Hilsa fish catch (Table 12) revealed that fishers spending almost 160 days per year with an average 10 hrs. per day. The price of hilsa found varies in different districts. Yield and net profit was low and expenditure for fishing was high may be due to increased tendency of juvenile catch, resource depletion and lack of timely information about the hilsa population. This may be increased by intensive care during information dissemination and implementation of law and order in time and

Table 10 Fishing technology-about the Craft

Sl. No.	Characteristics	Dimension	Overall
1	Name of Craft (%)	Dingi	58.17
		Donga	14.38
		Fishing boat	27.45
2	Type of craft (%)	Mechanised/	27.45
		Non-mechanised	72.55
3	Material Used (%)	Wood	81.05
		Tin	7.84
		Others	11.11
4	Ownership of craft (%)	Owned	51.39
		Hired/Other	48.61
5	Purchase Price (INR)	Min	500
		Max	1600000
6	Expected Life (Yrs.)	No.	4.76
7	Annual repairs cost (INR)	Min	100
		Max	6000

Table 11 Fishing technology-about the Gear

Sl. No	Characteristics	Dimension	Overall
1	Own fishing gear (%)	Owned	54.51
		Hired/ Other	45.49
2	Gear Used in Hilsa catch (%)	Gill net	96.08
		Traditional	3.92
3	O. A. L (Feet)	Avg	29.47
		Min	6
		Max	59
4	H. P.	Min	95
		Max	125
5	Type of nets (%)	Nylon	7.19
		Mono filament	87.58
		Others	5.23
6	Size of net (Kg)	Avg.	90.18
		Min	1.5
		Max	350
7	Dimensions (m)	Avg	3734.89
		Min	3
		Max	14000
8	Mesh size (mm)	Min	15
		Max	140
9	Purchase cost (INR)	Min	700
		Max	200000
10	Annual repairs (INR)	Min	50
		Max	120000
11	Go for fishing during monsoon (%)	Yes	100
12	No. of days in a week during June-August	Days	4.82

Table 12 Fishery activities during Hilsa fish catch

Sl No.	Characteristics	Dimension	Overall
1	Time devoted (Days Per year)	Avg	160.07
		Min	100
		Max	250
2	Time devoted (Hours Per day)	Avg	9.73
		Min	6
		Max	12
3	Price (INR/kg)	Avg	130.56
		Min	50
		Max	250

space, which can consequently increase the profit during catch. Fish catch and value addition of their catch can be integrated with other sectoral activities such as agriculture, animal husbandry, salt making and tourism in this state (Kutty, 2002, 2005; Chakraborty *et al.* 2004b; FAO/NACA, 1996). Problems encountered during Hilsa fish catch is depicted in Table 13. Resource decrease (92.81%), Credit facility (69.93%), Lack of coordination among fishers (59.48%), and Information on weather report (30.72%) were found main problems of hilsa fishers.

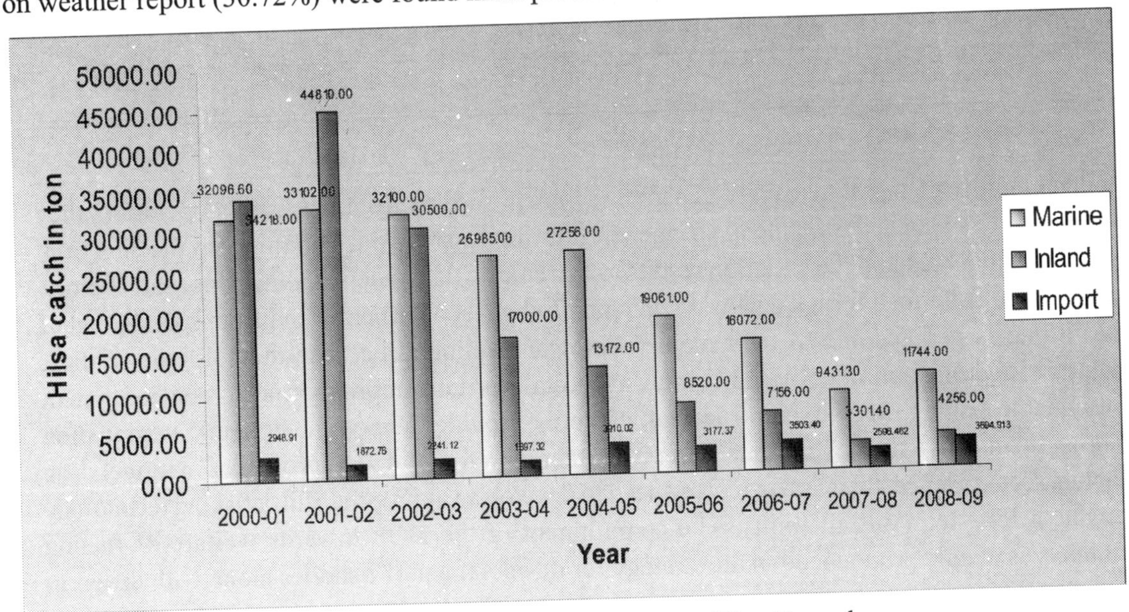

Hilsa Catch and Import Scenario in West Bengal

Hilsa fishing communities were found amongst the poorest groups with lower levels of literacy, poorer housing conditions and an ever deteriorating quality of life. As resources and returns decline, the operations of non-mechanised craft are becoming increasingly unviable. The artisanal fishing community is largely unorganized and lacks an effective voice to protect its livelihood. Since, the

Table 13 Problems encountered during Hilsa fish catch (In %)

Sl. No.	Problems	Overall
1	Resource decrease	92.81
2	Credit facility	69.93
3	Lack of coordination among fishers	59.48
4	Information on weather report	30.72
5	Middlemen interferences	30.07
6	Hamper of disposal by labour associations	28.45
7	Information on non removed sunk wooden/bamboo cages causing damage to gears	27.45
8	Confrontation with forest dept. /BSF	26.45
9	Decocting	23.53
10	Encroachment of foreign coast guard	22.83
11	Non availability of notice sunk ship/boat causing damage to the trawler	22.03
12	Craft related issues	9.15
13	Gear related issues	7.19

fisher community of this state cannot generate sufficient surplus from the fishing operation, they are compelled to face acute financial stringencies during lean period. Summing up the situation, they become indebted and the situation results to extreme critical when landing is not satisfactory. In general, the fishermen form a backward community both economically and from social point of view Fishing is perhaps one of their main sources of livelihood because the soil being saline, agriculture is getting unproductive day by day in coastal districts. Improved and diversified system of sustainable fishing method, if introduced, can only provide them with economic upgradation and bring them above the poverty line. Moreover, there is a trend of zero growth or marginal drop in capture sector. Hence, there is a perceived need to develop an appropriate hilsa fishing technology to enable a long-term sustainability. The commitment of the DOF towards welfare of fishing community is deeply pronounced in the strategy for development of fisheries along with different economic activities on production oriented programmes in inland, marine and brackish water sectors. The State Fisheries Department is pledged to implement a no. of welfare schemes for upliftment of the poor fishermen comprising of lower tier in the society both socially and economically. The following welfare schemes are worth mentioning viz. i) Pension Scheme—The scheme was commenced during 1991-92. Fishermen are provided pension @Rs. 300/- per month every year. ii) Fishermen Group Personal Accident Insurance Scheme—The scheme was commenced during 1984-85. Family members of deceased or permanently disabled fisherman have been benefited. iii) Fisherman Savings cum Relief Scheme Starting from 1998-99 provided with Rs. 300/- p.m. during lean period of four months. iv) Housing Scheme—A number of houses have been constructed for the fishing community. Apart from this the fishing community have been ably supported by the infrastructure development like minor fishing harbours, fish landing centres, diesel outlets, cold chain marketing units, flood shelter, community halls, tubewells, roads and fish markets.

Conclusion

For the development of the fisheries sector, proper database is very important. Base-line reliable data on resource inventory for planning and management for sustainable exploitation of resources are urgently needed. There are a large number of resources user groups and government regulatory and support agencies operating in the Coastal belt with disparate and sometimes conflicting interests. Both the goals for conservation and development can be achieved only through strategic planning and coordination among these various groups. The study clearly showed that hilsa fishing activities have no impressive effect on development of livelihood and standard of living of the fishers' households in the state of West Bengal. Many local fishers engaged in this sector are extremely poor. Formations of self help groups (SHGs) are also required to exchange technology achieve cooperation, market information and also to engage the women folk of fishers' family in post harvest activities. Fishers may be involved in integration and value addition with other sectoral activities to enable a long-term sustainability through employment generation, raising rural income and meeting internal market demands.

Plate 7 Hauling of nets and collecting of fishes in holds at Sankai, Katoa

Plate 8 Hilsa Fisher of Lalgola

Plate 9 Field Visit of Joint Director and departmental officers at Lalgola

Plate 10 Fishers ready for fishing at Sankai on river Ajoy

Plate 11-14 Mosquito net used at Balagar

Plate 15 Boat with gill net ready for harvesting hilsa at Lalgola

Plate 16 Bagnet in small boats for catching hilsa at Bugbazar

Plate 17 Hilsa fishing at Lalgola, Murshidabad

Plate 18 Hilsa fishing boats are at Farakka for harvesting

Plate 19 "Kona zal" used for harvesting Hilsa at Farakka Bindu gram

Plate 20 Mosquito nets are used at Jangipur for catching Hilsa

Plate 21 Naliapur, fishers hunting Hilsa using *donga* fishing boats.

Plate 22 Hilsa fishing with *vesal* nets, at Babughat, Kolkata

References

- Anon.1992. Participatory Rural Appraisal (PRA) for Agricultural Research at Aruppukottai, Tamil Nadu. Tamil Nadu Agricultural University and International Institute for Environment and development, London.
- Ayyappan, S., 2009. Indian Fisheries–The Way Forward. Presentation at National Conference of Fisheries Ministers, Bhubaneswar, 4-5 July, 2009.
- Bhaumik, U., Saha, S. K. and Mitra, P. M. 1993. Scenario of Brackish water Finfish and shell fish seed collection in the Sundarbans, West Bengal. *J. Inland Fish. Soc. India.* Vol 25 (1), p 44-50.
- Chakraborty Chinmoy, & Banerjee, S. 1995. Mass Mortality in Shrimp Farming, Fisheries World, p 30-35.
- Chakraborty Chinmoy and Banerjee, S. 2003. Production of *Penaeus monodon* from modified extensive method in Purba Medinipore, West Bengal. *J. Inland Fish. Soc. India.*Vol.35(1). p25-30.
- Chakraborty Chinmoy and Banerjee, S. 2004a. Economics of *Penaeus monodon* farming under modified extensive method. *Environment and Ecology*. Vol. 22 (1) 130-135.
- Chakraborty Chinmoy, S. Dutta and S. Banerjee, 2004b. Scope for *Macrobrachium rosenbergii* culture in coastal West Bengal An economical perspective. *J.* ***Environment*** *& Ecology.* Vol. 22 (Spl-2). p 159-162.
- CIBA. 1997. Final report Comprehensive survey of impacts of shrimp farming in coastal districts of Tamil Nadu. Mimeo.
- Gadhia Mohini, P.B. Patel, and P.K. Gadhia. 1999. Socioeconomic conditions of fisher community and status of fisheries around Korapur atomic power station. Fishing Chimes. Vol. 19 No. 9. p 49-51.
- Ghosh, A.K and Naskar, K.R.1991. The socio-economic status of rural villages of the Sunderbans engaged in estuarine fisheries- A case Study proc. Natl. sem. on conservation and management of Mangrove ecosystem.
- Ghosh, A.K, Pandit, P.K Karmakar, H.C and Banerjee, R.K. 1997. Ecology and bio diversity of different wetlands with their production potentiality. *In* K.K Vass and M.Sinha (eds) Changing Perspectives of

Inland fisheries proceedings of the national seminar, March 16-17, 1997. Inland Fisheries Society of India, Barrack pore, pp 221-227.

- Karmakar, H. C., Pandit, P.K. and Ghosh, A.K. 1999. A study on the present socio-economic status of the fishers engaged in three ecologically different zones of estuarine wetlands. p164-167. In M. Sinha, Dhirendra Kumar and P. K. Katiha (eds). Eco-friendly Management of Resources for Doubling Fish Production-Strategies for 21 st Century Proceeding of the National Seminar, December 22-23, 1999. Inland Fisheries Society of India, Barrack pore.
- Katiha, P. K. and Ravish Chandra, 1997. Income and employment generation from Indian inland waters. p 261-266. In K.K. Vass and M. Sinha (eds) Changing perspectives of Inland Fisheries. Proceedings of the National seminer, March 16-17, 1997, Inland Fisheries Society of India, Barrack pore.
- Kutty, M.N. 2002. Social Conflicts in Shrimp Farming-the rise and fall of advernturism in shrimp farming in India. Paper presented at the "World Aquaculture 2002" Meeting of the World Aquaculture Society at Beijing, China, May, 2002.
- Raman, R. P. and Gopalkrishna, G. 2002. Is shrimp farming detrimental to environment? A scientific study in coastal Andhra Pradesh. *Fishing Chimes*, Vol 22(9), p 34-35.
- Rao, G.R.M. 2000. Strategies for sustainable coastal aquaculture. Aquaculture Development in India Problems and Prospects. National Centre for Agricultural Economics and Policy Research, Workshop Proceedings. Ed. M. Krishnan and Pratap S. Birthal. p 74-78.
- Rao, A., Srinivasa, Krishnan, M. 2000. Coastal aquaculture issues of economics, equity and sustainability. *Fishing Chimes*. Vol.20 (5), p13-15.
- Schonhut, M. and U. Kievelitz, 1994. Participatory Learning Approaches– Rapid Rural Appraisal, Participatory Appraisal, an introductory guide. TZ-Verl-Ges., pp 183.
- Sen Neera, Dolanchampa Ghosh and K.R.Naskar, 1997. Socio-economic status of the rural population in the Indian Sundarbans and their role in the conservation of Mangroves, p. 267-272. In K.K. Vasds and M. Sinha (eds) Changing perspectives of Indian Fisheries. Proceedings of the National Seminar, March 16-17, 1997, Inland Fisheries Society of India, Barrack pore.
- Sinha, M. 1998. Wetlands and their implications on fisheries. *In* Meenbarta- Special issue on Wetland Department of Fisheries, Govt. of West Bengal pp 11-14.
- Srinath, K., Sridhar, M., Kartha, P.N.R., Mohanan, A.N. 2000. Group farming for sustainable aquaculture. Ocean & Coastal Management [Ocean Coast. Manage.], vol. 43, no. 7, pp 557-571.
- Townsley, P. 1993. Rapid Appraisal Methods for Coastal Communities. BOBP, Madras, pp 110.

4

FISHERIES IN WEST BENGAL, INDIA – PROBLEMS AND PROSPECTS

*C.S. Chakrabarti and K. C. Dora**

Abstract: Fisheries and aquaculture resources of West Bengal are extensive. Inland, brackish water and marine all types of fisheries are very important in the state of West Bengal possesses 2.76 hectares of ponds and tanks, 0.41 lakh ha of beels and boars, 0.16 lakh ha reservoirs, 1.72 lakh ha of rivers, 0.8 lakh ha of canals, 0.04 lakh ha of sewage fed fishery. Besides there are 2.1 lakh ha brackish water and 158 square kilometers productive coast line. Fish plays an important role in the nutrition, ecosystem, food security and livelihood system of the country. The fisheries sector provides direct support to 6 million fisher folk and indirectly provides almost the equal number of human being. Fishery sector contributes a substantial amount in National GDP. West Bengal has earned the reputation by producing 14.71 lakh tonnes of fishes (Inland and marine sectors together) in the year 2007-08. Still there is an annual deficit of about 6-7% which is approximately 259t/day in 2006-07. Although there is a big scope for the expansion of fishery sector in West Bengal but that is not happening up to the level of expectation for various reasons. The problems and prospects in Fisheries and Aquaculture in West Bengal would be discussed in the seminar.

Keywords: West Bengal, Aquaculture resources.

Introduction

Fish plays in important role in the nutrition, livelihood, ecology and health security of the country. The fisheries sector supports 6 million fisher folk directly and an equal number in ancillary activities. The sector contributes substantially to national Gross Domestic Product (GDP) and export earnings.

In West Bengal, the Fisheries and aquaculture is vast and varied. It possesses nearly three lakh hectares of ponds and tanks and approximately two lakh hectares of rivers, besides over two lakh hectares of brackish water fisheries. West Bengal also possess 158 square kilometers productive cost line-due to diversity of aquatic habitat, fishes adapted to cold water bodies of the northern

* West Bengal University of Animal and Fishery Sciences, 68, Kshudiram Bose Sarani, Kolkata- 700 037

Himalyan range to southern brakish water bodies of the Sunderbans, various types of fishes are available in West Bengal. Like the morphological variations among the fishes their genetic variations are equally interesting. There are 694 finfish species found in the water bodies of West Bengal out of the 2200 species available in India.

Although West Bengal is the highest producer of fish in the Country the State suffers a deficit of about 259 tones/day annually during the year 2006-2007 because more than 90% of the populations of West Bengal consume fishes in different forms.

Fisheries and aquaculture resources of West Bengal

Fisheries in West Bengal is broadly divided into inland and marine sectors. In the inland sector, about 30% of all India fish production comes from West Bengal. Though production from marine sector is going down to some extent, the Fisheries sector in West Bengal contributes about 3.50% SDP, which is next to only Agriculture. The table-1 shows the aquatic bodies available in West Bengal for Fisheries and aquaculture. The table-2 depicts the developmental activities in these sectors in West Bengal.

Table-1 Fisheries and aquaculture resources-West Bengal

Ponds & Tanks	2.76 lakh ha
Beels & Baors	0.41 lakh ha
Reservoirs	0.16 lakh ha
Rivers	1.72 lakh ha
Canals	0.80 lakh ha
Sewage Fishery	0.04 lakh ha
Brackish water Fishery	2.10 lakh ha
Coast Line	158 sq.km.

Table-2 Present status of development of fisheries in West Bengal (2007-08)

1	Total fish production (India)	6.57 million tones
2	Total fish production (West Bengal)	14.71 lakh tones
3	Export earnings from seafood	600 crores
4	Housing for fishermen	23,500 fishermen

5	Rural roads	2020 kms
6	Mechanized fishing vessels	4500 nos.
7	International Sea-food processing centre	1
8	Life and health insurance for fishermen	1.62 lakhs fishermen covered
9	Saving cum relief scheme	10,000 nos. benefited
10	Old age pension scheme	4500 fisher folks covered
11	Primary Fishermen Co-operative Societies	1053 nos.
12	Self Help Groups	6604 nos.
13	Average growth rate of fish production Inland	5% (Average for India is 4.29%)
14	Marine	0.11%

Fish consumption and requirement

According to ICMR, the minimum per capita requirement of fish for an average Indian is 11 kg per year. Which average per capita fish consumption year is 15.6 kg, in West Bengal, this is only 5 kg per capita/year for whole India.

Against India's total fish production of 6.57 million tones, West Bengal produced 14.71 lakh tones in 2007-08 and became highest fish producer in the country.

Though West Bengal in the highest producer of fish in the country, the State suffers an annual deficit of about 6-7%, which is in absolute tune of over 259 tones per day in 2006-07. (Table-3)

Table-3 Production figures (Inland and Marine sector combined) 2004-2008

1	2004-05	12.15 lakh tones
2	2005-06	12.50 lakh tones
3	2006-07	13.54 lakh tones
4	2007-08	14.71 lakh tones

Major constraints of development

- Gradual shrinkage water bodies due to urbanization and siltation, etc.
- Over exploitation of natural resources.
- Loss of natural habitat of fishes due to siltation caused by deforestation, multipurpose river valley projects, etc.

- Pollution due to industrial effluents, domestic sewage and agricultural wastes causing retarded fish growth, fish mortality and genetic disorders.
- Multi-ownership of water bodies in certain cases is a hindrance to development.
- Partial adoption or non-adoption of scientific methods of aquaculture and management of natural water bodies are some of the reasons for production less that optimum.
- Non-adoption or partial adoption of integrated farming system.
- Inadequate modern craft and gear for marine fisheries.
- Inadequate extension services including aqua input support.

Development for increasing production

The development for increasing fish production from 14.71 lakh tones in 2007-08 to 16.0 lakh tones during the 11th plan period in West Bengal would meet the immediate bare need of the people of the state.This can be achieved through focusing on Sustaining gains already made in the production technologies, Maintaining the productivity of the inland and marine fisheries, Increasing production through increased area of operation, New technological breakthrough, adding value to the produce, Ensuring quality, Creating increased livelihood avenues through remunerative and assured marketing opportunities, etc.

Recommendations

The entire developmental effort should be taken up not merely as a part of the assigned job but with right earnest and missionary zeal.

- The effort again should be in active participation with the stake holders.
- The co-operation of the people at large is also required which may be expected only through mass awareness.

Recommendations

For modern Fish Culture

Intensified pisciculture through adoption of modern polyculture or composite fish culture technologies with periodic multiple stocking, feed management, water replenishment, health monitoring and multiple harvesting, etc. should be ensured.

- Soil and water conditions should be periodically monitored.
- Intensification of aquaculture which should be ecologically sustainable.

Other viable avenues such as running water fish culture, cage culture in reservoirs and lakes are to be developed.

Organic fish farming should be given priority whereever conditions are suitable.

Stock improvement through diversification of species with locally available fishes of demand should be brought about.

Catfish and murrels culture should be propagated in seasonal water bodies.

Fish culture in small water bodies called 'Doba' should be taken up with hardy fishes like tilapia and other air breathing fishes.

Backyard cisterns should be used for tilapia culture for limited home consumption.

Integrated Fish Farming

Special emphasis for the extension of integrated fish farming especially poultry, duck, pig, dairy, paddy with horticulture, floriculture and seasonal vegetables on the embankment should be given.

Organic fish farming should be encouraged. Simultaneously utilization and treatment of organic waste including domestic sewage should also be encouraged enabling eco restoration and high production as well.

Fish preservation and product development

Recommendations

Modern preservation techniques should be popularized.

Post-harvest technologies should be adopted and popularized.

The traditional most popular method of beach drying of fishes should be compulsorily made hygienic, if necessary by adopting orders with strict enforcement.

Fish processing and export

Recommendation

Facilities available at the existing processing plants (except 5 units) do not meet the EU standards. Only few units meet national standards. Thus it is necessary to upgrade the standard of all factories.

Ice plant, cold storage, etc.

Recommendation

Adequate cold storage and ice plants at appropriate places/ locations should be encouraged all over the State.

Development of markets and marketing

Recommendations

For effective marketing, the basic requirement is adequate hygienic wholesale markets. The State Government should develop such infrastructure at the district level.

Retail markets are mostly in unhygienic condition. The State Government should convert them into hygienic ones and also establish modern markets accordingly.

The State Fisheries Department should ensure quality of fish in markets by introducing surveillance in retail markets through development of appropriate mechanism.

Fixing of minimum price of fishes both in wholesale and retail markets should be considered.

Marketing federations are required to equip themselves with market related information at per with 'e-chaupal' of some states of India. The Government should take initiative in this regard.

A cold chain should exist from producers to consumers for ensuring quality of fishes and thus the system may be ensured.

Fisheries development in West Bengal - A case study

With the sustained efforts, the fisheries development in the state has made considerable progress over the successive five year plan periods. The sixth five-year plan is the turningperiod in the development of fisheries in the state as the fish production has increased from 3.74 lakh MT during 1980-81 to 13.59 lakh MT in 2006-2007. In inland sector about 30% of all India fish production comes from West Bengal. Our success is attributable to the fact that the state tops in terms of per hectare productivity as well as fullest utilization of water bodies in the form of culture fisheries. Some major achievements are highlighted below.

Fish Production

The production of Fish during 2006-07 has reached 13.59 lakh MT of fish against the demand of 12.99 lakh MT. At the end of 11th plan the target for production of 15.58 lakh MT of fish has been fixed, which is expected to meet the domestic need.

Fish seed production

Our state is the pioneer in production of fish seed and contributes about 62% of the total production in the country. Fish seed production during 2006-07 has reached 13200 million whereas in the year 1981 the production was around 2300 million which records about 18% increase during the last 27 years.

Table 4 Year wise demand and production of fish in West Bengal from 1999-2000 to 2006-07

Year	Demand (lakh MT)	Production (lakh MT)	Deficit Or Surplus
1999-2000	11.15	10.450	(-)0.70
2000-2001	11.35	10.600	(-)0.75
2002-2003	11.68	11.200	(-)0.48
2003-2004	11.80	11.696	(-)0.104
2004-2005	12.09	12.150	(+)0.06
2005-2006	12.31	12.500	(+)0.19
2006-2007	12.99	13.590	(+)0.60

Source Annual Report, Govt. of West Bengal 2006-07.

Table- 5 State/Union Territory Wise Inland Fish Production

Sl. No.	State/Union Territory	2003-04		2004-05		2005-06	
		Inland Fish production	Percentag-e	Inland Fish production	Percentag-e	Inland fish production	Percentage
1.	Andhra Pradesh	648.70	19.00	642.32	18.25	672.25	17.901
2.	Arunachal Pradesh	2.65	0.08	2.70	0.08	2.754	0.073
3.	Assam	181.00	5.30	186.31	5.29	188.01	5.006
4.	Bihar	266.50	7.80	267.51	7.60	279.53	7.443
5.	Goa	3.60	0.11	4.23	0.12	4.04	0.108
6.	Gujarat	45.48	1.33	50.43	1.43	69.936	1.862
7.	Haryana	39.13	1.15	42.05	1.19	48.2	1.283
8.	Himachal Pradesh	6.56	0.19	7.81	0.22	7.295	0.194
9.	Jammu & Kashmir	19.75	0.58	19.90	0.57	19.15	0.510
10.	Karnataka	70.00	2.05	95.00	2.70	120.6	3.211

11.	Kerala	76.18	2.23	76.45	2.17	77.98	2.076
12.	Madhya Pradesh	50.81	1.49	62.06	1.76	61.077	1.626
13.	Maharastra	127.00	3.72	129.40	3.68	135.2	3.600
14.	Manipur	17.60	0.52	17.80	0.51	18.23	0.485
15.	Meghalaya	5.15	0.15	5.64	0.16	4.118	0.110
16.	Mizoram	3.38	0.10	3.68	0.10	3.75	0.10
17.	Nagaland	5.56	0.16	4.90	0.14	5.5	0.146
18.	Orissa	190.00	5.56	193.66	5.50	203.24	5.412
19.	Punjab	83.65	2.45	77.70	2.21	85.64	2.280
20.	Rajasthan	14.30	0.42	16.93	0.48	18.5	0.493
21.	Sikkim	0.14	-	0.14	0.00	0.015	0.000
22.	Tamilnadu	101.14	2.96	151.73	4.31	155.04	4.128
23.	Tripura	17.98	0.53	19.84	0.56	23.871	0.636
24.	Uttar Pradesh	254.00	7.44	277.00	7.87	289.58	7.711
25.	West Bengal	988.00	28.93	1035.60	29.42	1090	29.025
26.	Andaman & Nicobar Islands	0.09	-	NA	NA	0.043	0.001
27.	Chandigarh	0.08	-	0.08	0.00	0.087	0.002
28.	Dadra & Nagar Havelly	0.05	-	0.05	0.00	0.048	0.001
29.	Daman & Diu	0.00	-	0.00	0.00	0.069	0.002
30.	Delhi	2.10	0.06	1.41	0.04	0.7	0.019
31.	Lakshadweep	-	-	0.00	0.00	0	0.000
32.	Pondicherry	5.20	0.15	5.50	0.16	2.181	0.058
33.	Chhattisgarh	111.00	3.26	120.07	3.41	131.75	3.508
34.	Uttaranchal	2.56	0.07	2.57	0.07	2.789	0.074
35.	Jharkhand	75.38	2.21	NA	NA	34.27	0.913
Total		3414.72	100.00	3520.46	100.00	3755.4	100.001

Source: Extracted from the Agenda Note of TMC meeting (December 7-8, 20D6) for centrally sponsored scheme on Strengthening of database and information networking for the fisheries sector', DAHD, Ministry of Agriculture, GOI.

Brackish water Fish Farmers' Development Agency (BFDA)

The inland shrimp production has been raised from 28785 MT in 1990-91 to 80000 MT (approx.) at the end of the year 2006-07 under the efforts of B.F.D.A.

Table-6 Fish Seed Production in West Bengal (Unit in millions)

Year	Target	Achievement
2000- 2001	8850	8850
2001-2002	9000	9000
2002-2003	9100	9100
2003-2004	10000	11890
2004-2005	12500	12200
2005-2006	12500	12500
2006-2007	13000	13200

Source: Annual Report, Govt. of West Bengal 2006-07.

Inland Fisheries projects under Fish Farmers' Development Agency (FFDA)

During the year 2006-07, an area of 121373.37 hectare of water area has been covered under F.F.D.A. and the productivity has been raised from 600-800 kg/ha/annum to 3500-4000 kg/ha/annum. During 2006-07 a sum of Rs. 179.63 lakhs was spent on account of payment of subsidy and training cost to the farmers. In addition, an amount of Rs.706.5 lakhs was spent for payment of salary and wages to the employees and office expenses out of the state's non-plan budget. The following table illustrates the development of pond fisheries in West Bengal since 1996-97.

Marine Fisheries

About 10450 mechanized vessels are operating in the marine sector. The marine catch has been incredibly increased from 22000 MT in the year 1979-80 to 2.056 lakh MT. The construction work of fishing harbours at Sankarpur (stage-I & stage-II), Fregerganj, Sultanpur (Diamond Harbour), Hardwood Point near Kakdwip and Petuaghat along with 12 fish landing centers have been taken up and most of the schemes are on the verge of completion and the existing fishing fleets are utilizing this infrastructure in the specified area of Bay of Bengal.

Export

During 2006-07, 178000 MT of Marine fish and fish food products were exported whose export value was about Rs. 600.00 crore. It should be noted that in the year 1987-88 amount of foreign exchange earned by exporting marine products was only 50 crore. Within a span of less than 20 years it has been possible for this Department to increase export by 11 times in spite of so many limitations. The most important of which is that amongst the maritime states in our country West Bengal coastal region is very small, measuring approximately 158 km only. Still, West Bengal occupies the 4th position in the country in terms of export of seafood products.

Table-7 The development of pond fisheries in West Bengal since 1996-97.

Year	Pond Fishery (Cumulative in ha.)	Unit of Jhora Fishery (Cumulative one unit= 140 mt²)	No. of Fish Farmers trained (Cumulative)				
			SC	ST	Others	Total	Women out of Total
Upto 1997	101837.2	2576	50863	4954	129664	185481	4867
1997-98	103825.96	2691	52581	5090	132015	189686	5029
1998-99	105961.32	2841	53749	5195	133777	192781	5112
1999-00	107814.27	2983	58441	5454	136521	196816	5287
2000-01	110236.46	3102	55308	5504	7299	198111	5309
2001-02	112562.45	3273	56141	5563	139169	200873	5432
2002-03	114672.61	3416	56596	5586	139841	202023	5445
2003-04	116585.60	3538	57559	5631	142031	205221	5518
2004-05	118286.01	3692	57724	5651	142461	205836	5540
2005-06	119373.37	3810	57881	5657	142899	206437	5583
2006-07	121373.37	3960	58273	5677	143511	209742	5614

Source: Annual Report, Govt. of West Bengal 2006-07.

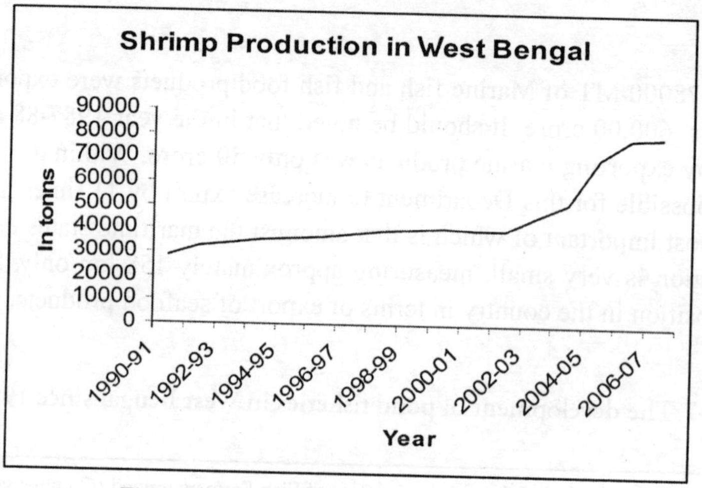

Fig.1 Shrimp production in West Bengal

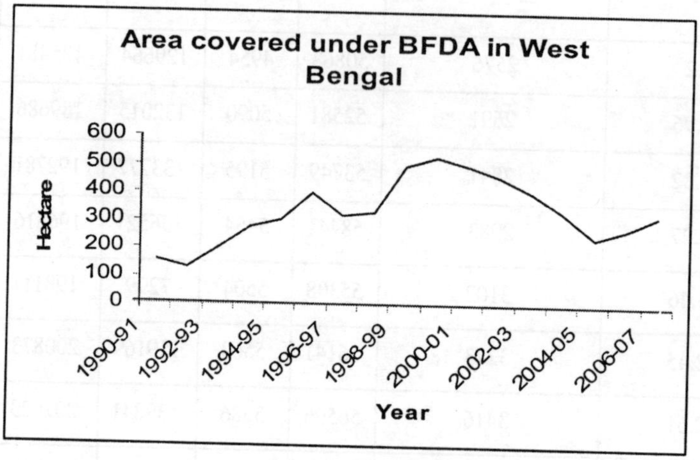

Fig. 2 Area covered under BFDA in West Bengal

Source: Annual Report, Govt. of West Bengal 2006-07.

Sewage fed fishery

It is a unique practice developed in the East Kolkata Wetland as low cost fish production in an eco-friendly manner coupled with recycling of organic waste. The department has also developed sewage fed fishery at Nabadwip and Sreerampur. Presently, an area of about 4083 hectares is being cultured yielding thereby approximately 21000 MT fish per annum.

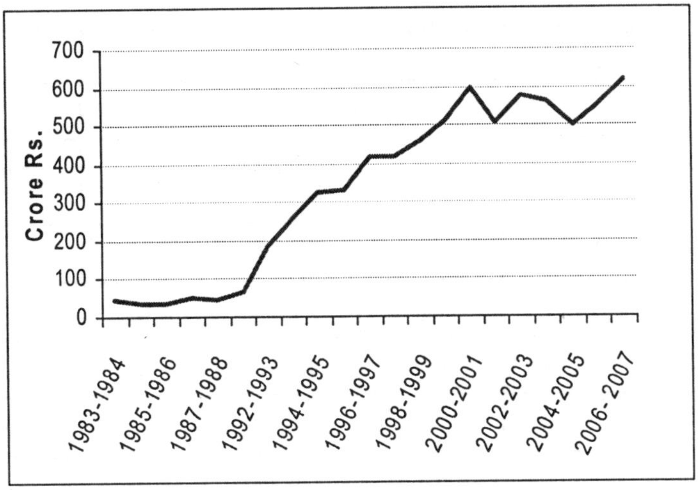

Fig.3 Year wise export of marine products through the ports of Kolkata and Haldia in terms value

Beel & Reservoir Fisheries

With the sustained efforts, productivity in the beels has been raised from 150-200 kg/ha/annum to 1200-1300 kg/ha/annum and in the reservoir from 50 kg/ha/annum to 800-1000 kg/ha/annum. Under NCDC assistance 5567 hectares of area have been developed.

River ranching

To supplement stock as well as to restore the breeding ground of indigenous carp, river ranching programme, a unique programme in the country has been taken up in the state. Some selected stretches of the river Bhagirathi and its tributaries are being ranched by liberating 20 lakh fingerlings every year.

Social Fishery

'Social Fishery' scheme was introduced in the year 1987-88 by the Fisheries Department aiming at stocking in the water bodies owned by the government department, public institutions etc by transferring management to group of persons selected by local panchayet. Under this programme more than 11352.13 ha of water areas have been developed up to 2004-05.

Tribal Sub Plan

For the socio economic upliftment of the tribal people Tribal Sub Plan (TSP) was formulated. The following schemes viz. "Economic upliftment of tribal people through operation of integrated Pisciculture Development", "Socio-economic upliftment of tribal people through operation of

Table-8 State/Union Territory Wise Marine Fish Production

Sl. No.	State/Union Territory	2003-04		2004-05		2005-06	
		Marine fish production	Percentage	Marine fish production	Percentage	Marine fish production	Percentage
1.	Andhra Pradesh	297.26	10.06	210.73	5.97	218.84	7.35
2.	Goa	90.00	3.04	91.61	2.60	100.91	3.39
3.	Gujarat	609.10	20.59	538.46	15.26	663.88	22.31
4.	Karnataka	187.00	6.32	171.23	4.85	176.97	5.95
5.	Kerala	608.53	20.57	601.86	17.06	558.91	18.78
6.	Maharastra	400.00	13.52	417.85	11.84	445.34	14.97
7.	Orissa	116.90	3.95	121.93	3.46	122.21	4.11
8.	Tamilnadu	373.00	12.61	307.69	8.72	307.99	10.35
9.	West Bengal	181.60	6.14	179.50	5.09	160	5.38
10.	Andaman & Nicobar Islands	31.06	1.05	177.65	5.04	12.053	0.41
11.	Daman & Diu	10.80	0.37	186.55	5.29	177.19	5.95
12.	Lakshadweep	10.00	0.34	207.91	5.89	11.964	0.40
13.	Pondichery	42.80	1.44	315.00	8.93	19.272	0.65
	Total	2958.05	100.00	3527.98	100.00	2975.5	100.00

Source: Extracted from the Agenda Note of TMC meeting (December 7-8, 2006) for centrally sponsored scheme on Strengthening of database and information networking for the fisheries sector', DAHD, Ministry of Agriculture, GOI.

Pisciculture development scheme in tribal areas by providing dwelling houses to the tribal people" are in operation.

Co-operative Sector

For the benefit of the fishermen community in the State a three tier system of constitution of co-operative societies from block level to state level has been developed in an organized manner. At

Table -9 Item wise Export of Marine products through the Ports of Kolkata
Q-Quantity in ton V-Value in Crore Rs.

Item Name		2003-04	2004-05	2005-06
Frozen Shrimp	Q	13970	12960	13600
	V	506.75	492.42	482.25
Frozen fish	Q	2644	2220	3624
	V	26.85	19.54	37.28
Frozen Cattle Fish	Q	558	1102	442
	V	4.26	8.64	3.46
Dried items	Q	233	416	208
	V	9.92	12.12	10.01
Live items	Q	359	440	435
	V	5.49	8.21	6.93
Chilled items	Q	408	418	456
	V	8.45	11.65	9.86
Others	Q	39	24	14
	V	0.13	0.02	1.41
	Q	18201	17580	18779
Grand Total	V	560.28	502.11	550.16

Source: Statistics of Marine Exports 2005, Govt. of India, MOC

the primary level there are 1053 Primary Fishermen's Co-operative Societies, 20 Central Fishermen's Cooperative Societies have been formed at the district level and at the state level there is an apex body for all such Fishermen's Co-operative Societies known as West Bengal State Fishermen's Co operative Federation Ltd. (BENFISH).

Special component Plan

Much emphasis has been given for poverty alleviation of Scheduled Caste (SC) people under Special Component Plan (SCP). A good number of potential need based schemes has been formulated and implemented for development of SC people. The following schemes viz. 'Grant and Managerial subsidy to Primary and Central Fishermen's Co-operative Societies', 'Group Personal Accident Insurance Scheme for Active Fishermen', 'Savings cum Relief scheme' and construction of 'Minor Fishing Harbours and Fish Landing Centers' etc. are in operation.

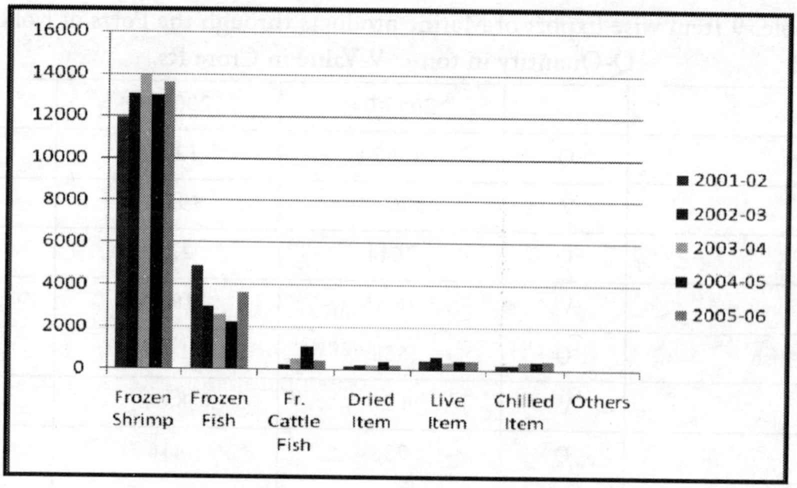

Fig. 4 Item wise Export of Marine products through the Ports of Kolkata

Source: Statistics of Marine Exports 2005, Govt. of India, MOC

Welfare Schemes for Fishermen

For the fishermen community following welfare schemes are being implemented

Housing Programme: This programme was taken up in the state for the benefit of poor fishermen for the first time in the country in the year 1984-85. Approximately, 23000 houses have so far been constructed for the fishermen under this scheme.

Pension scheme: The scheme started in the year 1991-92 to extent assistance to helpless old-aged fishermen. The Department has covered 4500 fishermen under the scheme during the year 2007-08.

Fishermen's Group Personal Accident Insurance Scheme: This programme under centrally sponsored schemes started in the year 1984-85 to provide financial assistance to the family members of deceased or permanently disabled fishermen. Financial assistance of Rs.50000/- in case of death and Rs. 25000/- in case of partial disablement due to accident is provided under the scheme.

Fishermen's Savings cum Relief Scheme: Under this scheme a fisherman deposits @ Rs.75/- for 8 months and equal matching grant is provided by the government. This accumulated amount of Rs.1200/- is given to the fisherman @ Rs.300/- per month during the lean period of 4 months.

Issue of Fishermen's Identity Card

Fishermen's Identity card has been issued to about 1, 84,974 fishermen.

Infrastructure Development

For increasing fish production as well as for upliftment of socio economic conditions of the fishermen community both inland and marine sector, emphasis is being given to develop the infrastructural facilities. Till now 5 Minor Fishing Harbours, 13 Fish Landing Centres, 4 Diesel Outlets, 2 Cold Chain Marketing Units, 1 Berge Jetty, 23 RCC Bridges and Culverts, 7 Boat Building Units, 24 Fishery Training Centres, 2 Auditoriums, 16 Flood Shelters, 78 Community Halls, 489 Tubewells,1854 km Village Road and 5 Fish Markets have been constructed.

Total number of mechanized and traditional boats operating in marine sector in West Bengal is about 5000 and 4800 respectively. In the marine sector out of 5000 boats about 1000 boats are owned by the members of co-operative societies. The construction of Minor Fishing Harbour at Harwood Point near Kakdwip is in progress. The project for construction of another Minor Fishing Harbour at Petuaghat in the district of Purba Medinipur started in the year 2004-05. Other infrastructure developmental schemes taken up by this department is given below

- Shore based wireless communication system.
- Potential Fishing Zone Forecast.
- Introduction of Integrated Fish Finder cum Navigational Guidance system (IFFNGS).
- Enforcement of the West Bengal Marine Fishing Regulation Act 1993 and Rules made there under.
- Setting up of Artificial Reef/ Fish Aggregation Device.
- Resource Development.
- Rural Electrification Programme.
- Distribution of fishing nets and fishery requisites.

Employment Generation

Fisheries sector has a significant role in employment generation, poverty alleviation and socio economic upliftment in the rural areas of the state. During the last 20 years since 1982 about 20 lakh fisher families and fish farmers have been directly benefited and about twice the number have been indirectly benefited from the various activities undertaken by the department. Implementation of the scheme under S.G.S.Y. has benefited many women Shelf-Help groups, comprising of women below poverty line to become economically independent. Formation of Fish Production Group and leasing out of vested water body to this group not only elevatated their socio-economic condition but also improve revenue earning from lease rent for the Government and Local Bodies.

Extension

The findings of research stations of this department for the development of fisheries and fish farmers involved are regularly reached to the target group through extension services viz. through meeting/ training etc. upto the panchayet level in the districts.

Training

So far about 210000 nos. of fishermen/ fish farmers and 10800 fisher women have been trained up on scientific pisciculture technology.

Research & Technology

The aim to meet up increased demand of fish and better utilization of limited water resources call for refinement of advanced technology to produce more fish for which Fresh Water Fisheries Research Station, Kulia, Kalyani in the district of Nadia and Microbiology and Parasitology Research Centre at Captain Bhery near Eastern Bypass have been working on various aspects in this direction. The Microbiology and Parasitology Research Centre has been set up with special emphasis for rendering assistance towards control of fish diseases. District level laboratories have been organized in all districts for extending technical help to the pisciculturists (Source of data Annual Report 2006-07, Department of Fisheries, Government of West Bengal).

References

- Advances in Fish Processing Technology- D. P. Sen. Published by Allied Publishers Private Limited, New Delhi, India, 2005.
- All India Seminar on Sustainable Aquaculture for Augmentation of Export with Special Reference to Environment, Engineering and Value Addition, Kolkata, India, September 03-04, 2004.
- Annual Report, Department of Fisheries, Government of West Bengal, India, 2008.
- Fish and Fisheries- B. N. Yadav. Published by Daya Publishing House, Delhi, India, 1999.
- Fish for All National Launch, Kolkata, India, December 18-19, 2003.
- Fish Stock Assessment Manual, FAO 393, 2007.
- State Level Workshop on NABARD Scheme for All Fisheries Officials in West Bengal, 20th March, 2007.
- Status of Centre-State Coordination in Agricultural Research Education and Extension, In Region II West Bengal, Assam, March, 2000.

5

LIVE FEED – THE LIFE LINE OF ORNAMENTAL FISH INDUSTRIES IN WEST BENGAL, INDIA: A LIVELIHOOD STUDY

Madhumita Mukherjee, Shivendra Singh And Sandip Kumar Mondal*

Abstract : A large number of youths and housewives have been engaged themselves in the trade of wild collection of live feed from the sewage for their livelihood specially in the periurban area of Kolkata such as Howrah, Hooghly, North & South 24 parganas and Nadia, to meet the huge demand of live feed especially in ornamental fisheries sector. A study was performed with an objective to study the status of the livelihood of these people, their collection procedure, preprocessing and the hazards involved. The study revealed that the net daily income for a group of 2-3 persons is about 200 to 300 rupees. They suffer some health problems due to their involvement in this business. There is a scope tc develop a backyard hatchery where mass culture may be made using the target species being isolated from the wild collection. Owing to high demand of live feed in ornamental fish market, isolation & mass culture as well as organization concept may focus an avenue for income generation and livelihood development.

Key Words: Live feed, Backyard hatchery, Isolation

Introduction

A variety of live feed consisting of different species of algae, rotifers, copepods, nauplii of brine shrimps, molluscan larvae, cladocerans, polychaetes, nematods etc. have been used individually and in combination with artificial diet for larval rearing in many fish and shrimp hatcheries. Live feed is extremely essential for rearing the spawns and the brooders in ornamental fishery sector. They are extremely important for health and colour formation due to their contents of protein and carotenoid pigments. Nowadays, the ornamental fish producers are being appreciating the importance of live feed in aquarium culture system due to the flourish of commercial and scientific culture of

* Aquatic Resource Health Management Center, Pailan, P.O. Pailan Hat, P.S. Bishnupur, Kolkata- 700104, E-mail:arhmc_pailan@yahoo.co.in, madmita_mukh@yahoo.co.in

ornamental fish. Live feed are not very readily available in the market, mostly they are procured from the natural sources like creeks, ponds, canals and rivers by hand nets during early morning hours when Moina and Daphnia are available. Natural abundance of live feeds mainly in the sewage canals contribute to the wealth of our fishery resources. In the periurban area of Kolkata such as Howrah, Hoogly, North & South 24 parganas, Nadia, a large number of youths and housewives have been engaged themselves in the trade of wild collection of live feed for their livelihood. Approx 7000 poor people are engaged in collection of sludge worm from urban sewage channels & drain and sell their collection to the ornamental and aquarium feed suppliers & seller. As the people are realizing necessity of live feed for the aquarium fish industry there is a rising marketing scope for the live feed producers. In West Bengal, many people both man & women are engaged in live feed collection from natural resources.

The main objective of this work was to study the status of the livelihood of the people engaged in wild collection of live feed, collection process and preprocessing and its hazards. The survey of livelihood study was conducted in the adjoining areas of South 24 parganas and Howrah, West Bengal, India. The reason behind selecting this two district is that these two districts hold the leading position in ornamental fish farming in West Bengal as well as the whole country. Besides, these two districts of West Bengal rank the first in India in the live feed collection from natural sources.

Methodology

The field survey data were collected from the people engaged in wild collection of live feed from the Sewage Canal of different locations of the districts of Howrah and South 24 parganas, West Bengal. The collection spots were Sewage Canal of Hooghly River & Budge Budge (Chodia), sewage canals from Khidirpur to Kalighat and other large sewage drain located at Santoshpur (Akra) and Alipur. Studies on different aspects of livelihood were made like place of collection, collection period and time, procedure of collection and pre-processing, numbers of person involved in collection, sex and age group of the collectors, cost involved in collection, marketing place and market price of the products, net earning (income) per head and the common hazards faced by the collectors during the process. The collected samples were analysed and types of live feed were recorded. The target species of the live feed collected from local water bodies were identified and the quantitative and qualitative analyses of the collected samples were made. Different types of microalgae collected from the natural water bodies were washed with the filtered water. 1 ml. water sample was taken and watched under the microscope, the number of target organism was then counted and isolated by serial dilution and streak plate mainly based on the technique reported by Gopinathanan, 1996. The stock culture were maintained in the laboratory in sterilized 15 ml culture tubes and 100 ml conical flasks plugged with cotton containing 'Conway' or Walne's media. (Gopinathanan, 1993) maintained in 25° C temperature and 12L/12D irradiation (1000 lux illumination). Cladocerans (*Moina micrura* and *Ceriodaphnia cornuta*) were also cultured in the laboratory by feeding with microalgae at regular interval. The ovigerous females of the copepod

(Cyclopoid, *Oithona* sp.) and harpacticoid (*Euterpina acutifrons*) were also cultured in the laboratory in 10 ml vilas and 100 ml beakers and fed twice a week with micro-algae. The larvae of non biting midges of the family Chironomidae (Blood worms) were processed by the new freeze dry technique-removal of moisture under vacuum at sub zero temperatures. Flash frozen natural fish foods in blister packs were also developed with mosquito larvae, daphnia, moina, cyclopes, rotifers (brachionus) micro worms and diatoms.

Result

Collections are being made early in the morning and in the time depending on low tide of Hooghly river from different Sewage Canal connected to the Hoogly river and other large sewage drains in the different areas of South 24 Parganas and Howrah district. The collected samples were composed of Microalgae, Water Fleas/Daphnia, Tubifex, Moina, Mosquito larvae, Earthworm, *Paramecium caudatum,* Cyclops, Rotifers, Nematodes Infusorians, Whiteworms and Bloodworms under the family Chironomidae. (PLATE 1) The collections are being done throughout the year but maximum collection recorded in the rainy season. Live feeds are collected by mosquito net. (PLATE 2) The collected macro live feed are segregated and being pre-processed by cleaning thoroughly by repeated washing with fresh water (PLATE 3). They are supplied to Ornamental fish shop, Ornamental fish Culture units and local markets. It has also been reflected that in most cases male individuals are involved for such wild collection but in some cases women and child are also involved. It has been revealed from the study that the initial cost involved was initially about Rs. 2000/- to 2500/- towards the cost of Bicycle, Mosquito Net, Plastic Bucket, Plastic Bag and Plastic sheet. The amounts of collection were recorded from 5 to 8 Kgs per day and the net income was Rs. 200 to 300/- per day per individual. The initial cost involved was initially about Rs. 2000/- to 2500/- towards the cost of Bicycle, Mosquito Net, Plastic Bucket, Plastic Bag and Plastic sheet. The amounts of collection were recorded from 5 to 8 Kgs per day and the net income was Rs. 200 to 300/- per day per individual.

During the study, the problems and hazards faced by the collectors recorded were Skin disease, Injury of limbs, troublesome movement of long distance for planktons due to its scarcity, frequent fluctuation of market price, lack of organization concept and standard price limit. (Table-1). *Paramoecium* sp. (infusorions) and microworm (*Anguilula silusae*) are being isolated from the wild collection of live feed and cultured in the laboratory of Aquatic Resource Health Mangement Centre, Pailan *Paramoecium* sp. and microworm is a good source of nutrients for ornamental fish larvae. Beer are being used as a food medium for the microworm culture. Microworms were harvested wiping a stiff brush in the mass of microworms moving along the walls of the cultivation pots. The other species of zooplankton cultured here are *Daphnia* sp. and *Moina* species.

The Cladocerans and ovigerous females of the copepod cultured in the laboratory provided good result. Flash frozen natural fish foods in blister packs containing mosquito larvae, daphnia, moina, cyclopes, rotifers (brachionus) micro worms and diatoms were also developed in the laboratory (PLATE 4).

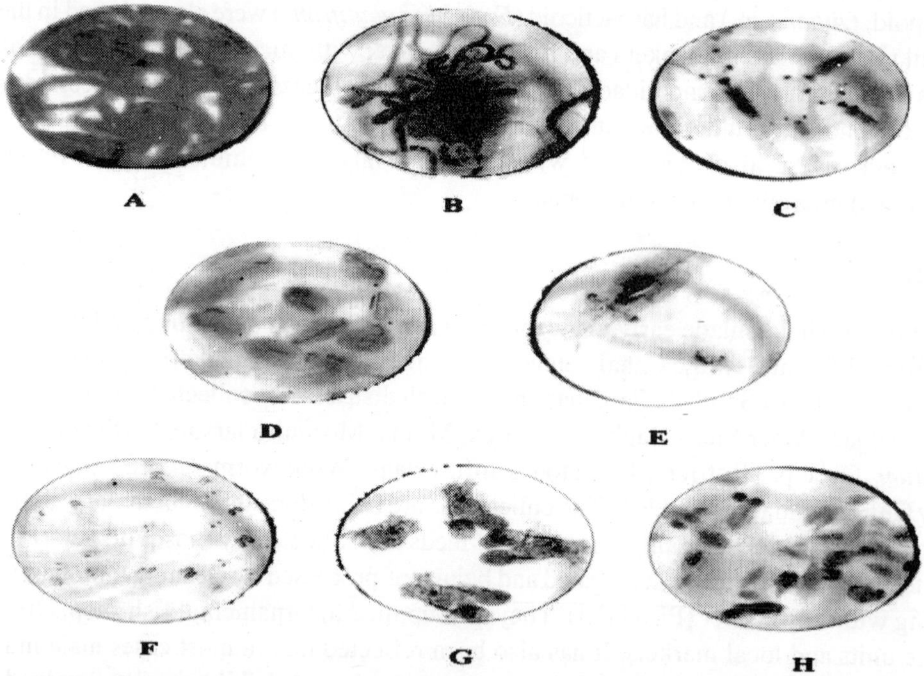

PLATE 1. Collected Live Feed (A. Red mosquito larvae; B. Tubifex; C. Mysis; D. Red water fleas; E. White shrimps; F. White mosquito larvae; G. Rotifer; H. Cyclops)

PLATE 2. Collection procedure of sludge worm and earthworm from wild source

PLATE 3. Washing and cleaning of collected sludge worm

PLATE 4. Flash frozen natural fish food in blister packs (Practiced in many countries)

Table-1 Observation during livelihood study for the people engaged in wild collection of live feed

Sl No.	Aspects of Study	Observation
1.	Place of collection	Sewage Canal of Hooghly River & Budge Budge (Chodia), sewage canals from Khidirpur to Kalighat and other large sewage drain located at Santoshpur (Akra) and Alipur.
2.	Collection Period	Throughout the year Low tide), but maximum collection recorded in the rainy season
3.	Collection Time	Depend on low tide of Hooghly river.
4.	Quantity of collection	Generally 5 to 8 kg per day, prepared from 35 to 40 kg of raw material.
5.	Market price	Rs. 30 to 35 per kg (But cost of increased during Summer season)
6.	Collection Procedure	Live feed are collected from by mosquito net.
7.	Processing procedure	By frequent washing.
8.	Marketing place	Ornamental fish shop, Ornamental fish Culture, Local Market (Galif street market, Howrah market).
9.	Cost involved in collection	First involve in collection initially it is about Rs. 2000-2500/- (Bicycle, Mosquito Net, Plastic Bucket, Plastic Bag, Plastic sheet etc.)
10.	Net earning (income) per group of 2-3 persons.	200 to 300/- Rupees per day.
11.	Nos. of person involved in collection	Near about 300 person in each location
12.	Sex and age group of the collectors	Male and Female collectors (20 to 45 ages), Child (8-10 years), Male contributed approx 60%, Female-25% and Child-15%
13.	Problems if any	Generally they faced the problems of Skin disease Injury of limbs Troublesome movement of long distance for planktons due to its scarcity Frequent fluctuation of market price Lack of organization concept and standard price limit.

Discussion

Live feed is the natural feed for fishes-very essential for proper growth of fish and up-keeping quality fish production- particularly normal color of fishes. The emergence ornamental fish culture has been creating more opportunity to the unemployed youths and women's of the society. Similarly with the growth of aquarium fish market it is also focusing the need of continuous and steady supply of live feed. This may create a new avenue of an ancillary aquarium industry. The floating microscopic plants or phytoplankton are the micro algae, which form the basic food for most of the animals in aquatic eco-system. Plankton is of immense value as food and plays an important role in the disposal of sewage and in the natural purification of polluted waters. However, some plankton form a harmful bloom that may cause high mortality among the aquatic organisms and pose a serious hazard in the water supply for domestic and industrial use. Bloodworms are a special diet for all tropical fish, gold fish, small fish, and fry. Rotifers are truly planktonic, mostly free swimming and also solitary. Majority of them inhabit freshwater bodies (e.g., *B. calyciforus* and *B. rubens*), while some are found in marine and estuarine ecosystems (e.g. *B. plicatilis*). Rotifers are most suitable live foods for early larval (zoeal) stages of crustaceans, due to their small size (0.1-1.0 mm long) slow swimming habit and nutritional value. Rotifers are referred to as 'biocapsules' in hatcheries since they contain and transfer macronutrients, micronutrients and also vitamins. Hino, A., 1993, studied the successful use of rotifer, *Brachionus calyciforus* in larval rearing of some freshwater ornamental fishes, The application of freshwater rotifers in larval rearing of fish has not been fully exploited and its application is limited to only some of the freshwater food fish species like Gudgeon and perch (Lubzens, 1987, Snell and Carrillo 1984). The cladocerans (both neonates and adults) are also used as live food in culturing fish fry (Alikuni *et al.* 1955; Ivleva, 1975; Chen, 1979) and penaeid larvae (Muthu, 1983). The high densities of Chironomids and Oligochaetes were reported to be found in the sewage fed bheries of east Kolkata wetlands specially near the inlet and in the middle. The growth rate of fishes can only be ascertained after the food and feeding habits of stocked fishes are studied. (Mukherjee and Begum 2002) The use of *Moina micrura*, in the culture of *Cyprinus carpio* juveniles were examined and significant higher production was obtained by Jana and Chakraborty 1993. As regards tropical aquarium fishes Shim Kum Fah (1988) and Rani Mary Jacob (1995) have pointed out the value of *Moina* sp. as an important live food item. The food organism used during the larval period are vitally important in determining the success behind ornamental fish industry and it is dependent on the availability of live organism. (Rahaman & Mukherjee 2005). Cultivation of live feed like Paramoecium and micro microworms is necessary to keep the business running for the aquaculture farms if natural food resources are absent or due to climatical conditions become unpredictable. Aquatic Resource Health Management Centre, the state level laboratory under the Department of Fisheries, West Bengal, maintains the stock of different live feed like *Chlorella* sp., *Phormedium tenuii, P. val, Anabaena, Ulva, Enteromorpha, Microworm, Paramoecium,* Blood worm etc. The technique for mass culture has been translated to different fish farmers by different training programmes and workshops.

A major problem encountered by the aquaculture industries is the non-availability of the right kind of food, especially the live food for rearing larvae and juveniles of finfish and shellfish under controlled systems. Ornamental fish farming is mainly dependent on the availability of live organisms of suitable size for feeding different stages of their larvae. In early life stages, different aquarium fishes are zooplankton-feeders, i.e. they prey on small free-living planktonic animals. As no artificial larval diet can at present totally fulfill their nutritional requirements, their successful rearing still depends on an adequate supply of high quality live feeds, usually in the form of Daphnia, Brachionus, Moina and brine shrimp (*Artemia* sp.). It has been reported that Blood worm breeding are in practice in Vietnam specially in shrimp ponds. The study indicated that a large number of people are engaged for wild collection of live feed but it is quite insufficient as per vast demand for the aqua industry. There is an ample scope to solve the problems with a view to organization concept and better livelihood for poverty alleviation. Availability of native strains will be a boon to aquaculture as this will help in increasing the quality and bringing down the cost. Moreover there is a need for developing live feed culture units involving native strains for both stock and mass culture for the supply of proper nutritional supplement for the scientific culture systems.

The potential natural feed can easily be brought into flash frozen natural fish food in blister packs of Bloodworms, Mosquito larvae, Daphnia, Moina, Cyclopes, Rotifers (brachionus) Microwarms, Diatoms etc. It can be processed by the new freeze dry technique and removal of moisture under vacuum at sub zero temperatures. This process preserves the excellent nutritional factors of live bloodworms and also preserves their flavor just like real worms without all the fuss and bother. In the process of blister packing the maintenance of temperature is an important factor. This industry will ultimately open a new avenue of employment generation.

In order to meet the huge demand of live feed especially in ornamental fisheries sector, there is a scope to develop a Backyard hatchery where mass culture may be made using the target species being isolated from the wild collection. It may give a new avenue of income generation and livelihood development for the poor people specially the live feed collectors from the wild source. It needs to be standardized and popularized the scientific technique for isolation and mass culture of the live feed species. The National Fisheries Development Board (NFDB), a body under Government of India, introduced different subsidy oriented schemes. There is a need to formulate and initiate a model project by NFDB on different aspects of live feed for its future translation for popularization.

References

- Gopinathan, C.P. 1993, Micro-algal culture, Live Feed, Hand Book on Aqua Farming 7, published by The Marine Products Export Development Authority, Kochi, India, 61 pages.
- Hino, A., 1993. Present culture systems of the rotifer *Brachionus plicatilis* and the function of micro-organisms. In C. S. Lee, M. S. Su. and I. C. Liao (eds). Finfish Hatchery in Asia Proc. Finfish Hatchery in Asia'91. TML Conf. Proc. 351-59, Tungkang Marine Lab.,Taiwan Fisheries Research Inst., Tungkang, Pintung, Taiwan.

- Jana,B.B and Chakraborty, R.1993. The effect of management protocols for juvenile carp (*Cyprinus carpio)* culture on live history responses of a zooplankton food source, *Moina* sp. (Kurz). Aquaculture, 110; 285-300.
- Lubzens, L., 1987. Raising rotifers for use in aquaculture. Hydrobiologia **147** 245-255
- Rahaman, M. & Mukherjee, M., 2005. Live feed culture a new path for Aquarist, Coloured fish in Water Garden, published by Marine Products Export Development Authority & Department of Fisheries, Aquaculture, Aquatic Resources and Fishing Harbour, Govt of West Bengal, pp 176-200.
- Mukherjee, M. & Begum, S., 2002. Macrobenthic faunal diversity in sewagefed bheries of east Kolkata wetlands, Madhumita Mukherjee & Kunal Chattopadhyay (ed) pp 128-138
- Snell, T. W., K. Carrillo, 1984. Body size variation among strains of the rotifer Brachionus plicatilis. Aquaculture **37** 359-367.

6

WATER QUALITY MANAGEMENT PRACTICES IN FRESHWATER CULTURE AND HATCHERIES OPERATION

*Chandra Prakash, S.N.Dwivedi * and K.D. Raju*

Abstract: The quality of water plays a very important role to ascertain the quantum of aquatic productivity in different aquatic ecosystems. Higher productivity oriented results need the essential water quality parameters under desirable ranges. This operational management practice is a three tier system; analysis of water quality parameters, knowledge of their favorable ranges and the chemical and mechanical control techniques to make the water quality parameters conducive for culture and hatchery practices. In the present study, the molecular weights, atomic weights, combining weights and valences of principal chemical compounds, elements and radicals are discussed coupled with their applicability for control measures of undesirable limits of water quality parameters. Further, the ideal values of various parameters are discussed in relation to fresh, brackish and sea water. In the last tier of chemical and mechanical control techniques, the various control interventions like liming, coagulation, oxidation, chlorination, ozonation coupled with mechanical, air lift, biological and chemical filters and their application technique are mentioned.

Keywords: Productivity, Coagulation, Sea water

Introduction

The importance of water quality control is two fold, first in the context of burgeoning aquatic pollution problem and secondly with reference to intensive rearing and culture per unit volume of water. Good water quality provides better growth and survival rates to culturable organisms. The water quality management is a three tier system comprising the analysis of various water parameters, to have the knowledge of the favourable ranges of these parameters and thirdly the chemical treatment and control techniques involved to bring the parameters under desirable limits. Aquaculture is

Central Institute of Fisheries Education (Deemed University), Versova, Andheri(W), Mumbai-400 061; * E-1/106, Arera Colony, Bhopal-462016 practiced in fresh, brackish and marine waters. Therefore, based on the nature and quantum of solids available in water medium, they can be classified as follows.

Freshwater, i.e., such water in which concentration of dissolved solutes is lesser than 0.5 ppt. Freshwater can be further sub-divided into two categories.

(a) Soft water, among ionic composition the dominances is of carbonate and bicarbonate ions.

Hard water, among ionic composition the dominances is of sulphate ions. Saline water, i.e., such water in which concentration of dissolved solutes is more than 0.5 ppt. Saline water can be further sub-divided into three categories:

Brackishwater, salinity range is 0.5 ppt to 30 ppt.

Seawater, salinity is 30 ppt. to 37 ppt.

Metahaline water or brine, salinity is >37 ppt.

Acidic water, it relates to such water, which is dominated by the ionic concentration of iron, aluminium and sulphur.

Wastewater or polluted water, there are two categories under this type.

Sewage, it is the combined liquid wastes discharged from domestic and industrial sources within a given area. Effluent, it refers to industrial wastewater.

First Tier–The Water Analysis

Chemical analysis of water plays a significant role in aquaculture. It helps to diagnose the constraints in the aquatic system and to explore management options for optimum economic returns. At this stage of starting the analytical work, the points to be taken in to consideration, are laboratory safety, proper sampling and appropriate storage of the samples collected and the requisite chemicals and solutions. To prepare the solutions of known strength, the normality, atomic weight, molecular weight, equivalent (combining) weight and valency of common elements, radicals and compounds should be known. These are shown in the table 1 and 2.

Second Tier–The Favourable Ranges

Proper water quality management has become a watch word for achieving maximum production in aqua farming. After completion of analysis of desirable parameters, the next step is the knowledge of ideal values or favourable ranges of these water parameters.

Third Tier–Chemical Treatment and Mechanical Control

The last step involved in the water quality management practice is the chemical treatment and mechanical control to bring the parameters under suitable limits. The important techniques are described as follows.

Table 1 Molecular weights and combining weights of principal compounds involved in water treatment

Substance	Formula	Molecular weight	Combining weight	Applicable to control
Aluminum hydroxide	$Al(OH)_3$	78	26	Turbidity, Alkalinity and Phosphate
Aluminum sulphate	$Al_2\ (SO^4)_3$	342	57	
Calcium bicarbonate	$Ca(HCO_3)_2$	162	81	Acidity
Calcium carbonate	CaO_3	100	50	Fluoride
Calcium hydroxide	$Ca(OH)_2$	74	37	Phosphorous and Pathogens
Calcium oxide	CaO	56	28	
Calcium sulphate	$CaSO_4$	136	68	pH 4.0-6.5 and Above 9.5
Ferric chloride	$FeCl_3$	162.5	54.2	
Ferric hydroxide	$Fe(OH)_3$	106.84	35.61	The zone for turbidity & oil
Ferric oxide	Fe_2O_3	159.70	26.61	pH 4.0-10.0
Ferric sulphate	$Fe_2(SO_4)_3$	400	66.6	pH 9.5 and above
Ferrous oxide	FeO	71.84	35.92	Iron and sulphate
Ferrous sulphate	$FeSO_4$	152	76	
Hydrochloric acid	HCl	36.5	36.5	Heavy metals and acidity
Magnesium bicarbonate	$Mg(HCO_3)_2$	146	73	
Magnesium carbonate	$MgCO_3$	84	42	
Magnesium hydroxide	$Mg(OH)_2$	58	29	
Magnesium sulphate	$MgSO_4$	120	60	
Sodium bicarbonate	$NaHCO_3$	84	84	Hardness, Pathogens Acidity and oil
Sodium carbonate	$NaCO_3$	106	53	
Sodium chloride	NaCl	58.5	58.5	
Sodium hydroxide	NaOH	40	40	
Sodium silicate	Na_2SiO_3	122	61	
Sodium sulphate	Na_2SO_4	142	71	
Sulphuric acid	H_2SO_4	98	49	Iron, Chlorides and alkalinity

N.B. When ferrous salts are to be used, addition of lime is necessary for hydrolysis of ferrous ions. Ferrous salts do react with natural alkalinity, but it is a delayed reaction. Therefore, lime is to be added first.

Table 2 Atomic weights, valencies and combining weights of elements and radicals involved in water treatment

Element/ Radical	Atomic symbol	Atomic weight	Valency	Combining weight
Aluminum	Cl	26.98	+3	8.99
Calcium	Ca	40.08	+2	20.04
Carbon	C	12.01	+4	3.00
Chloride	Cl	35.46	-1	35.46
Copper	Cu	63.54	+2	31.77
Hydrogen	H	1.008	+1	1.008
Iodine	I	126.91	-1	126.9
Iron	Fe	55.85	+3	18.22
Magnesium	Mg	24.32	+2	12.16
Nitrogen	N	14.01	Several	-
Manganese	Mn	54.93	Several	-
Oxygen	O	16.00	-2	8.00
Potassium	K	39.10	+1	39.10
Sodium	Na	23.00	+1	23.00
Sulphur	S	32.06	Several	-
RADICALS				
Bicarbonate	(HCO_3)	-	-1	61.01
Carbonate	(CO_2)	-	-2	30.00
Hydroxyl	(NO_3)	-	-1	17.00
Nitrate	(NO_2)	-	-1	62.01
Nitrite	(NO_2)	-	-1	46.01
Phosphate	(PO_4)	-	-3	31.66
Silicate	(SiO_2)	-	-2	38.05
Sulphate	(SO_4)	-	-2	48.03

A. Chemical Treatment

Liming

Coagulation

Oxidation

Chlorination

Ozonation

B. Mechanical control

Mechanical filter

Air-lift filter

Bio-filter

Chemical filtration

i. Liming

Lime is frequently applied in aquaculture practices to improve water quality and acidity of soils. There is a close relationship between the base saturation of bottom soils and total hardness of water. Total hardness of waters exceeded 20 mg/1 when bottom muds were 80% or more base saturated. Fertilization of ponds seldom produces good plankton growth if either the total hardness or total alkalinity is lesser than 20 mg/l. Some pond waters which contain more than 20 mg/1 total hardness or total alkalinity, but are dark coloured because of the presence of humic substances, may also need lime for other important following factors:

Lime creates a buffer system to prevent marked diurnal changes of the water from acidic to alkaline conditions.

- It neutralizes the soil acidity.
- It precipitates colloidal matters such as clay suspended in water.
- It kills pathogens and promotes bacterial breakdown of organic matter.

It helps phosphorus fixation in the pond soil and water, essential for the assimilation of nitrogen into cellular matter.

Table 3 Ideal values of various physico-chemical parameters for aquaculture in three types of water bodies

Sl.No.	Parameter	Fresh water	Brackish water	Seawater
1.	Colour (colour units)	Clear water with greenish hue <100 colour units	Clear water with greenish hue <100 colour units	Clear water with greenish hue <100 colour units
2.	Transparency (cm)	20-35	26-35	26-35
3.	Clay Turbidity (mgl^{-1})	<30	<30	<30
4.	Solids (mgl^{-1})			
	a. Total	<500	>500	>500
	b. Suspended	30-200	25-200	25-200
5.	Temperature (^{O}C)			
	a. Tropical Climate	25-32	25-32	25-32
	b. Temperate climate	10-12	10-12	10-12
6.	pH	7.5-9.5	8.0 -8.7	8.0-8.5
7.	Hardness (mgl^{-1})	50-280	>50	>50
8.	Alkalinity (mgl^{-1})	50-300	>50	>50
9.	Chlorides (mgl^{-1})	31-70	>500	>500
10.	Salinity (ppt)	<0.5	10-25	>30
11.	Dissolved Oxygen (mgl^{-1})	5-10	5-10	5-10
12	Total dissolved free carbon dioxide (mgl^{-1})	<3	<3	<3
13.	Ammonia Nitrogen (NH_3^- $Nmgl^{-1}$)			
	a. Unionized (NH_3)	0-0.1	0.0-1	0-0.1
	b. Ionized (NH_4^+)	0-1.0	0-1.0	0-1.0
14.	Nitrite Nitrogen (NO_2N mgl^{-1})	0-0.5	0-0.5	0-0.5
15.	Nitrate Nitrogen (NO_3^-N mgl^{-1})	0-1.3	0-1.3	0-1.3
16.	Total Nitrogen (mgl^{-1})	0.5-4.5	0.5-4.5	0.5-4.5
17.	Total Phosphorous (mgl^{-1})	0.05-0.40	0.05-0.5	0.05-0.5
18.	Potassium (mgl^{-1})	0.5-10	>0.5	>0.5
19.	Calcium (mgl^{-1})	75-150	>75	>75
20.	Magnesium (mgl^{-1})	20-200	200-1350	>1350
21.	Sulphate (mgl^{-1})	20-200	200-885	>885
22.	Silica (mgl^{-1})	4-16	>5	>5
23.	Iron (mgl^{-1})	0.01-0.3	0.01-0.3	0.01-0.3
24.	Manganese (mgl^{-1})	0.001-0.002	0.002-0.02	0.002-0.02
25.	Zinc (mgl^{-1})	0.002-0.01	0.002-0.01	0.002-0.01
26.	Copper (mgl^{-1})	0.003-0.005	0.003-0.005	0.003-0.005
27.	Cobalt (mgl^{-1})	<0.003	<0.003	<0.003
28.	Biochemical oxygen demand (B.O.D.) (mgl^{-1})	<10	<15	<15
29.	Chemical oxygen demand	<50	<70	<70

30.	Hydrogen sulphide (mgl^{-1})	<0.002	<0.003	<0.003
31.	Residual chlorine (mgl^{-1})	<0.003	<0.003	<0.003
32.	Primary productivity (mg C/m^3/day)	1000-3000	1000-2500	1000-2500
33.	Plankton (ml/100 litre)	2	1	1
34.	Chlorophyll – a (μgl^{-1})	20-275	20-250	20-250
35.	Redox – potential (volts)	0.40-0.52	0.40-0.52	0.40-0.52
POND BOTTOM SEDIMENT				
36.	Carbon Nitrogen ratio (C N)	101	101	101
37.	Organic carbon	0.5-2.75	0.5-3.00	0.5-3.00
38.	Total nitrogen ($mg100g^{-1}$)	>50	>50	>50
39.	Phosphorus ($mg100g^{-1}$)	>6	>6	>6
40.	Potassium ($mg100g^{-1}$)	5-10	5-10	5-10

$$FeSO_4.7H_2O+Ca(OH)_2 = Fe(OH)_2+CaSO_4+7H_2O$$

It supplies calcium needed for bone formation in fish and for plankton growth.

Table 4 Kinds of lime

Chemical formula	Chemical name	Common name	Relative neutralizing value (%)
$CaCO_3$	Calcium carbonate	Lime stone	100
CaO	Calcium oxide	Quick lime	150 - 175
$Ca(OH)_2$	Calcium hydroxide	Hydrated or slaked lime	120 - 135

Application of lime dose (Fig. 1)

Alkalinity or hardness more than 20 mg/l needs lime @ 5 mg/l.

Generally lime dose should not exceed more than 10 mg/l as beyond this limit, it may act as overdose.

Coagulation

Raw water with 30 mg/l turbidity can be treated by filtration without any pre-treatment. If turbidity exceeds more than 30 mg/l that too if it has become undesirable for specific culturable organism, the alum can be added as per jar test method observing out the optimum dose per litre of sample water. Commercial alum containing combination of $A1_2(SO_4)_3$ $18H_2O+A_2(SO_4)_3$ $9H_2O$ is used as a

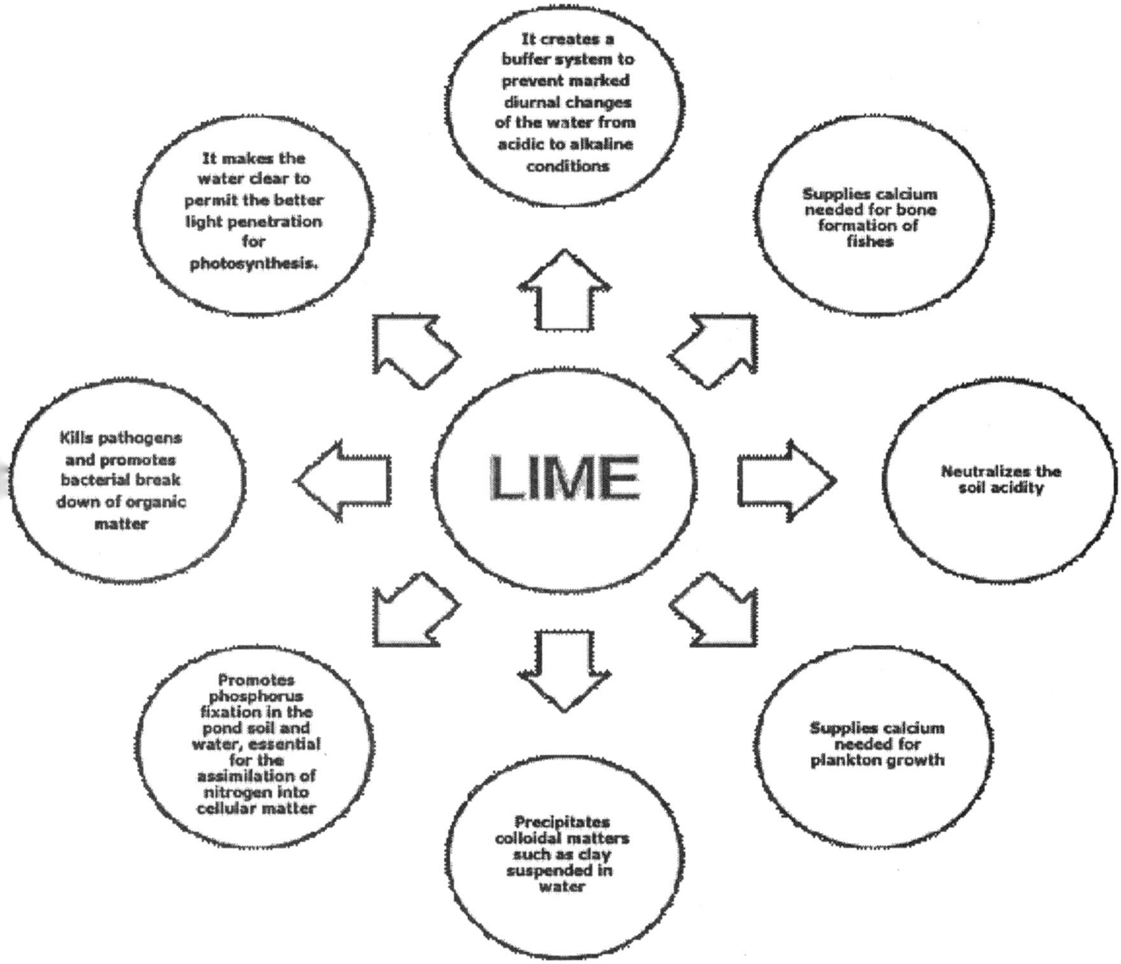

Role of lime application in aquaculture

oagulant. Alum reacts with natural alkalinity and produces a precipitate of aluminium hydroxide long with suspended or colloidal impurities due to agglomeration. So, by the application of oagulant alum, suspended impurities can be controlled.

1. $A1_2(SO_4)_3\ 9H_2O + 3Ca\ (HCO_3)_2 = 2A\ I(OH)_3 + 3CaSO_4 + 9H_2O + 6CO_2$
2. $A1_2(SO_4)_3\ 9H_2O + 3MgCO_3 = 2A\ I(OH)_3 + 3MgSO_4 + 6H_2O + 3CO_2$

)xygenation / Aeration

)xygen is the most important factor in fish production. A minimum acceptable level is considered) be 5 ppm. At levels below 4 ppm, fish may thrive but will not feed and grow normally. Low xygen levels are frequently met during the summer in fish ponds with heavy algal blooms. Typically, e fish will be found dead or under severe stress at dawn when the oxygen levels may be considerably depleted or approach zero. Low concentrations of potassium permanganate (2 to 4 ppm) can be used either as algaecides or as an oxidising agent to enhance the oxygen levels under emergency conditions. It is known that 1 mg of $KMnO_4$ can generate 4 mg of oxygen after getting dissolved in water. Air is almost having 20% oxygen by volume.

Chlorination

Sodium hypochlorite solution (NaOCl) is easily available in glass bottles, 5 gallon carboys and 30 gallon drums and is generally known as javelle water, chlorine bleach or bleach liquor. The quantities of hypochlorite required for chlorinating known volumes of water at the required levels vary in case of different samples. The materials best suitable for handling hypochlorite solution are rubber, ceramics, glass and plastics. The exact quantities of sodium hypochlorite to be taken in case of samples of different concentrations to chlorinate known volumes of water and dosages of chlorine required in the treatment of water are Chlorination and de-chlorination practices can be to the tank contents in the absence of cultureable organisms. Water in bulk quantity can be chlorinated with bleaching powder ($CaOCl_2$), which contains approximately 33% available chlorine. Dechlorination can be made with sodium thiosulphate ($Na_2S_2O_3.5H_2O$). One ppm de-chlorination needs 7 ppm sodium thiosulphate.

Ozonation (Fig. 4)

In the aquaculture, a lot of organic wastes is introduced into the water which cannot be removed by normal filtering or by settling tanks. The ozonisation of this water effectively not only eliminates these organic components by neutralization and oxidation but also kills bacteria in the circulating water. Ozone concentrations of 0.6 - 1.0 mg/l and contact time of 1-3 minutes are sufficient to kill most of the pathogens in aquaculture systems. Safe permissible exposure of ozone to aquaculture 0.002 mg/l.

The merits of ozonation in aquaculture systems are as follows:

- Prevents occurrence of water borne diseases.
- Increases survival in packing and transport of fish and prawn seed.
- Reduces turbidity created by organic matter.
- Maintains optimum pH and dissolved oxygen.
- Converts toxic nitrite into non-toxic nitrate through oxidation as nitrite is toxic to fishes in concentrations as low as 0.5 ppm.
- Neutralizes obnoxious gases like ammonia, hydrogen sulphide, methane and carbon dioxide etc.
- Favours increased growth of plankton.
- Reduces BOD and COD loads to trace level.

The Mechanical Control

The process of removing undesirable impurities from a fluid by passing it through a porous material is called filtration. It consists of passing the water through a thick layer of sand and gravel which act as strainer. Suspended and colloidal matter in the water and also a large number of bacteria are caught in the interstices of the sand during its passage. Central Institute of Fisheries Education, Mumbai has designed and developed different types of filters for management of water quality in aquaculture practices. Filtration devices also hold good for recirculatory system of aquaculture, particularly at places where there is shortage of water. In Israel for example, where there is scarcity of water, recirculatory system is the most widely practiced device of aquaculture.

Types of Filters

Mechanical Filter

It is the physical separation of concentration of suspended particulate matters from circulating water. It is accomplished by passing the water through suitable substrate that trap the particles. The trapped materials are then removed by various methods depending upon the type and design of the filter. Gravel reduces turbidity in water by trapping particulate matter and removing it from suspension. This is accomplished in two ways. First, suspended matter is physically trapped in the interstices among gravel grains. Second, the electrostatically charged surfaces of the gravel grains attract oppositely charged particles or colloids and remove them from solution. Rough and angular gravels are best for mechanical filtration. Their numerous surfaces increase the electrostatic potential of the filter bed.

The mechanical filtering efficiency of gravel increases with decreasing grain size of the individual granule. Smaller granules have more surfaces exposed to the water for electrostatic attraction of particulate matter. The smaller interstices facilitate removal of finer particles. This results in a greater percentage of suspended matter removed per volume of filtered water. Even distribution of each grade of gravel is very important when initially filling a sand filter. This assures well distributed circulation through the gravel column. No gravel bed should be lesser than 6 cm no matter how small the culture system is. Systems lesser than 50 m^3 rarely need gravel more than one meter deep if they are well designed and managed. These filters are applicable on large-scale water demand, e.g., in carp hatcheries and in aquafarms more particularly of slow sand mechanical filtration system.

Bio-filter (Fig.2)

Contrary to other devices of filtrations, bio-filter is a device in which filtration is effected by involving micro-organisms. It restores the quality of culture water by mineralization of organic nitrogenous compounds, nitrification and denitrification by bacteria suspended in the water and attached to the gravel in the filter-bed. Bio-filter should be kept away from sunlight to avoid algal

Fig.1 Application of lime in pond

Fig.2 Biofilter in aquaculture

Fig.3 Air-lift filter in aquaculture

growth, which may affects its working efficiency. Types of bio-filters used in Aquaculture are tricking, submerged, updraft and rotating disc filters. Heterotrophic and autotrophic bacteria are the major groups present in aquaculture systems. Heterotrophic species utilize organic nitrogenous compounds excreted by the animals as energy sources and convert them into simple compounds, such as ammonia.

The mineralization of these organics is the first stage in biological filtration. It is accomplished in two steps, ammonification, which is the chemical breakdown of proteins and nucleic acids producing amino acids, and organic nitrogenous base and deamination in which a portion of organic and some of the products of ammonification are converted to inorganic compounds. Once organics have been mineralized by heterotrophs, biological filtration shifts to the second stage, which is nitrification, it is the biological oxidation of ammonia to nitrite and nitrite to nitrate by autotrophic bacteria. These organisms, unlike heterotrophs, require an inorganic substrate as energy source and utilize carbon dioxide as their only source of carbon. *Nitrosomonas* sp. and *Nitrobacter* sp. are the principal nitrifying bacteria in aquaculture systems. *Nitrosomonas* oxidizes ammonia to nitrite whereas *Nitrobacter* oxidizes nitrite to nitrate. The third and last stage in biological filtration is denitrification. This process is a biological reduction of nitrate or nitrite to either nitrous oxide or free nitrogen. Denitrification can apparently be carried out by both heterotrophic and autotrophic bacteria. Bio-filter removes dissolved impurities upto a limited extent and they are more applicable in intensive fish spawn and fry and prawn larval rearing systems.

Airlift filter (Fig.3)

It is the most trouble free appliance of filter. In this device, water is strained through a synthetic sponge layer by pumping the water with airlift. In small aquaculture tanks, e.g, aquaria, lift pipe extends below water level and filter chamber rests above the water surface. When air is supplied in the vertical pipe of the filter, the air-mixed culture water comes up and gets filtered through the filter chamber. The suspended impurities up to the size of 0.002 mm can be filtered out through this device. By pumping 5 cm^3 air/sec., 2 liters of water per minute can be filtered when the diameter of the lift pipe is 1 cm. The airlift filters are applicable on small-scale water demand, such as in aquaria and glass jars etc. By doubling the diameter of lift pipe the water flow will get increased by 5.6 times more.

Chemical filtration

It is the removal of substances (primarily dissolved organics but also nitrogen and phosphorus compounds) from solution on a molecular level by absorption on a porous substrate, or by direct chemical fractionation or oxidation. The removal of dissolved organics by absorption is done by activated carbon or charcoal, which is a porous substance, and its degree of efficiency or absorptive capacity is measured by the total surface within the pores that are available to chemically attract organic molecules. In aquaculture, it is usually the granular type that is used. One kg of a high grade powdered variety may contain several million square meter of surface area, but granular types are both chemically efficient and easy to handle. After the removal of colloidal and suspended solids, the soluble refractory organics may be efficiently removed by contact with activated carbon granules. Such carbon absorbs up to 20-30 percent of its own weight in mixed organics from wastewater. It is better to place glass wool on top of the carbon to reduce colloidal surface coating of the granules. Granular activated carbon is best kept in a separate container or contactor. Scattering

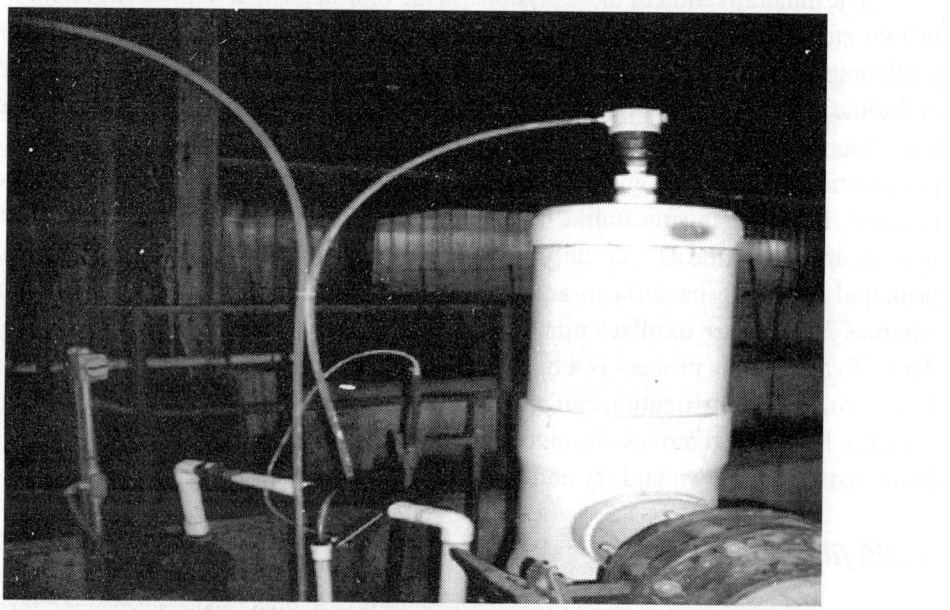

Fig. 4 Ozone application in aquaculture

it on the surface of a filter bed is impractical because once the material is saturated; it must be separated from the gravel grains and removed. Chemical filtration can make almost complete removal of dissolved impurities and they are more applicable in prawn hatcheries, as they need very clean water.

Fig. 5 Village pond for aquaculture

Precautions

When the resistance in the filter (due to clogging), i.e., head loss, is equal to the total depth of water on the filter, the operation will stop. The loss of head should not be greater than depth of filtering sand; when it becomes excessive and before a negative head is formed, the filter should be cleaned. Further, whenever any of the following conditions appear, filter should be cleaned. Formation of heavy surface mats, the slimy, gelatinous filter skin. Reduced flow rate. Dissolved oxygen below saturation in the filter effluent.

References

- APHA, 2005. Standard methods for the examination of water and wastewater, 21st edition, APHA-AWWA-AEPA, Washington D.C.
- Banerjea, S.M., 1967. Water quality and soil condition of fish ponds in some states of India in relation to fish production, *Indian J. Fish.* 14(1&2) 115-144.

- Boyd, C.E., 1978. Water quality in warm water fish ponds. Agril.Expt. Stn. Auburn, Univ, USA. p 359.
- Boyd, C.E., 1984. Water quality in worm water fish ponds, *Auburn University*, Alabama, USA, 34 pp.
- Boyd, C.E., 1990. Water quality in ponds for aquaculture, Alabama Agricultural Experiment Station Auburn University, Auburn, USA.
- Boyd, C.E., and Tucker, C.S., 1998. Pond aquaculture water quality Management, Kluwer Academi Publishers, London , pp.187.
- CIFE, 1997. Manual on recent advances in management of water quality parameters in aquaculture CIFE, Mumbai.
- CIFE, 1997. Manual on hatchery and grow-out technology of giant freshwater prawn, CIFE, Mumba
- Dwivedi, S.N. and Parkash. C., 1985. Lime application in aquaculture. Training manual on fres water farm management, CIFE, Mumbai.
- Egna, H.S., and Boyd, C.E., 1996. Dynamics of pond aquaculture, CRC Press, New York.
- Lin, C.K and David, R., 1997. *In* Dynamics of pond aquaculture. (ed. Egna, H.S. and Boyd, *C.E.).CR Press, New York.* 73-107 pp.
- Garg, S.K., 1984. Sewage and waste disposal engineering, Khanna Publishers, Delhi.
- Garg, S.K.,1985. Hydrology and water resources engineering, Khanna Publishers, Delhi.
- Garg, S.K., 1986. Water supply engineering, Khanna Publishers, Delhi.
- Prakash, C; Sinha, P.S.R.K and Reddy, A.K. 1990. Economic viability of aquaculture in sewage, *J Environ. Biol* 11(1).
- Prakash, C; Reddy, A.K. and Thakur, N.K. 1996. Water quality manipulation in aquaculture. *In* Proceedings of the National seminar on water quality issues in aquaculture systems (ed. Santhanam R., Ramdhas, V. and Gopalakrishnan, P.), Tamilnadu Veterinary and Animal Sciences University Tuticorin.
- Tebbutt, T.H.Y. 1983. Principles of water quality control, University of Birmingham, UK.

7

THE STATE OF ART OF FISHERIES OF UTTARAKHAND STATE, INDIA

*H.S.Raina and R.K.Langer**

Abstract: About 800 million people – 20 percent of the world's population – are food insecure, lacking economic and physical access to the food required to lead healthy and productive lives. Much of the poverty, food insecurity and malnutrition that exist in developing countries occur in rural areas with limited and unreliable rainfall and fragile soils. These areas often referred to as 'low productive areas' because current agricultural productivity tends to be low, due to much degradation and deforestation. Failure to address the problems effectively in the low-potential areas themselves will accelerate degradation of natural resources and increased poverty, food insecurity and at the same time out-migration. In areas of low-productivity but significant agricultural potential, public policy and public sector investment should promote sustainable use of existing natural resources to enhance the productivity of agriculture, fisheries and other rural enterprises. The interaction between agricultural and non-agricultural rural activities should be explicitly considered in efforts to accelerate growth in rural areas. Dr. Norman Borlaug the Nobel Orient, had observed that "The only way for agriculture to keep pace with population and alleviate hunger is to increase the intensity of production in those ecosystems that lend themselves to sustainable intensification, while decreasing the intensity of production in the more fragile ecosystems". This observation, if we look from the angle of prevalent practices of hill farming and land-use patterns, it is found that due to environmental sensitivity of the region; this concept is very much relevant to hills. With regard to crop diversification, fishery can play an important role provided proper policy initiatives are taken at respective levels. The Agriculture alone cannot be sustainable on the long run, it needs to be integrated with other commodities; therefore, concept of Integrated Agriculture and Aquaculture (IAA) needs to be seriously considered.

Keywords: Malnutrition, Crop diversification, Initiatives

* Shiv Ram Enclave, Post office Sidhra, Sidhra –Jammu 180 019 (JK); * Project Directorate for Farming Systems Research, Modipuram – Meerut 250 110 (U.P.)

1. State of Uttarakhand

The Uttarakhand was constituted as 27th State of the Indian Republic on November 9th 2000.The state is spread over a geographical area of 53,204 sq.km. in the altitudinal range from 198 m to 7,816 m above the sea level in the Central Himalaya. It is diverse in its geological, geographical, climatic, biotic and social characters. It receives the highest rainfall ranging from 245 cm to 260 mm and records low temperatures in majority of regions. Nature has endowed this State with wealth of natural flora, fauna and minerals. The aquatic resources are present in the form of mighty rivers, serpentine rivulets-streams, beautiful lakes and reservoirs. Many hill stations of the state are world famous for their pristine environment, landscape, salubrious climate and mighty mahseer since British rule. But the mountainous character and rugged terrain and adversity of climate have stood in the way of the development and prosperity of this beautiful region. It is very poorly developed state, less accessible, thinly populated, nearly urbanized and physically isolated region in India. But the continuous efforts of the state and the central government, the hilly areas are getting the impetus for the economic development for which some infrastructural facilities are being provided and some new central places/service centers are being developed to generate and to sustain the economic growth and for diffusion of such growth to the remote rural area.

2. Physiography

The Uttarakhand is carved out from 13 northern districts of the erstwhile hilly region of Uttar Pradesh state. This is spread between the Latitudes 28° 47′ to 31° 20′ N and Longitudes 77° 35′ to 80° 55′ E in the Central Himalayas. Out of its total geographical area of above 51,122 sq.km. 88.9 % area is mountainous with population of 48,35,912 as per 1981 Census. The length and breadth of the state is 320 km. and 250 km., respectively. The state comprises four longitudinal physiographic sub-divisions Trans Himalayan domain; Great Himalayas; Lesser Himalayas and Tarai-Bhabar – Siwalik ranges. The highest mountain peak in the territory - Nanda Devi (7,816 m asl) is situated in the Trans Himalayan Zone. The southern Tarai belt (198 m asl) is located in the lowest altitudes in the state of Uttarakhand. The region embraces a great geographical diversity stretching from "Tarai-Bhabar and Don' belt in the south to Trans Himalayan range in the north in between Siwalik, Lesser and Greater Himalayas ranges. This narrow foothill strip of Tarai-Bhabar and Dun make a remarkable dent on its overall physical and economic status. Except Tarai belt, the proportion of area irrigated in hilly parts does not exceed 15 %. Further, augmentation of land under cultivation is restricted due to physical constraints as well as economic and social cost involved in it. Whatever land is available for cultivation is mostly rainfed as the percentage of net irrigated area to net sown area in 1980-81 was recorded only 8.6. The pressure of population in Hilly regions of Uttarakhand is exorbitantly high. The increasing trend in population growth at an annual rate of above 2.72% after 1970 has surpassed the limits of available natural resource base. Obviously, the people are encroaching upon forestland and steep slopes for bringing extra land under cultivation to meet food supply, thereby causing a great damage to the entire ecosystem of the region. Under these conditions agriculture is bound to be a subsistence activity for most of the people in hilly regions of this state.

3. Climate

The altitudinal and geographical variation, mountain slopes, expansion of river valleys and vegetation cover results in varying climates in different parts of the Uttarakhand State. It is hot and humid (tropical) in the Tarai-Bhabar and in the river valleys below 600 m asl. At higher elevations the climate becomes sub-tropical upto 1,200 m asl. It has Cool temperate upto 1,800 m and cold temperate upto 2,500 m. at still higher altitudes the climate is extreme cold. While, the vegetation in the state may be categorized in four groups, tropical, mountain (sub-tropical), temperate and typical alpine.

4. Natural Resource Base

4.1 Forest Cover

As per the recent study the state has only 28.7 % of the region is under forest with only 4.4% of the area has a forest with greater than 60% crown cover. The high variation in the altitude (250-3,090 m) has influenced the distribution of large number of plants. The Deciduous forests of *Shorea robusta* (Sal), *Embica officinalis* (Anola), *Termanalia tomentosa* (Sain), *T.belerica* (Behar) and *T.chebula* (Marr) etc. are confined to lower valley area (200-300m) while miscellaneous deciduous forest of *S.robusta* (Sal), *Cedrela toona* (Toon) are confined to the lower valley areas (800-1,400 m). The *Rhadodendron arboreum* (Burns), *Myrica sapida* (Kaphal), *Pyrus pashia* (Mehal), *Pinus roxburghii* (Pine) from the subtropical Himalayan forests between 1,200 and 2,100 m. In the middle elevations the mixed forest of *Quercus leucotrichophora* (Banj) and *Rhododendron* and pine are quite common. At the higher elevations *Quercus* and *Abies pindrow* (silver fir) are quite common and form the moist temperate forests in the region. The oak forest shows grow due to excess precipitation while in pine areas rockslides and landslides are common.

4.2 Aquatic Resources

Uttarakhand is blessed with innumerable aquatic resources both lotic (rivers, rivulets and streams) and lentic waters (freshwater lakes and reservoirs). The region on the whole harbours four major river systems; the Yamuna, the Ganga, the Kali and the Ramganga-Kosi systems. A number of glaciers are present in the state laying above 3,600 m asl. The important glaciers are Gangotri, Yamunotri, Pindari, Kafini, Sunderdhunga, Nakuri, Baldhunga, Poling, and Balati etc. The region-wise major resources in this state are described as under.

5. Garhwal Region

The Garhwal Himalayas encompass seven districts of Uttarakhand, *viz.* Dehradan, Tehari, Pauri, Chamoli, Uttarkashi, Rudraprayag, Hardwar lying between latitudes 29° 26′ – 31° 28′ N and longitudes 77° 49′ – 80° 6′ E. It has an area of about 30090 sq. km, spreading 220 km in the east west and 235 km in the north south. On the whole, the region constitutes more than a half (59%) of Uttarakhand state and about 7.01% of the entire Himalaya. The Tons, a tributary of the Yaumna separates it from Himachal Pradesh in the west and the district boundaries of Nainital, Almora and Pithoragarh separate it from Kumaon in the east and starting from the south, the region extends through the snow clad peaks of the Himadri, making the Indo-Tibetan border northwards. Highest snow covered ranges are located in the northern part of the region while the altitude gradually decreases following southward. The region has some very high peaks of which Gangotri, Yamurati, Nadadevi (7,817 m), Nandakot (6,861 m) and Trisul (7,120 m) are important. These high peaks of the region are covered with glaciers and form part of the Great Himalayas. Barring a narrow strip of terai–bhabar and duns, Garhwal is largely a rugged mountainous country where the altitude dramatically fluctuates between 300 m (Shivaliks) to 7,817 m (Nanda Devi Peak). This abrupt altitudinal variation has obviously resulted in a complex but interesting diversity in topography, climate, flora and fauna. This land has the privilege of being the source region of the two mighty river systems- the Ganga and the Yamuna and superb natural surroundings The super Himalayan resorts of Kedarneth, Badrinath, Gangotri, Yamunorti and Hemkund along with their numerous satellites are witnessing a spontaneous flow of pilgrims every season and number is increasing every year. The aquatic resources holding coldwater fishes of the region may be grouped as under:

5.1 Riverine Resources

The main water resources of Garhwal Himalaya are the rivers and their tributaries. Snowmelt waters feed most of these rivers while a few have their catchments in the high ridged without snow. Main rivers like Bhagirathi, Alaknanda, Yamuna, Bhilanganga, Pinder, Nandakini, Mandakini are perennial resources of water from ice melt, while Nayar and Ranganga originating from the highest reaches have either no snow or receive little snow only during winter season. The total length of important rivers in these districts of Garhwal Himalaya is estimated to be 831 km. Table -1 depicts the district-wise important riverine resources.

5.2 Lacustrine Resources

There are numerous (about 37) high altitude natural lakes in Chamoli, Uttarkhasi, Pauri and Rudraprayag districts of Uttaranchal state having a total area of approx. 100 ha. They are small in area ranging from 0.12 ha to 4.0 ha. The Chamoli district has maximum number of lakes as high, 25 out of 37 recorded in the Garhwal region. These are typical coldwater lakes located above the elevation of 2100 m asl are shown in Table-2.

Table - 1 The Riverine systems holding coldwater fisheries in Garhwal region.

istrict	Major River System	Total length (Km.)
hamoli	Alaknanda, Birhi, Nandakini, Mandakini, Vishnuganga, Amritganga, Madmeheshwari and Heylang	102
Jttarkashi	Bhagirathi, Asiganga, Yamuna, Tons, Gadugad, Peber, Rupin and Supin.	220
'ehri	Bhagirathi, Bhilangna and Alaknanda	215
'auri	Ranganga, East Nayar, West Nayar, Nayar, Alaknanda, and Ganga.	115
)ehrudan	Yamuna, Tons, Ganga, Asan, Song, Suswa and Bediagad	17

3 Reservoir Resources

xcept Tehri dam presently, there is no reservoir in the region, but a number of hydro-electric ojects are under construction where reservoirs of various dimensions will be created. The prominent nong them is one at Maneri Uttarkashi District. The reservoir of 46.0 sq.km area was created by hri Dam which is the Asia's biggest dam, the resultant impoundment will is ideal for fisheries velopment.

4 Ponds

n account of typical mountainous topography, the pond resources in the districts as per records e very limited. Village ponds of different dimensions having an approx. area of 100 ha are confined ly in the Terai region of Dehradan which are used, both for irrigation and fisheries. In the hilly gion concrete ponds and hilly tanks are available. There is possibility of common carp culture in se water bodies. In the plain of Haridwar, Udhamsingh Nagar and Dehradun districts, there is ssibility to grow to major carps culture in these ponds. The coldwater fishes are not present in se ponds, particularly in highland zones.

Kumaon Region

e Kumaon region is blessed with scenic beauty and varied natural resources. Lying between the itudes of 28° 44′ and 30° 49′ N and longitudes 78° 45′ and 81° 01′ E, Kumaon is situated at the junction of Nepal, Tibet and India. It constitutes a distinct geographical entity of great strategic nificance and is spread over 23,000 sq.km area. The altitudinal range varies from 200 m to 7436 asl. The Kumaon Division comprises five districts, out of these –Nainital, Almora, Bageswar, am Singh Nagar are very important. The altitudinal variations and the slope aspects have given e to varying climates in different parts of Kumaon. It is hot and moist in sub-mountain region below 400 m (21.0 to 40.0°C at Tarai and Bhabar belts) and cold temperate in highlands, above 2500 m (6.0 to 20.0°C for higher ranges).

The southwest monsoon, the cyclones and the local thunderstorms are the main source precipitation in Kumaon region. Nearly 80% of the total precipitation is attributed to the southwest monsoon, while 15% is from time to time cyclones and others through local thunderstorms. The winter precipitation, associated with cyclones, is in the form of rainfall at lower reaches and snowfall on higher elevations. The summer precipitation in this region of Uttarakhand is usually associated

Table -2 Main high altitude lakes in Garhwal region

District	Lakes	Altitude (m asl)
Chamoli	Panya Tal	4880
	Vasundhara Tal	4800
	Roop Kund	4760
	Kashni Tal	4680
	Nandi Kund	4450
	Dev Tal	4400
	Hemkund	4329
	Gandhi Sarovar	3970
	Braham Tal	3280
	Bhekl Tal	3010
	Gaurikund	3000
	Nag Tal	2200
Tehri	Shahastra Tal	4600
	Tamba Kund	3440
Uttarkashi	Kedar Tal	3970
	Dodi Tal	3255
	Masar Tal	2890
	Rishi Tal	1880
Pauri	Tara Tal	2000
	Airoli Tal	1800
Rudraprayag	Bhenti Tal	-

with local thunderstorms caused by atmospheric instability owing to strong diurnal temperature variations. The areas above 4,000 m receive perpetual snow. The notable glaciers of this region are Pindari, Kaphini and Sundardhunga. The melting of snow and ice from the glaciers significantly contributes to the water of the rivers emerging in the Great Himalayan domain i.e. the Pindar, Eastern Ramganga, Gori, Dhauli and Kali.

6.1 Riverine Resources

The watershed drainage systems of Kumaon are mainly – the Alaknanda, the Western Ramganga, the Kali, the Kosi and Gaula. Both in the volume of water discharge and nature of course length, the Kali is the largest river in the region, it forms a natural boundary between Nepal and India in Kumaon region. The catchments of the Kali and the Ramganga rivers draining the whole Kumaon extend between those of the Karnali and the Ganga. The Alkananda river system includes the drainage areas of Girthi-Kisogad in the north and the Pindar in the northwest. Details of different rivers in this region is given on Tables 3 & 4.

Table-3 Principal river systems and their tributaries in Kumaon

River system	Major tributaries
The Kali	Dhauli, Kali, Gori, Saryu, Ramganga (East), Gomti, Panar, Ladhiya
The Ramganga-Kosi	Ramganga (west), Kosi, Fenka, Dabka, Baur, Nandhau

6.2 Lacustriane Resources

6.2.1 Lakes

A number of natural freshwater lakes constitute very valuable resources for recreation and fisheries activities. Lakes Nainital, Bhimtal, Naukuchiatal, Khurpatal, Sattal, Gurualtal are situated at an altitude ranging from 1290 to 1937 m asl in the latitude of 29° N within a 25 km radius of the town Nainital. All lakes are small in size, with the largest one Bhimtal, covering 72 ha. Lake Naukuchiatal is the deepest, with a maximum depth of 41.0 m. The tectonic movements along faults and the diversified lithology are reported to be responsible for their origin. Among the natural lakes outside the Nainital district, the Tarag Tal in the valley of Western Ramganga, the Shyamala Tal near Sukhidhang in Champawat district, the Jolingkong group of lakes in the Kuti valley and the Anchari Tal in the Dhauli valley are also important. These lakes have originated as a result of damming of rivers by landslide debris.

In Kumaon, the growing demands of local residents to meet the challenge of increasing tourism and other developing programmes have increased pressure on all most all the lakes particularly on lake Nainital. This lake has been witnessing a dramatic increase in recreation.

Table-4 Major Kumaon Rivers and their Tributaries

River -Master Trunk	Origin River	Stream Length(Km.)
1. Kali	Fluvial (glacial)	220
Tributaries		
Gori	Glacial	100
Dhauli	Glacial	170
Kuti	Glacial	45
Master stream		
Saryu	Fluvial (glacial)	120
Tributaries		
Ramganga	Fluvial (glacial)	85
Tributaries		
Mehar Gad	Fluvial (Erosion)	35
Bhujpatri, Kalpani	Fluvial (Erosion)	32
Gomati	Fluvial (Erosion)	100
Saryu	Fluvial (Erosion)	48
Jaingan, Godi, Panar	Fluvial (Erosion)	87
2. Ganga	Glacial	95
Tributaries		
Pindar, Girthi	Fluvial (Erosion)	95
Kosi		150
Master streams		
Suyal, Simi Gad etc.	Fluvial (Erosion)	94
West Ganga	Fluvial (Erosion)	95
Master streams		
Gagas, Ramganga, etc.	Fluvial (Erosion)	132
Phika, Dhela	Fluvial Deposition)	257

Amongst the Himalayan lakes, Nainital is perhaps the most ecologically damaged system by the input of sewage that it cannot fully recover even if the pollution discharge into the lake is diverted. Once the water quality of the lake is improved, right fishery management policy is in place, scientific norms are followed for stocking and harvesting, desired hatchery is established to produce required seed of indigenous species. In the years to come the species which have declined will start establishing themselves, but sustained stocking is needed because there are no natural breeding grounds of these species within the lake. This restoration of biodiversity will be a major achievement.

Amongst the Kumaon lakes, Bhimtal lake is well suited for mahseer fishery from the various bio-ecological viewpoints as the fish has well established and many big specimens have been reported in the past from this lake so the glory of this wealth can be further restored.

Table -5 Important freshwater lakes of Kumaon region

Features	LAKES					
	Nainital	Bhimtal	Sattal	Naukhuchiatal	Khurpatal	Shayamtal
Altitude (m asl)	1938	1345	1286	1220	1670	1510
Maximum length	1457	1975	875	1050	-	750
Maximum width	467	532	-	880	-	210
Maximum depth (m)	25.7	24.8	18.0	41.24	15.0	10.0
Area (ha)	45.7	72.0	-	37.0	13.4	4.0
Catchment Area (Km^2)	5.5	11.2	5.0	2.6	-	-
Shoreline (m)	3630	4023	-	3560	-	-

Though the mahseer has auto spawning and production capacity in the lake itself but on account of increased anthropogenic stresses and burgeoning angling pressure, the fish production is presently at a very low level. The Bhimtal lake can be developed primarily for mahseer fishery, it will boon for the Uttarakhand since it may be recognized as angler's paradise both from home and abroad eventually boosting the tourist industry. Sattal lake along with a few small lakes located within its vicinity can also be developed for mahseer fishery. Possibilities of raising mahseer and snow trout fingerlings for stocking these lakes using 'Pen' and 'Cages' could be exploited which have a vast potential. The stock would be fed with highly nutritive and efficient cheap feeds.

There are innumerable high mountainous lakes in Garhwal Division of Uttarakhand State, located in Central Himalayas. Amongst these, the Hemkund Sahib lake is very clean and utra-oligotrophic in nature. The lake is a large glacial lying on the cirque floor of the Saptsring glacier. Normally the lake remains under ice cover more than six months in a year. It is fed mainly by the glacier located on the eastern slopes of Saptsring peak. It is surrounded by towering snow-clad peaks of the central Himalayas. Water of the lake is transparent having secchi reading about 3.0 m. So far there is no record of any fish in this lake. Deorital lake is known to have been stocked with the fishes of Himalayan mahseer (*Tor putitora*), Common carp (*Cyprinpus carpio)*, and brown trout (*Salmo trutta fario*). Some lakes in this region are important from religious point of view.

6.2 Reservoirs

A number of man-made lakes (reservoirs) have been created in the Tarai region of Kumaon from time to time for storage of surface water for irrigation and also developing fisheries in them as a by product. Notable amongst them are Baigul (2200 ha.), Tumaria (2681 ha.), Dhaura (1200 ha.), Nanak Sagar (4662 ha.), Haripura and Kalagarh. These reservoirs contribute major portion of fish production of the state.

7. Fisheries Resources

Both the regions Garhwal and Kumaon of the Uttarakhand state are bestowed with valuable indigenous fish germplasm (Table-7 & 8) and pristine water resources with tremendous range in their thermal regime. These systems hold large population of both indigenous and exotic cultivable and non-cultivable fish species. The biological productivity in high altitudes is constrained by number of factors; however, in the foothills the biological productivity is quite high which can suitably be harnessed into fish biomass. The per unit production of coldwater fisheries to the total inland production catch is not very significant but it has the natural resources/potentials to contribute to the economy of hill state. The present status of fishery of the state is briefed under following heads:

7.1 Capture Fisheries

The major components of capture fisheries in this state are snow-trouts, mahseers and trouts. Minor carps and loaches also contribute to the fish biodiversity of the region.

7.1.1 Snow-trout – *Schizothorax richardsonii* is endemic to Himalayan and known to occur in almost all streams and rivers of Uttarakhand region and constitute the main fishery. There is only one species in the snow-trout group *Schizothorax richardsonii*, which is widely distributed from 600 m – 2000 m asl in Central Himalayan. The middle reaches of all the watersheds of this region have a great potential of this fishery. The fish is caught in good number and consumed locally and plays a vital role in supplementing nutrition requirement to poor people living in remote and hilly area of Uttarakhand. In hills its fishery also provides some income to a section of society who because of limited resources in terms of cultivable land is unemployed. In winters with the onset of winters, the species migrates down streams traveling a considerable distances and constitutes a significant fishery during winters. It is generally caught through conventional methods and has a local market. Winter fishery in uplands is mostly dominated by snow-trout. But of late it has been observed that its population is declining and needs some measures to conserve to sustain its population, particularly in open waters.

7.1.2 Mahseers –*Tor putitora* and *Tor tor* are one of the finest groups of food and sport fishes in this region. Among the two, *Tor putitora* is the dominant species in both Kumaon and Garhwal waters. In addition to the above *Tor chelynoides* is also recorded from these regions. In rivers and streams, the species during spring and summer months migrate upward and ascend to the smaller tributaries for feeding and breeding. Ramganga, Saryu and Kali in Kumaon and Alaknanda, Bhagirathi, Pinder, Asiganga etc. from Garhwal region are important mahseer waters. Besides the above streams, Kumaon lakes are also important mahseer resources.

7.1.3 Trouts – Brown (Salmo *trutta fario*) and rainbow-trout (*Oncorhychus mykiss)* are being reared in a few trout farms in Garhwal region and have been stocked in a number of snow-fed rivers, rivulets and other lotic waters of the same region. The regions situated above 1500 m asl in both the divisions of the State, have a good potential for trout both for culture and sport. In order to harness this potential, the suitable areas need special attention to promote and support trout fishery activities.

7.1.4 Others miscellaneous fishery- of the local hill streams, constitute minor carps such as *Labeo dero, L.dyochelius, Raiamas bola,* and species of *Garra, Barilius, Balitora, Botia, Brachydanio, Danio, Glyptothorax, Neomachelius, Pseudocheneis etc.* Though these fishes do not grow in big size and have less commercial value but do contribute in the catches of local fishermen to meet their day-to-day requirement and generate income for them.

7.2 Sport Fishery

The important fishes in the Uttarakhand state with good sport value are –golden mahseer (*Tor putitora*), deep-bodied mahseer (*Tor tor*), Indian trout (*Raiamas bola*) among the indigenous species and brown trout (*Salmo trutta fario*) and rainbow trout (*Oncorhychus mykiss*) from exotics. Gooach (*Bagarius bagarius*) typical migratory species has also sport value in this region.

7.3 Culture Fishery

Though attempts for fish culture in the state were initiated in the beginning of Twentieth Century and trout farms were constructed at Kaldyani (Uttarkashi) and Bhowali (Nainital) in the year 1910. Since the State till November 2000 was part of Uttar Pradesh, the fish farming actually in the hills restricted basically for seed production of brown and rainbow trouts in trout farms located in Garhwal region and of mahseer in Bhimtal of Kumaon region. The aquaculture of major carps and exotic carps was restricted to foothills and other parts of the Terai region of erstwhile U.P. The details of the cultivable species are as under

Indigenous species	Golden mahseer - *Tor putitora*
	Snow-trout - *Schizothorax richardsonii*
Exotic trouts	Brown trout - *Salmo trutta fario*

	Rainbow trout - *Oncorhynchus mykiss*
Chinese carps	Silver carp - *Hypophthaimichthys molitrix*
	Grass Carp - *Ctenopharyngodon idella*
Other exotics	Common carp - *Cyprinus carpio*
	Indian species Indian major carps - *Labeo rohita, Catla catla, Cirrhinus mrigala*

Table 7. Piscine diversity in hill region of Uttaranchal
(* Transplanted within the country; ** Exotic species)

Order Cypriniformes

Family Cprinidae

Subfamily Cyprininae

Carassius auratus (Linnaeus)**

Carassius carassius (Linnaeus)* *

Catla catla (Hamilton-Buchanan)

Chagunius chagunio (Hamilton-Buchanan)

Cirrhinus mrigala (Hamilton-Buchanan) *

Ctenopharyngodon idellus (Valenciennes)**

Cyprinus carpio Linnaeus * *

Labeo boga (Hamilton-Buchanan)

Labeo dero (Hamilton-Buchanan)

Labeo dyocheilus (McClelland)

Labeo rohita (Hamilton-Buchanan)*

Puntius chola (Hamilton)

Puntius conchonius (Hamilton-Buchanan)

Puntius dukai Day

Puntius gelius (Hamilton-Buchanan)

Puntius phutunio (Hamilton-Buchanan)

Puntius sarana (Hamilton-Buchanan)

Puntius sophore (Hamilton-Buchanan)

Puntius ticto (Hamilton-Buchanan)

*Tor chelynoides (*Hamilton)

Tor putitora (Hamilton-Buchanan)

Tor tor (Hamilton-Buchanan)

Hypophthalmichthys molitrix (Valenciennes)**

Subfamily Rasborinae

Barilius barila (Hamilton- Buchanan)

Barilius barna (Hamilton- Buchanan)

*Barilius bendelisis (*Hamilton- Buchanan)

*Barilius shacra (*Hamilton- Buchanan)

Barilius vagra (Hamilton- Buchanan)

Brachydanio rerio (Hamilton- Buchanan)

*Danio aequipinnatus (*McClelland)

Danio devario (Hamilton- Buchanan)

Esomus danricus (Hamilton- Buchanan)

*Parluciosoma daniconius (*Hamilton- Buchanan)

Raiamas bola (Hamilton-Buchanan)

Subfamily Schizothoracinae

Schizothorax richardsonii (Gray)

Schizothorax kumaonensis (Menon)

Subfamily Garrinae

Crossocheilus latius latius (Hamilton-Buchanan)

Garra gotyla gotyla (Gray)

Garra lamta (Hamilton-Buchanan)

Subfamily Garrinae

Crossochelius latius latius (Hamilton–Buchanan)

Garra gotyla gotyla (Gray)

Garra lamta (Hamilton–Buchanan)

Family Balitoridae

Balitora brucei (Gray)

Bhavania australis (Jerdon)

Subfamily Nemacheilinae

Noemachelius beavani Gunther

Noemacheilus botia (Hamilton-Buchanan)

Neomachelius corica (Hamilton-Buchanan)

Noemachelius montanus (McClelland)

Noemachelius multifasciatus (Day)

Noemacheilus rupecola (McClelland)

Noemachelius savona (Hamilton-Buchanan)

Noemachelius scaturigina (McClelland)

Family Cobitidae

Subfamily Cobitidae

Lepidocephalus guntea (Hamilton-Buchanan)

Subfamily Botinae

Botia almorhae Gray

Botia dario (Hamilton-Buchanan)

Botia dayi Hora

Order Siluriformes

Family Bagridae

Mystus vittatus (Bloch)

Family Schilbeidae

Subfamily Schilbeinae

Clupisoma garua (Hamilton-Buchanan)

Family Amblycipitidae

Amblyceps mangois (Hamilton-Buchanan)

Family Sisoridae

Bagarius bagarius (Hamilton-Buchanan)

Euchiloglanis hodgarti (Hora)

Glyptothorax alaknandi Tilak

Glyptothorax brevinnis (Hora)

Glyptothorax cavia (Hamilton-Buchanan)

Glyptothorax conirostre conirostre (Steindachner)

Glyptothorax dakpathari Tilak & Hussain

Glyptothorax garhwali Tilak

Glyptothorax pectinopterus (McClelland)

Glyptothorax telchitta (Hamilton-Buchanan)

Glyptothorax trilineatus Blyth

Nangra nangra (Hamilton-Buchanan)

Pseudocheneis sulcatus (McClelland)

Family Salmonidae

Oncorhynchus mykiss (Walbaum)**

Salmo trutta fario Linnaeus **

Family Belonidae

Xenentodon cancila (Hamilton-Buchanan)

Family Pocciliidae

Gumbusia affinis (Baird & Girad)**

Order Perciformes

Suborder Channoidei

Family Channidae

Channa orientalis Bloch & Schneider

Suborder Mastacembeloidei

Family Mastacembelidae

Mastacembelus armatus (Lacepede)

7.4 Existing Fish Yield

There is hardly any fish-landing centre in the uplands from where the exact contribution of hill fishery to the total production of this region could be ascertained. The major contribution in this sector is from the reservoir and pond fishery of the Terai region. However, the production from all resources including reservoirs, lakes etc. is estimated about 900 t/Yr. Based on the experimental fishing and field observations made by many workers from the local markets of different rivers revealed that the fish catch per unit effort from the highland open waters ranged between 125-425 g using cast net method. The dominant fishery in rivers was dominated by *Schizothorax richardsonii* 60.0-95% in total catches in all most all the rivers. The other important species recorded was *Tor putitora*, which 2.0-10.0% in the catch. Apart from these species, brown trout (*Salmo trutta fario)* was also recorded in lesser magnitude in the river Bhagirathi, Bireihi and Pinder.The other important species encountered were *Labeo dero, Garra gotyla, Barilus bola, B.bendelisis,* and species of *Glyptothorax, Neomachelius*. Normally the upper reaches of these rivers support fewer species due to altitude, stream gradient, temperature and current variations. A total of 24 fish species have recorded from both the Kumaon and Garhwal regions. Majority of the fishes is caught individually by local fishermen in the rivers and streams and does not form the fisheries of commercial catch. The average catch per person with their conventional gears range from 500 –1500 g/day. The local market and village shops act their sale centers, which normally fetches them a price of Rs.60-80 per kg.

8. Existing Constraints in Fisheries Development

In the past many organization have done commendable work on coldwater fisheries and developed technologies for both capture and culture of exotic and indigenous species. But the impact of those technologies in this state due to various constraints was very marginal. It is obvious that resources in Uttarakhand are quite valuable for the development of hill fishery for food and sport, but there is an urgent need for proper management of these resources and in this direction certain steps of substantial importance need active consideration. In order to manage these ecosystems ecologically sustainable for contributing to overall development of remote hilly regions, focus has to make on the following issues.

Ownership of resources Practically all the water resources suitable for hill fishery in the state are owned by the forest department. For implementation of fishery development programme these need to be placed under the management of fishery department.

Infrastructural development in the form of construction of coldwater fish farms and hatcheries has been taken on top priority to achieve the target set for action plan.

Due to lack of feed mill in this region the production of both exotic trouts and native species could be achieved to the national level.

Up-gradation of working facilities in the farms & labs.

Introduction of better exotics for higher production and sport.

Strategies for the management of lake fishery.

Survey of upland regions to know the status of hill fishery and frame polices for efficient harvest of fish.

Development of fishing angling spots for tourists in the uplands both for mahseer and trout for sport fishing is of utmost importance.

Conservation of native and exotics in natural ecosystems need enforcement of protective legislation and adopting aquaculture practices on a large scale.

Development Needs

There is immense potential for development of hill fishery in the state. Hence, an integrated approach is urgently required for proper and sustainable development of the available diverse resources. The action plan is proposed for both culture and capture of exotics and endemic fisheries, which are as under:

Table-8 Details of important Food, Sport and Ornamental fishes of Uttarakhand State

Food fishes	***Sport fishes***	***Ornamental***
Indigenous fishes		
Tor putitora	*Tor putitora*	*Brachydanio rerio*
Tor tor	*Tor tor*	*Danio devario*
Schizothorax richardsonii	*S.richardsonii*	*Botia almorhae*
Labeo dero	*Raiamas bola*	*Barilius bendelisis*
Labeo dyocheilus	*Tor chelynoides*	*Bagarius bagarius*
Exotic fishes	Cyprinus carpio	Salmo trutta fario
Carassius carassius	*Oncorhynchus mykiss*	*Carassius auratus*
Hypophthalmichthys molitrix	*Ctenopharyngodon idella*	*Indian major carps*
Catla catla	*Labeo rohita*	*Cirrihina mrigala.*

8.1 Open water Fisheries

In the past, many organizations have done commendable work on coldwater fisheries and developed technologies for both capture and culture of exotic and indigenous species. But the impact of those technologies in this state due to various constraints was very marginal. It is obvious that resources in Uttarakhand are quite valuable for the development of hill fishery for food and sport, but there is an urgent need for proper management of these resources and in this direction certain steps of substantial importance needs active consideration. In order to manage these ecosystems ecologically sustainable for contributing to overall development of remote hilly regions, focuses have to make on the following issues Ownership of resources practically all the water resources suitable for hill fishery in the state are owned by the forest department. For implementation of fishery development programme, these need to be placed under the management of fishery department.

For resource base inventories, survey of both the divisions of the Uttarakhand to collect detail resource base information on fish and fisheries and other related aspects to formulate polices/ strategies to develop and improve hill fishery in potential zones.

Development of angling spots for tourists in the uplands both for mahseer and trout for sport fishing is of utmost importance.

Conservation of native and exotics in natural ecosystems need enforcement of protective legislation and adopting aquaculture practices on a large scale.

8.2 Management of riverine ecosystems

Production enhancement in lotic systems through stocking of most suitable species.

Installation of various hatchery units for mass scale seed production of snow-trouts, golden mahseer and other important fish species at the suitable sites along the river stretches.

Fabrication of modern suitable fishing gears for harnessing fish resources properly for the betterment of hill communities.

Cumulative participation by various agencies-ichthyologists, environmentalists, planners, law forcing agencies, angling associations/ clubs, NGO's and other on-line authorities at higher level, besides local and voluntary organizations need to join hands sincerely to save and conserve the natural fish diversity in the hills.

Creation of mass awareness especially bank dwellers and fisher folk as regards to abuse of unscrupulous and warden fishing besides the fisheries conservation through press, electronic media and arranging exhibitions, skits, frequent personal interactions.

Habitat Improvement

Along with the ranching of sport and food fish, programmes of stream improvement to maintain ideal ecological conditions for the well being of existing stocks and of transplanted fishes is equally important. Hardly any attention is paid to this aspect in any region in the country. In this direction following points needs special attention for habitat improvement:

a) Poaching and illegal fishing should be totally banned;

b) Removal of sand, pebbles and stones and other stream material from the main streams should be totally checked;

c) Artificial pools, with central spillways may be created in certain suitable stretches of the stream holding mahseer fishery;

d) Certain productive stretches having deep pools and gorges may be declared as sanctuaries for mahseer and snow-trouts;

e) Certain waters should be declared purely for recreational fishing (reserve solely for game fishing), and such waters should not be allowed for commercial or any other type of fishing;

f) Proper watch and ward of waters holding endangered species like *Raiamas bola*.

Provision of spawning and survival facilities, creation of shelter and feeding range and augmentation of food supply are some of the physical conditions, which could be accomplished by human efforts to improve streams for fish.

8.4 Sport fishery

Development of sport fishery especially trouts in Garhwal and golden mahseer in Kumaon regions.

In-situ conservation of native species and declaring certain sanctuaries and national parks would also provide opportunities for tourism promotion.

Legislation

The present day fishery legislation is based on the Indian Fisheries Act.1897. The Indian Fisheries Act defined the functions and responsibilities on the states to protect fish resources. The state fisheries departments were created to protect and increase fish stocks and to assure that good fishing would continue and improve. The fishery regulations are intended to take several steps, like creel limits (control on over exploitation); size limits (impairment of reproduction); closed season (total ban on fishing during spawning season) etc. In this direction NRCCWF has already provided its impetus for framing and finalizing the fisheries Policy and Act of Uttaranchal State in 2003-04.

8.6 Infrastructural Developments

Development of physical facilities likes fish hatcheries/farms with adequate numbers of nursery, adult and brood stock pond/raceways, feed mill and other facilities for intensive seed production of desirable species required for ranching and for overall development of hill fishery in Central Himalayas.

8.7 HRD support on coldwater fisheries

Hill fishery is a specialized sector as compared to warm water fisheries, so expertise in this sector is required to develop it for food and sport purposes. State should make efforts to train staff of Fisheries Department belonging to different categories through various programmes conducted within and outside the country. The staff needs to be trained in related fields, which will improve the working and overall management and fish production of the hill fishery.

8.8 Aquaculture

- Trouts
- Intensive culture of rainbow trout.
- Mass scale seed production of brown trout for ranching for sport.
- Commercial culture of trouts (rainbow trout) in Govt. farms.
- Carps
- Intensive Common carp culture in ponds and running water systems.
- Intensive culture of Chinese carps in water harvested and mud ponds.
- Himalayan mahseer and snow-trout
- Brood stock management of golden mahseer and snow-trouts in captivity.
- Intensive seed production of mahseer and snow-trout.

9. Future Perspective

The Hill Area Development is now being focused as priority in the country. But up-till now the economic up-liftment of hill states including Uttarakhand, due to various climatic, geomorphologic and resource constraints has been very insignificant in comparison to states in the plains of India. Through adequate financial support and institutional building and strengthening R&D support this imbalance can be removed. Both the hill divisions of Uttarakhand region of the country are bestowed with valuable indigenous fish germplasm and pristine water resources with tremendous range in their thermal regime. The fish can play a vital role in supplementing protein requirement to poor people located in remote Himalayan region and providing source of income to a section of people who because of resource constraints in-terms of cultivable lands in hills. Therefore, adequate research and development support in fisheries sector is very important and critical for this state. Infrastructural development in this region is the most important aspect for achieving success in this sector and various issues related with the development of both exotic and indigenous fisheries in each division should be strongly looked into.

10. Action Plan to Achieve the Objective

Depending upon the natural aquatic resources in the hills of Uttarakhand, species variability, technology support and scientific database available. Fishery if developed on scientific lines will go a long way in contributing to the rural economy in remote hilly zones of the state. It is suggested that fishery development should be initiated based on the available resource for both exotic and indigenous fishes. Higher reaches where water temperature and other conditions are suitable should be explored for trout culture to promote sport fishery and culture of rainbow trout on commercial lines. While at comparatively lower altitudes, Himalayan mahseer and snow-trouts seed production centres have to be established to cater to the demand for stocking natural water bodies for food and sport fishery view point. Establishment of carp hatcheries and farms at lower altitudes is another pre-requisite of the state to meet to the demand of local farmers who are engaged in small-scale fish farming. In addition to the above, the regions situated between 1,000 to 1,800 m asl has a tremendous potentials for exotic carp culture, this should be encouraged by providing know-how and seed to the farmers will, indeed, supplement the income of farmers earned by him through other agricultural activities. To promote all these activities especially in hilly regions of Uttarakhand state, besides re-modeling or up scaling the existing facilities, there is immediate need to create more infrastructural facilities in the form of hatcheries, farms with tanks and raceways. There is urgent need for establishing one feed mill in Garhwal division of the state to produce feed on a pilot scale for trout both for departmental use and subsequently for the use of farmers or the beneficiaries.

11. Coldwater Fishery Resources Survey

Resource database inventory of Uttaranchal state, this will involve survey of fishery resources and their status both in Kumaon and Garhwal regions and output will be a complete document along-with the information on the aspects which are at present lacking. This document will help in formulating management and conservative polices for native as well as for exotic food fishes, sport fishery and open water fishery. The activities will be covered on the following aspects.

11.1 Coldwater Aquaculture

One of the priority areas, in the state is to promote culture of exotic trouts. To grow fast growing strain of rainbow trout (*Oncorhynchus mykiss*) in the farms for raising food fish and intensive seed production of brown trout for ranching in suitable natural waters. To achieve these twin objectives there is an immediate need to equip and remodel the existing hatchery and farm facilities located in both the divisions of the state to carry out the proposed activities.

Both the indigenous fishes (mahseers and snow-trouts) are very critical to the Uttarakhand fish biodiversity. The mahseer is world famous and its linkage with angling can give boost to State tourism. This is only fish that is presently propagated on scientific lines at Bhimtal. Therefore, the improvement of existing facilities at Bhimtal is very essential and critical as far as its development and conservation of mahseer fishery in Kumaon region is concerned. The other equally important fish in the state is snow-trout which can also be attempted if adequate facilities are developed at Bhimtal and Bhowali. The Bhowali unit located close to Nainital can be very handy to develop a sale centre / small rearing unit for rainbow trout in the Kumaon region.

11.2 Running water aquaculture system for farmers

In Uttarakhand the resource base for aquaculture ponds / tanks is limited but hill stream diversion offer a good scope for developing running water culture system. In this system water is used only as medium of fish growth and water as such is not consumed in the process. The experiments conducted so far in this culture system have shown promising results. This culture system is based on combination of exotic carps (Common carp, Silver carp, & Grass carp). The fishes have to be stocked @ 2-5 advanced fingerlings per m^2. The running water culture system requires that stocks should be fed adequately for which a combination of conventional and artificial balanced diet will be needed. To initiate this activity in the State the selected farmers have to be identified as beneficiary and running water system / ponds to be established / constructed under the proposed activity.

Depending upon the site specific approach the size of a running water unit should be in the range of 50 – 100 m^2 with a depth range of 0.5 – 0.8 m with proper screen to control the escape of stocks. There should be adequate water inlet and outlet mechanism. The recommended management practices need to be followed by the farmer. Depending upon the site selected on scientific norms in both the divisions ponds / tanks will be constructed.

11.3 Trout rearing raceways for farmers

The rainbow trout farming has very good potential in the State of Uttrakhand. The trials conducted on its rearing to table size and seed production in Garhwal region by the State Department and in the Kumaon region by NRCCWF have given very promising results. Further, the sale of this fish to the local people in the Champawat district @ Rs.150/- per kg was initiated by initiated by NRCCWF. There was a good response. The fish fetches a good price and the species has excellent aquaculture qualities. Therefore, its culture should not remain confined to government farms but technology be extended to willing farmers. Further, the seed production capacity in State owned trout farms have to be properly utilized. This is possible if the farmers are involved in the rearing of trout. Since the trout rearing requires specialized tanks with running, high oxygenated water and proper balanced diet apart from supply of required seed material. These aspects have to be looked into by the State under this specialized project / scheme. In this connection as a first step trout rearing raceways have to be constructed at selected site to be provided to registered farmers both in Garhwal and Kumaon regions.

11.4 Establishment of Feed Mill

In case the rainbow farming technology has to be extended to the farming community in hills and also the State trout farms have to sustain their trout stocks the assured supply of trout feed to these requirements is very essential. The trout feed is a specialized item and has to be prepared in bulk at the farm. In this connection establishment of one feed plant each in Garhwal and Kumaon region is a re-requisite for success in trout farming. Therefore, one feed mill is to be established in the Grahwal region preferably at Biarnagna Trout farm. Amongst the coldwater fishes, the fast growing strains of rainbow trout being the high value species has a tremendous demand.

11.5 Input for supply of feed and seed

Regular supply of stockable fish seed of both exotic carps and trouts to the farmer has to be assured for culture of stocks in running water ponds and raceways. In addition regular feed has to be provided to farmers identified under this fish arming scheme in both the divisions of the state.

11.6 Technologies for Highland Reservoirs

In Uttarakhand the highland small and large rivers are impounded into huge reservoirs under hydro-power projects. The number of these projects is likely to double in the next two decades. As a result of these projects, there has been considerable interference with the riverine environment, inevitability affecting both the habitat and the indigenous biota, including fishes. Moreover, these water bodies can be used to enhance the overall fish biomass production of the state effectively by employing modern techniques. This is a big challenge to the hill fishery sector of this state, can only be achieved only if techniques/fish species combination available with other Himalayan states on the reservoir fishery can be scientifically tried, and thus needs very special attention. These systems can also be used for fishery enhancement effectively by employing the modern cage or pen culture tools. For the proper implementation of fishery development programmes, these ecosystems should be place under the management of fishery department.

The capture fisher sector of this state particularly aquaculture of freshwater fishery has been given lower priority in the overall fishery planning. This has contributed top a gradual decline on the fish stocks in the majority of waters of the state. There has been slow progress in promoting aquaculture of important candidate fish species and little has been done in enhancing the riverine and lacustraine stocks of both native and exotic fish varieties. Even fishery of reservoirs has not been scientifically managed, leading to poor fish biomass production from these systems. The awareness among the local people about protecting catchments to prevent further deterioration of ecosystems holding native fishes has been lacking. It is suggested that due priorities should be given to protect and enhance highland fish stocks. It should allocate sufficient financial resources for adequate creation of infrastructure and training of manpower in this sector to improve and manage both food as well as sport fisheries in the state.

8

GENETIC VARIABILITY OF *MYSTUS CAVASIUS* (HAMILTON, 1822) POPULATIONS USING RANDOMLY AMPLIFIED POLYMORPHIC DNA

R.K. Garg, P. Sairkar, N. Silawat, N. Batav and R. Sharma

Abstract: Random amplified polymorphic DNA (RAPD) markers were used to assess genetic diversity of *Mystus cavasius* (Hamilton), a lower risk near threatened species collected from two locations of Madhya Pradesh i.e., Gandhisagar reservoir and Bandsagar reservoir. High level of genetic polymorphism was observed in the collected individuals. Ten RAPD primers were primarily scored in 3 individuals from each of the two locations. Five primers, which gave polymorphism, were selected for polymerase chain reaction (PCR) and used in the final RAPD analysis. Using these primers, 64 scorable bands were generated in the fishes of Gandhisagar reservoir, of which 55 (84.54%) were polymorphic; whereas, 108 bands were found in the fishes of Bandsagar reservoir, out of which 103 (93.07%) were polymorphic. 30 individuals formed two major clusters, Gandhisagar reservoir population forming one cluster and Bandsagar reservoir population in another cluster. The present study yielded elucidating the usefulness of complementary approaches to make diversity analysis more explanatory and purposeful for optimum genetic amelioration and effective conservation of its genotypic variability.

Key words: Status of fishes in M.P., Genetic Diversity, RAPD, Polymorphism, monomorphism.

Introduction

National Bureau of fish Resources (1998) has published a list of threatened fishes of which about 30 species were categorized as endangered from central India. Some of these are *Tor khudree, T. putitora, T. tor, Ompok bimaculatus, O. pabda, Bagarius bagarius, Pangasius, Labio dero, L. dussumeiri, L. dyocheilus* etc. 700 freshwater fishes reported from Indian inland waters out of

Centre of Excellence in Biotechnology, M.P. Council of Science and Technology, Vigyan Bhawan, Nehru Nagar, Bhopal-462003 (M.P.)

which, 320 were assessed in CAMP workshop jointly organized by NBFGR and Zoo Outreach organization during 1998. Of these 320 species assessed, about 166 fish species were from central region and among these 21 were categorized as endangered, 27 vulnerable, 50 at lower risk or near threatened, 8 at lower risk of least concern and 58 species were not evaluated (CAMP, 1998).

According to Islam *et al.* (2002) information on genetic variation is essential for conservation and stock improvement programs. Genetic polymorphism is classically defined as the simultaneous occurrence of a trait in the same population of two or more discontinuous variants or genotypes. Although DNA sequencing is a straightforward approach for identifying variations at a locus, it is expensive and laborious. A wide variety of techniques have, therefore, been developed in the past few years for visualizing DNA sequence polymorphism.

Mystus cavasius (Hamilton) is a scale less big sized fish and distributed in Bangladesh, India and Burma (Bhuiyan, 1964). The fish is usually found in freshwater and are mostly available in rivers, canals, beds, ponds, ditches and inundated fields. It is very excellent as food and source of protein to the peoples of India. Molecular studies on this fish has not done by any worker in Madhya Pradesh, therefore, present study has been carried out on two populations of *Mystus cavasius.*

Materials and methods

Collection of fish samples from sampling sites

The fish samples were collected from two water bodies of Madhya Pradesh i.e., Bandhsagar reservoir and Gandhisagar reservoir. Bandsagar is constructed on Son river in district Rewa, whereas, Gandhisagar reservoir is constructed on Chambal river in district Neemach. The distance between both the reservoirs is more than 500 km so, chances of genetic diversity within the fish species may be more. 15 individuals from each reservoir were collected with the help of castnet and local fishermen brought to the laboratory for their morphometric and molecular study.

Fish taxonomy on the basis of its morphology

3.2.1 *Morphometric measurements*

The morphometric analyses of the fishes were done on the basis of certain parameters. Total 15 parameters were taken for taxonomic identification of fishes. They are Total length, Standard length, Body weight, Body width, Body depth/height, Head length, Head length excluding snout, Width of head, Snout length, Eye diameter, Length of caudal Peduncle, Length of anal fin, Length of pelvic fin, Length of pectoral fin and Height of caudal peduncle. All the parameters were taken with a flexible 30 cm scale and a graduated thread as shown in figure 1. A physical balance took the weight of all fishes.

- Total length (A-H). Measured from the tip of the snout to tip of caudal fin.
- Standard length (A-G). Measured from the tip of snout to the base of the caudal fin.

- Length of head (A-D). Measured from the tip of the snout to the most posterior bony extremity of the operculum.
- Depth of body (I-J). Measured along the vertical line at its deepest part.
- Length of caudal peduncle (F-G). Measured posterior base of anal fin to the base of caudal fin.
- Height of caudal peduncle (K-L). Measured along the vertical line at its narrowest part.
- Length of head excluding snout (B-D). Measured from the anterior margin of the orbit to the most posterior bony extremity of operculum.
- Width of head. Measured along the broadest part of the head.
- Diameter of eye (B-C). Measured from margin to margin of the bony orbit.
- Snout (A-B). Measured from the tip of the mouth to anterior margin of the orbit

Molecular Study

In the present study, most of the morphometric characteristics of fishes has been done and often overlapped with populations. These morphometric data were not enough to support the established genetic structure of the populations that often leads to taxonomic uncertainty. Therefore, we have performed the RAPD analysis for molecular study. For this purpose, DNA was isolated from tissues using protocol (Janarthanan and Vincent, 2007). Then, isolated DNA was introduced to polymerase chain reaction (PCR) using arbitrary random primers. After which, electrophoresis of PCR product has been done, then, obtained the random amplified polymorphic DNA bands on the gel.

Genomic DNA extraction from tissues

For the isolation of total genomic DNA, a short procedure was applied according to a modified protocol than reported (Wu *et al.,* 1995). Tissues (200-500mg) were placed in a 1.5 ml micro-centrifuge tube and homogenized by using Eppendorf micro-pestle. In the homogenized tissue, 0.5 ml of lysis buffer (4mM NaCl, 0.5 mM EDTA, 0.1% SDS and 0.02 NP 40) and 0.01% proteinase-K were added, mixed gently and incubated at 55°C on dry bath (Genei, Model-SLM-DB-120) for 45-60 minutes for complete lysis of cells. After incubation, 250μl chloroform and 250μl phenol were added, mixed gently, and centrifuged at 10,000 rpm at room temperature (High Speed Brushless Centrifuge, MPW-350R, Poland) for 5 minutes.The supernatant was then transferred to a new micro-centrifuge tube and further chloroform (250μl) and 250μl of 7.5 M ammonium acetate were added, mixed well and centrifuged at 10,000 rpm for 5 minutes at room temperature. The DNA was precipitated from supernatant with two volumes of 99% ethanol (ice-cold). The DNA pellet was then washed with 70% ethanol, dried and dissolved in a Tris-EDTA buffer (10mM Tris HCl, 1mM EDTA, pH 7.6). UV-VIS Spectrophotometer (ND-1000, USA) was used to check quality as well as quantity of isolated DNA. The concentration of extracted DNA was adjusted to 50 ng/μl for PCR amplification.

In this work, ten commercially available oligonucleotide primers (ten to twenty bases long) of Bangalore Genei, India were used to initiate PCR amplifications. Primers were randomly selected on the basis of GC content and annealing temperature for RAPD-PCR amplifications. After initial screening with all the ten primers, the five which were used in final study were RAn-1, RAn-3, RAn-4, RAn-5 and RAn-6.

The reaction mixture (30 µl) for PCR was composed of 3 µl of 10X Taqpolymerase buffer, 1.2 µl of 10mM dNTP, 1.2 µl of RAPD primer, 0.6 µl Taq DNA polymerase (3U/µl), 23.4 µl sterile distilled water and 0.6 µl template DNA. A negative control, without template DNA was included in each round of reactions. After pre-heating for 5 minutes at 94°C, PCR was run for 45 cycles. It consisted of a 94°C denaturation step (0.45 minutes), 37°C annealing step (1 minute), and 72°C elongation step (1.5 minutes) in a Thermal Cycler (Corbet Research, Australia). At the end of the run, a final extension period was appended (72°C, 7 minutes), and then stored at 4°C until the PCR products were analyzed.

The amplified DNA fragments were separated on 1.2 % agarose gel and stained with ethidium bromide. A DNA size marker was run with each gel (100, 200, 300, 600, 1000, 1500, 2000, 2500 and 3000 bp DNA ladder from Bangalore Genei, Bangalore, India). The amplified pattern was visualized on an UV Transilluminator and photographed by Gel Documentation System (Alpha-Innotech, USA). The RAPD fragments were scored for the presence and absence of fragments on the gel photographs, and RAPD fragments were compared among the populations. RAPD banding patterns were recorded on spreadsheets, which were used to determine gene diversity, gene flow, number of polymorphic loci and genetic distance through a construct by an un-weighted pair group method of arithmetic mean (UPGMA-Nei, 1978).

Statistical analysis Reproducibility of RAPD-PCR was tested by gel documentation system. The bands were visualized under UV-light and the gel was photographed. The DNA profile or fingerprint of each fish sample was recorded. The data was scored for the presence or absence of the amplified fragment for all individuals. The data matrix was generated and each individual profile constructed using the following criterions if a given amplified fragment was present in an individual, then it was assigned as '1' and when the fragment was absent it was assigned as '0'. These generated data called as binary matrix which was used for phylogenetic relationship between two populations of *Mystus cavasius* using Clustar vis (version.1.4.1) and Tree me software.

Results

Fifteen samples from each reservoir i.e., Bandsagar and Gandhisagar were collected for detail study of intra-specific diversity between *Mystus cavasius*. The fish specimens from both reservoirs were collected with the help of local fishermen (fig. 1). All the fishes were brought to the Biotechnology Core Laboratory, MPCST. The morphometric measurements of all the fishes were done.

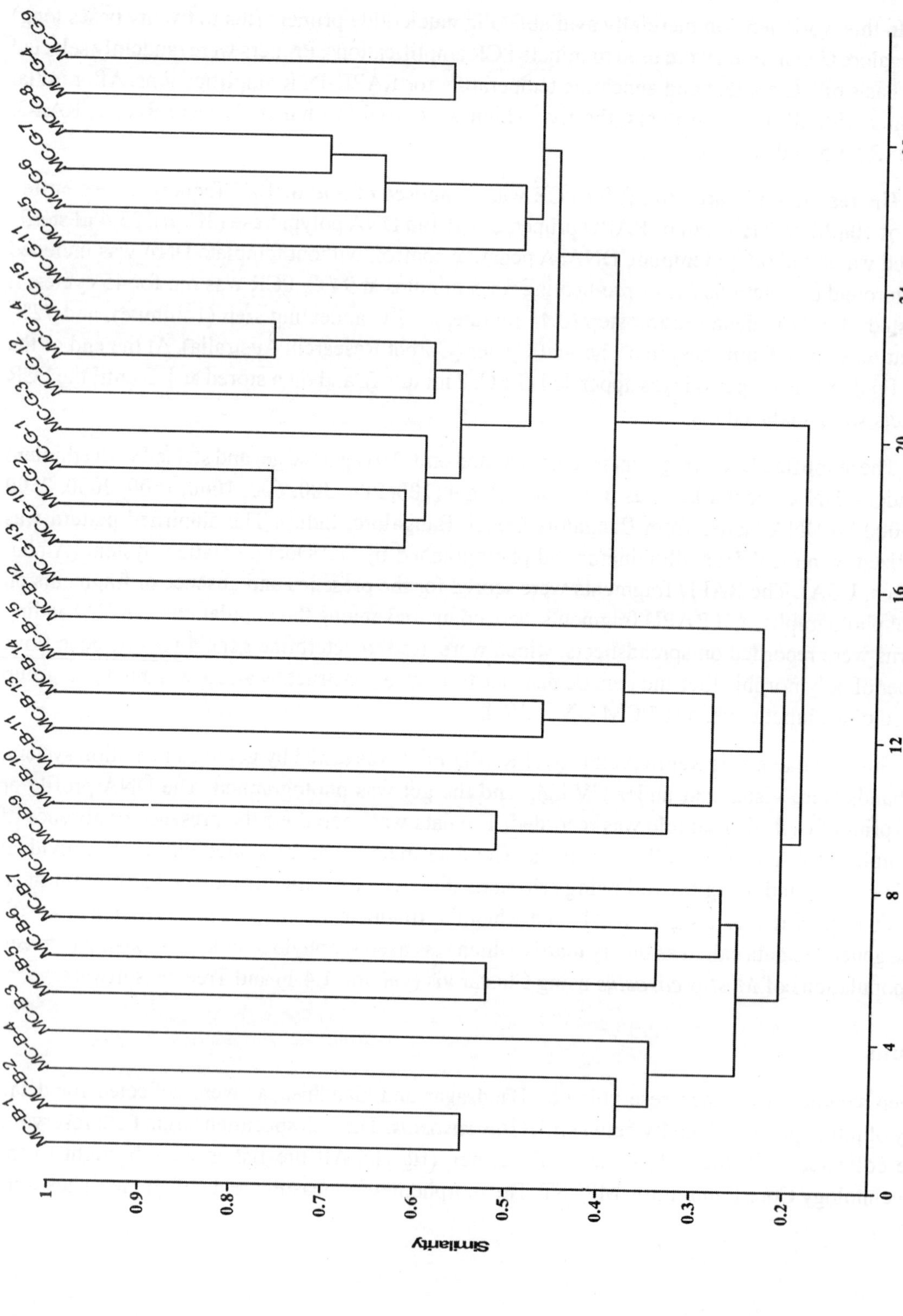

Figure 1. A dendrogram based on UPGMA clustering of 30 individuals of *Mystus cavassius* using genetic distance

4.1 Morphometric measurements of fishes

The morphometric measurements of all the fishes were done in the laboratory. Following parameters Total length, Standard length, Body weight, Body width, Body depth/height, Head length, Head length excluding snout, Width of head, Snout length, Eye diameter, Length of caudal peduncle, Length of anal fin, Length of pelvic fin, Length of pectoral fin, Height of caudal peduncle were taken. Most of the morphometric characteristics of fishes are similar and often overlap with population. These morphometric data are not enough to support the established genetic structure of the population often that leads to taxonomic uncertainty (Daniel, 1997; Panniah and Gopalakrishnana, 2000). The Fin formula of *Mystus cavasius* was found as D.1/7/0; P.1/9; V.6; A.11-12 (3/8-9); C.19-21; Barbles 4 pairs. In recent studies some scientists have given name instead of *Mystus cavasius* range of variation, mean and standard deviation in morphometric measurements of *Mystus cavasius* were also calculated.

4.2 Qualitative and quantitative analysis of DNA

Isolated DNA was qualitatively and quantitatively estimated by UV-Spectrophotometer (ND-1000). The detailed results of quality and quantity of DNA of fishes of both reservoirs have been given in tables 4 and 5. The samples of Bandsagar Dam, Rewa were coded as MC-B-01, MC-B-02, MC-B-03, MC-B-04, MC-B-05, MC-B-06, MC-B-07, MC-B-08, MC-B-09, MC-B-10, MC-B-11, MC-B-12, MC-B-13, MC-B-14, and MC-B-15 and quantity of DNA of these samples were recorded as 248.67, 260.34, 526.96, 473.98, 435.81, 314.86, 575.29, 651.90, 65.11, 385.95, 203.28, 562.70, 220.56, 1043.05, 578.75 ng/µl, respectively. The ratio of O. D. between 260nm and 280nm were recorded as 1.98, 1.98, 1.97, 1.99, 1.83, 1.81, 1.64, 1.97, 1.75, 1.75, 1.87, 1.78, 1.74, 1.56, and 1.67 respectively. DNA yield and quality were checked by Spectrophotometer from the absorbance data of the sample DNA at 260 & 280nm. The purity of DNA sample was calculated from OD 260/OD 280 ratios (Sambrook and Russell, 2001). DNA samples were diluted with sterile Milli-Q water to 50 mg/µl for further use in RAPDs.

DNA amplification

RAPD markers have been found to have a wide range of application in gene mapping, population genetics, molecular evolutionary genetics and plant and animal breeding (Bardakci, 2001; Ertas and Seker, 2005). After checking the quality and quantity of isolated DNA, eight RAPD primers of animal (RAn-1 to RAn-8, make Genei, Bangalore) were screened on all the samples collected from different sampling stations. Amplifications were performed according to the protocol (Williams *et al.*, 1990) with some modifications in the reaction volume. Amplifications were performed in a Corbet Research Thermal Cycler (PCR). Only five primers (RAn-3, RAn-4, RAn-5, RAn-6 and RAn-8) generated a large number of bands.

The PCR protocol as adopted in the study resulted in reproducible pattern of amplicons using specific combination of accession and primer. Only the primers which displayed reproducible, scorable and clear bands were considered for analysis. The image profiles of banding patterns were recorded and molecular weight of each band was determined by Alphaview software. The banding pattern was scored based on the presence or absence of clear, visible and reproducible bands (fig. 2-11). The results were analyzed based on the principle that a band is considered to be 'polymorphic' if it is present in some individuals and absent in others.

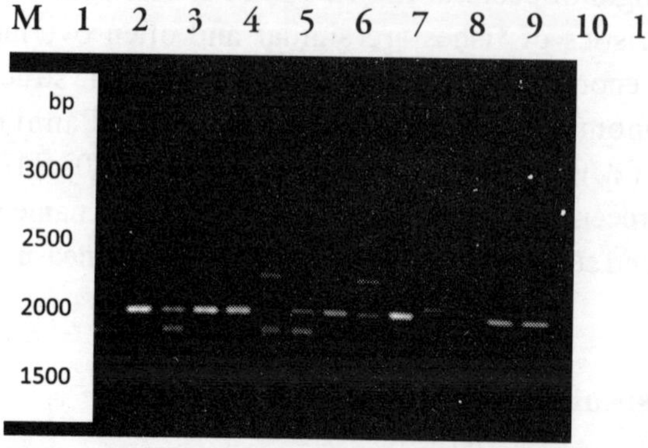

Figure 2. Primer RAn-3

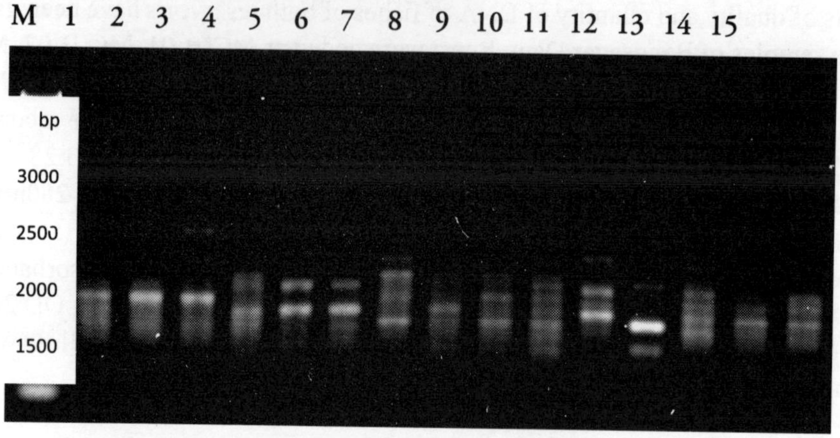

Figure 3. Primer RAn-4

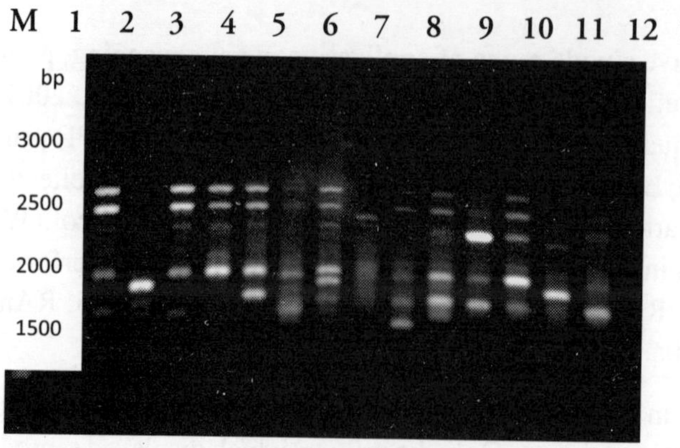

Figure 4. Primer RAn-5

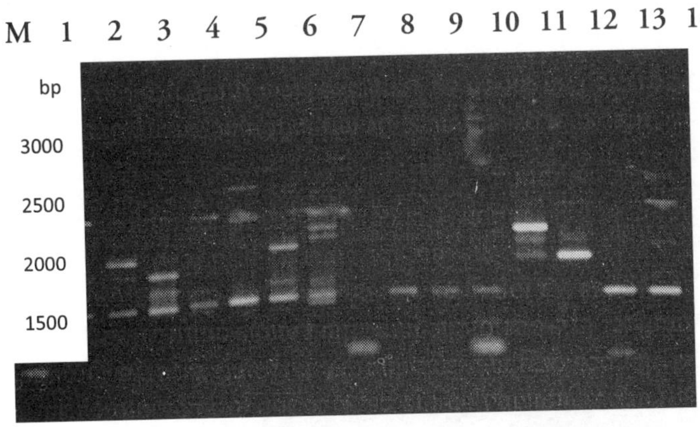

Figure 5. Primer RAn-6

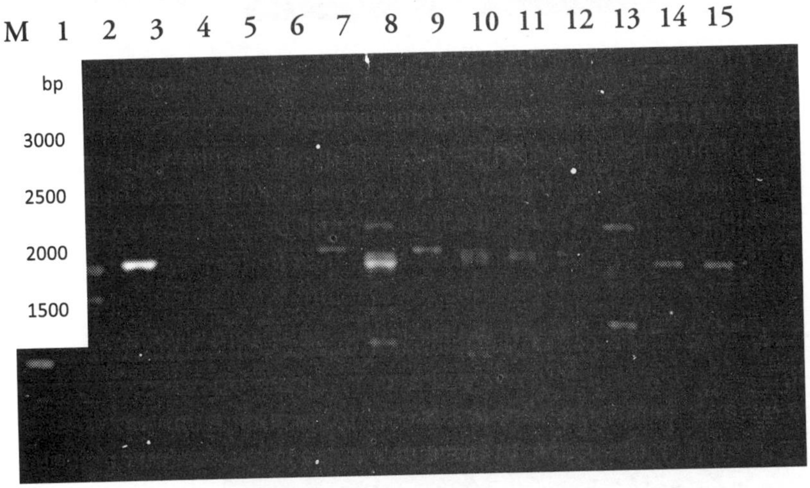

Figure 6. Primer RAn-8 Bandsagar

Figure 2-6 DNA fingerprints of 15 individuals of obtained with primer. 1-15 individuals of Bandsagar reservoir and M is molecular weight marker

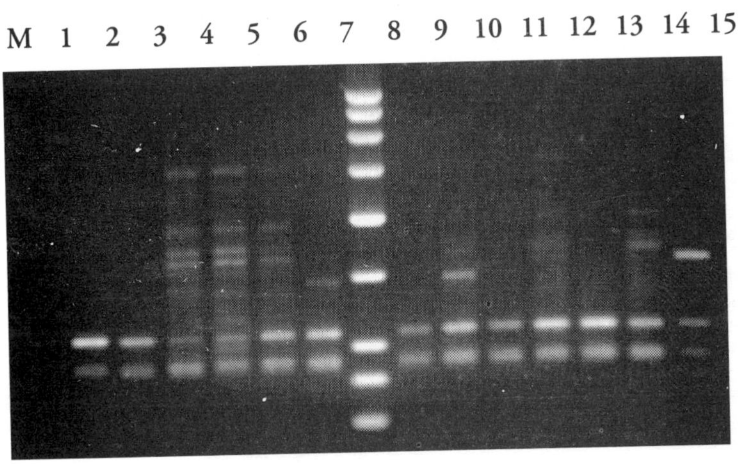

Figure 7. Primer RAn-3

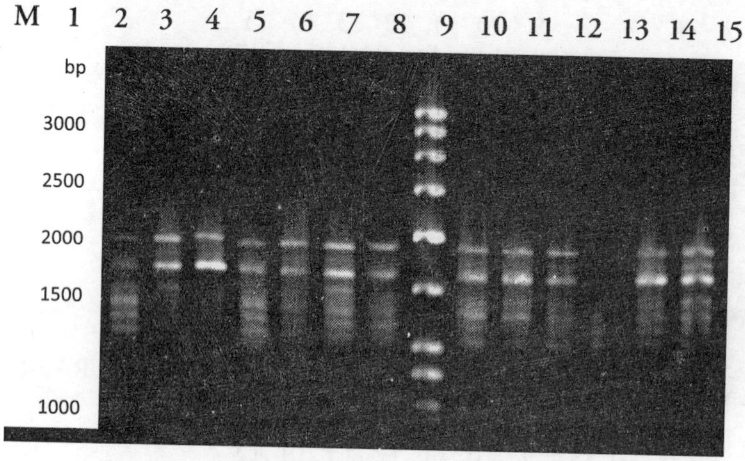

Figure 8. Primer RAn-4

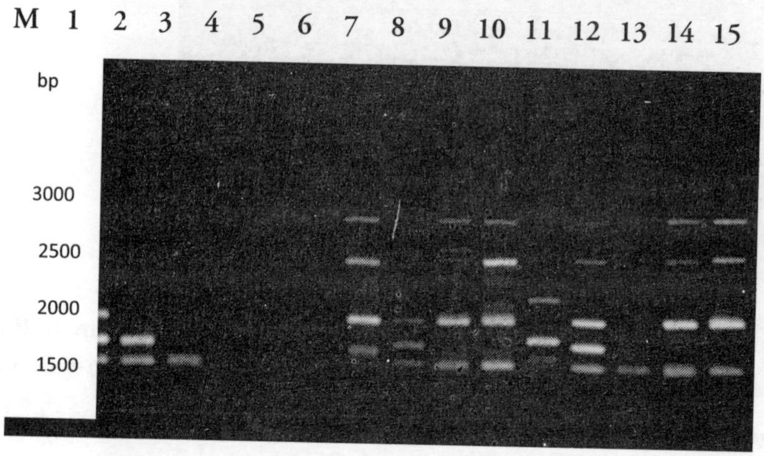

Figure 9. Primer RAn-5

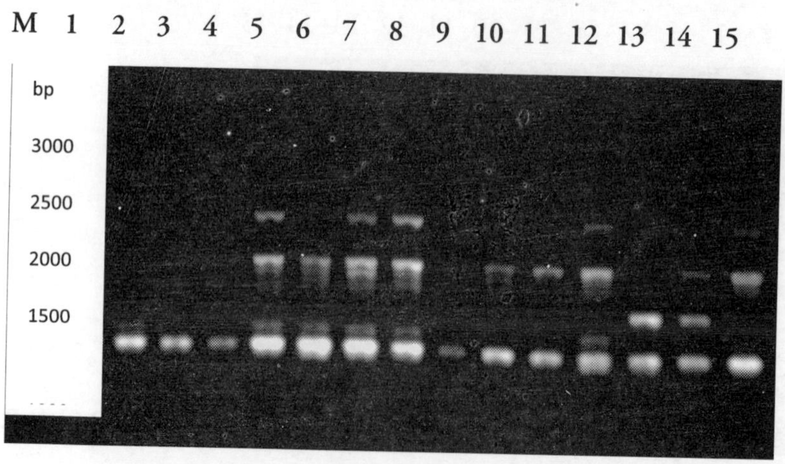

Figure 10. Primer RAn-6

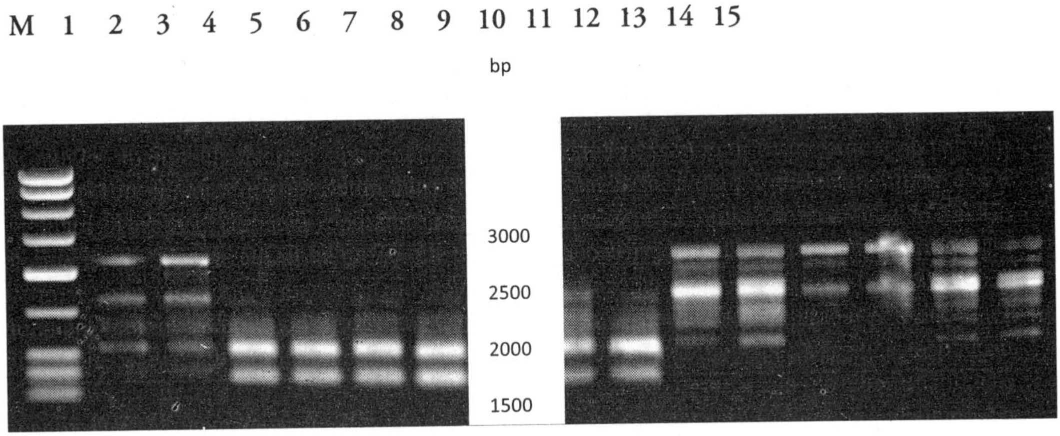

Figure 11. Primer RAn-8

Figure 7-11 DNA fingerprints of 15 individuals of obtained with primer. 1-15 individuals of Gandhisagar reservoir and M is molecular weight marker

Polymorphism

Since RAPD is a dominant marker, we assumed that each band represented the phenotype at a single biallelic locus. Therefore amplified fragments were scored for the presence (1) and absence (0) of homologues bands.

Polymorphism assessments were done in 30 individuals of *Mystus cavasius* collected from 15 individuals from each reservoir (Gandhisagar and Bandsagar). Total 64 RAPD scorable bands were recorded in the 15 individuals of Gandhisagar reservoir out of 64 bands 55 bands were polymorphic which shows 84.54% polymorphism. The average amplification of primers in individuals of Gandhisagar reservoir was 12.8 bands while average polymorphic banding pattern was 11 bands (Table 18). Primer wised polymorphism in fishes of Gandhisagar reservoir were obtained as Primer RAn-3 gave total 16 bands out of which 14 were polymorphic (87.50%). The range of molecular weight of DNA fragment length was recorded as 257-2545 base pair. Primer RAn-4 gave total 9 bands out of which 7 were polymorphic (77.77%) and range of molecular weight of DNA fragment length was as 271-900 base pair. Primer RAn-5 gave total 17 bands out of which 16 were polymorphic (94.11%) with range of molecular weight of DNA fragment length as 209-1135 base pair. Primer RAn-6 gave total 10 bands out of which 8 were polymorphic (80.00%). The range of molecular weight of DNA fragment length was recorded as 192-1100 base pair. Primer RAn-8 gave total 12 bands out of which 10 were polymorphic (83.33%). The range of molecular weight of DNA fragment length was recorded as 180-1758 base pair (Table 18).

In similar way total 108 RAPD scorable bands were recorded in the 15 individuals of Bandsagar reservoir out of which 103 bands were polymorphic which shows 93.07% polymorphism while average banding patter was 21.6 and average polymorphic bands were 20.6 (Table 19).

Primer wised polymorphism in fishes of Bandsagar reservoir were obtained as Primer wised polymorphism in fishes Bandsagar reservoir were obtained as Primer RAn-3 gave total 19 bands out of which 18 were polymorphic (94.73%). The range of molecular weight of DNA fragment length was recorded as 223-1675 base pair. Primer RAn-4 gave total 15 bands out of which 13 were polymorphic (80.66%). The range of molecular weight of DNA fragment length was recorded as 210-1265 base pair. Primer RAn-5 gave total 32 bands out of which 32 were polymorphic (100%). The range of molecular weight of DNA fragment length was recorded as 240-2631 base pair. Primer RAn-6 gave total 22 bands out of which 22 were polymorphic (100%). The range of molecular weight of DNA fragment length was recorded as 145-1700 base pair. Primer RAn-8 gave total 20 bands out of which 18 were polymorphic (90%). The range of molecular weight of DNA fragment length was recorded as 150-1400 base pair.

The similarity indices were also calculated with each primer and a matrix was developed. Finally, all matrices were analyzed as an average. The graphic phenogram of the genetic relatedness among the four accessions was produced by means of UPGMA (Un-weighed pair group method with arithmetic average) cluster analysis of averaged similarity index. UPGMA dendrogram clearly distributed all the specimens in two clusters (fig. 2). The fishes of Gandhisagar reservoir are in one cluster and fishes of Bandsagar reservoir in second cluster.

Shifat *et al.* (2003) have also reported RAPD Jaccards dissimilarity coefficient in 34 individuals of *Tenualosa ilisha* and observed 20.41% polymorphism. Rahman *et al.* (2009) studied on genetic variations of wild and hatchery populations of *Catla catla* revealed by RAPD markers and found overall 54.55% polymorphism. We found that 88.80 % of the loci in our study (average of both studies) were polymorphic as compared to the 75% reported by Islam *et al.* (2005) in *Catla catla*, 55.76% in *Oreochromis niloticus* (Zaeem and Ahmed, 2006).

Discussion

The protocol for RAPD technique may serve as a reliable, simple, easy to handle and elegant tool in molecular diagnosis of different accessions/variants of an important animal species. In essence, the RAPD method used in this study displayed appreciable intra–population variations or molecular polymorphism, which pre-existed in different locations. In spite of their morphological identity, substantial polymorphism was observed among the variants using RAPD. The percentage of polymorphism as 93.07% was observed in Bandsagar reservoir which shows that the species genetic diversity by itself is very much high, which might be due to favorable environment conditions. Whereas, 84.54 % polymorphism was recorded in the fishes of Gandhisagar reservoir which is also high if compared with other studies on various fish species.

References

- Bardakci, (2001). Random Amplified Polymorphic DNA (RAPD) Markers. *Turk. J. Biol.*, 25 185-196.

- Bhuiyan, AL. (1964). Fishes of Dhaka. Asiatic Society of Pakistan, Dhaka. 148p
- Daniel, R.J. Taxonomic uncertainities and conservation of the western Ghats. *Curr.*
- *Sci.*, 73, 169-170 (1997).
- Nei, M. and Li, W.H. (1979). Mathematical model for studying genetic variation in terms of restriction endonuclease. *Proc. Nat. Acad. Sci., USA*, 76, 52699-3263.
- Ponnian, A.G., A. Gopalakishnan (2000) Cultivable, ornamental, sport and food fishes endamic to peninsula India with special reference to western Ghats *In* Ponnian, A.G. Gopalakishnan, A. (Eds). Endemic fish diversity of western Ghats. NBFGRNATP Publication, National Bureau of Fish Genetic Resources, Lucknow, India, pp. 13-32.
- Rahman, S.M.Z, M. Khan, S.A.S. (2009) Islam Genetic variation of wild and hathery population of the catla India major carp (*Catla catla* Hamilton, 1822 Cypriniformes, Cyprinidae) revealed by RAPD markers. *Genet. Mol. Biol.*, 32, 197-201.
- Rahman, A.KA. (1989). Freshwater fishes of Bangladesh Zoological Society of Bangladesh, Dhaka, 364p.
- Sambrook J, Fritish EF and Maniatis T (1989). Molecular Cloning a Laboratory Manual. 2nd edn. Cold Spring Harbor Laboratory Press, New York.
- Shifat, R., B. Anwara, H. Khan (2003) Use of RAPD Fingerprinting for Discriminating Two Populations of Hilsa shad (*Tenualosa ilisha* Ham.) from Inland Rivers of Bangladesh. *Biochem. Mol. Biol.*, 36, 462-467.
- Williams, J.G.K., Kubelik, A.R., Livak, K.J., Rafalski J.A. and Tingey, S.V. (1990). DNA polymorphisms amplified by arbitrary primers are useful as genetic markers. *Nucl.* acids *Res.* 18, 6531-6535.
- Wu, Q., M. Chen, M. Buchwald and R.A. Phillips (1995) A simple, rapid method for isolation of high quality genomic DNA from animal tissues. *Nucleic Acids Res.*, 23, 5087-5088.
- Zaeem, S.Y.E. and M.M.M. Ahmed (2006) Genetic differentiation between sex reversal and Normal of Full-Sib Nile Tilapia, *Oreochromis niloticus* Based on DNA Fingerprinting. *Res. J. Fish. Hydrol.*, 1, 1-5.

9

ZOOPLANKTON DIVERSITY IN CHANGING ENVIRONMENT OF RIVER GOMTI IN LUCKNOW, UTTAR PRADESH, INDIA

P. K. Varshney, R. K. Agrahari, S. K. Singh, A. K. Yadav and W. S. Lakra*

Abstract: To evaluate the zooplankton diversity of river Gomti in relation to the prevailing environmental conditions four stations Maa Chandrika Devi, Daliganj, Ambedkar Park and Aquaduct were identified from upstream to downstream along the course of river in Lucknow. DO was low on many occasions at all the stations except Maa Chandrika Devi and COD values were high. There was a gradual increase in mean nitrite and phosphate values from up to down stream. Planktonic density was poor. Copepoda and cladocera were the most pre dominant groups followed by rotifers, planktonic oligochaetes, diptera, ostracoda, protozoa and hydra at all the stations. Relatively low density of zooplankton inspite of high nutrients budget may be a function of great infestation of water quality. Seasonally high density was reported during monsoon while low in post-monsoon. Increasing number of rotifers next to copepodes and cladocerans indicate an aquatic regime with substantial load of organic matter. Planktonic diversity was poor. Abundance of a particular group of organisms in a deteriorated and polluted environment may not culminate the energy at higher trophic level. Filthy condition with foul smell throught the length of the river coupled with poor water quality and appearance of indicator organisms.

Keywords: Environmental Condition, Planktonic Density

Introduction

The disposal of wastes into the aquatic environment has long been an accepted fact but it is being done without considering the possible consequences it may pose. The aquatic environment has a good deal of assimilation capacity but when the waste load exceed absorbing capacity pollution becomes a serious problem. Today aquatic pollution assumes a major challenge to man because the living resources form an essential source of animal protein. The rivers have become seriously

National Bureau of Fish Genetic Resources (ICAR), Aquaculture Research & Training Unit, Chinhat, Lucknow - 227105;
*CIFE, Rohtak Centre, Lahli, Via Anwal, Rohtak-124 411

vulnerable to pollution in industrialized cities leading to serious limitations to our harvesting the food from the polluted environments. Lucknow city is an example of the conditions posed by increasing urbanization and industrialization. It has a population of about 3.68 million. The quantity of domestic sewage and industrial waste produced in Lucknow is about 325 mld (Jal Nigam,1993). At present there is only one treatment plant of 42 mld capacity located at Gaughat to treat the sewage from Sarkata, Pata, Nagaria as well as from Gaughat itself. Therefore, a major portion of the sewage is discharged untreated at various places through 5 sewage pumping stations in the river Gomti. Usefulness of zooplankton in pollution monitoring programme to ascertain the river health environment is well established.

Material & Methods

Four sampling stations were identified along the course of river in Lucknow. The selection of two sampling stations was made in specific discharge dominated areas while one station in up stream presuming a stress free unpolluted environment and the other one station in down stream considering recovery zone were located (Fig. 1).

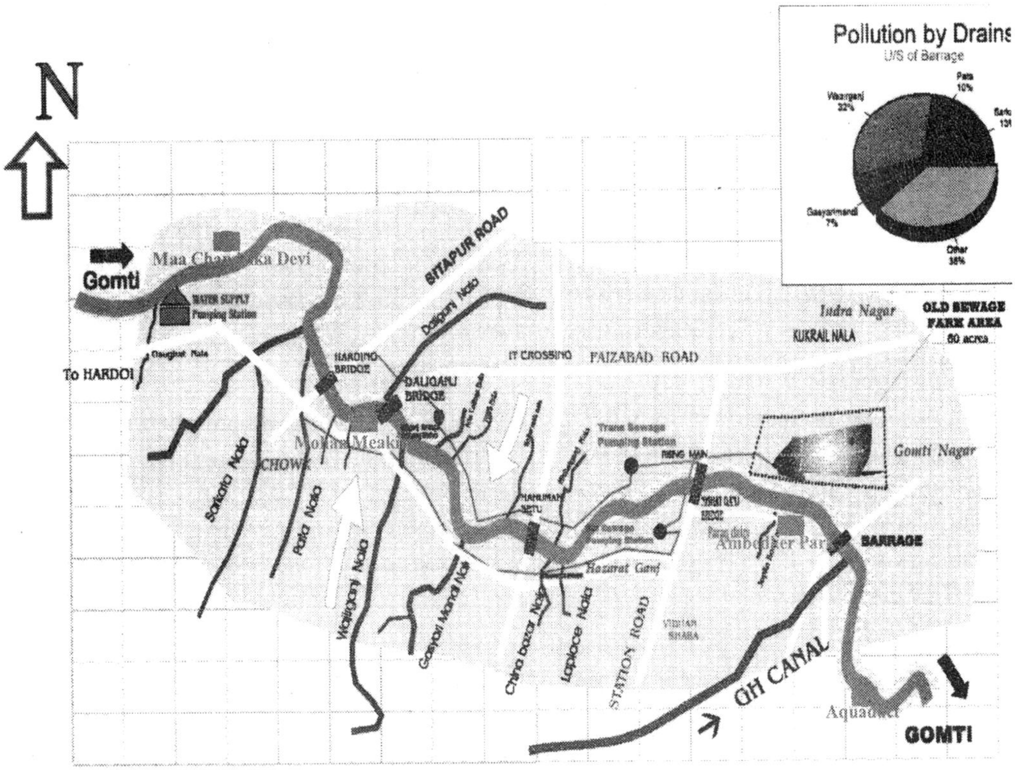

Fig 1: Location of Stations, main drains & sewers draining Gomti at Lucknow

DRAIN

TRUNK SEWER

Maa Chandrika Devi. It is one of the most sacred temple in the city and is situated about 35 km from Lucknow on Sitapur road. On auspicious day of "Amavasya" every month huge crowed gathers in the form of a mela to worship Maa Chandrika Devi and offer prayer. The station is situated in the upstream of river Gomti behind the temple. On this site the width of the river is less and comparatively the depth is more. The people are coming for picnic on this site on various occasions.

Daliganj. The station Daliganj is situated in the vicinity of the city close to Mohan Mekin a distillery plant discharging the waste directly into the river Gomti. The quantum of the discharge is around 9.00 mld. It is a fishermen locality. Washermen wash the clothes here on both sides of the bank of river. On this spot the river bed is black in colour because of the decomposition of organic wastes discharged by the distillery, other adjoining plants and sewage hence, the area is badly affected by fowl smell. This locality is also utilized for night soil. Hence, this station has been identified as dominated by industrial effluent apart from sewage.

Ambedkar Park. Ambedkar Park is situated on the bank of the river Gomti opposite to the five star hotel "Taj Residency". The Ambedkar Park is recently constructed by the Government of Uttar Pradesh and dedicated to the nation. It is one of the important and beautiful picnic spot. In the upstream of this park at a distance of about 0.5 km there is a cremation ground known as "Bhainsa Kund". Close to this Bhainsa Kund city drains opens into the river. Gomti receives around 73 mld of sewage from Jopling road adjoining at this station. Hence, the area is dominated with domestic sewage discharge and was identified for sampling.

Aquaduct. It is one of the peculiar spot on the river Gomti in down stream in the out skirt of the city Lucknow. A barrage is constructed on the river. Over the river there is Sharda canal flows. On the top of the Sharda canal there is a bridge for pedestrians. It gives a beautiful architectural view. On the side of the river bank there is a guest house of irrigation department surrounded by beautiful garden. This area is utilized for recreation by lucknowites during holidays. This station has been identified as a recovery zone in down stream.

Samples were collected monthly for assessment of environmental parameters (water & soil) and biological parameters. Water sample include temperature (air & water), pH, DO_2, CO_2, alkalinity, hardness, and nutrients (NO_2, NO_3 & PO_4). Temperature, pH, CO_2 and DO_2 were analyzed immediately in the field itself, while for rest of the parameters the water samples were collected in half liter narrow mouth polythene bottle and were analyzed following the Standard methods (APHA, 1998). Soil analysis was done for pH, $CaCO_3$, water Retention Capacity and organic carbon (APHA, 1998).

Zooplankton, the microscopic free floating animal component of aquatic system are endowed with many remarkable features and are often armored with spines, which hamper their predation by higher organisms. The ability of movement not only provide them an effective defense measure but also enable them to actively search and feed upon the phytoplankton. Their high & rapid rate of parthenogenetic reproduction usually overcomes the predation losses and enables them to exploit algal blooms. These constitute an important link between primary producers and consumers of higher order in aquatic food web. Therefore, the population dynamics of zooplankton with reference to system energetic provides key information for management practices.

The zooplankton samples collected following the Standard methods (APHA, 1998). The samples thus collected were preserved in 5% formaline. Identify the zooplankton in the sample in laboratory using keys & monographs and then proceed for qualitative & quantitative analysis (APHA, 1998; Edmondson, 1959, Needham & Needham, 1962 and Ward & Whipple, 1959).

Results and Discussion

The results of environmental parameters are presented in tables 1 and 2 and that of zooplankton in figs. 2-5. The water temperature follows the trend of atmospheric temperature. The mean pH shows a gradual decrease from upstream to down stream stations except Daliganj with exceptionally low value (7.20). DO were low at both sewage dominated (Ambedkar Park) and industrial waste dominated (Daliganj) stations. On many occasions DO was less than 1.00 mg l^{-1} (0.20). Mean DO value was higher (6.15 mg l^{-1}) at Maa Chandrika Devi. Perhaps the high organic load might have causes the depletion of oxygen. There was a gradual increase in CO_{2--} from upstream to downstream. High mean alkalinity (266.96 mg l^{-1}) at Daliganj and hardness (288.07 mg l^{-1}) at Maa Chandrika Devi were observed. Mean values of phosphate and nitrite show increasing trend from upstream to downstream Higher mean nitrate was noticed at Daliganj and Ambedkar Park may be because of high organic load. COD was also high in this river almost through out the stretch. Increasing trend of CO_{2--}, PO_4 and NO_2 from upstream to downstream as well as poor DO values represents the poor river condition due to influx of sewage and industrial effluents. The water quality of the river indicate that the river is under severe pollution stress both by sewage as well as industrial wastes.Comparatively mean sediment pH was low at Daliganj and Ambedkar Park. Increasing trend in mean value of $CaCO_3$ was noticed from upstream to downstream. Water Retention Capacity and Organic Carbon (32.74%) were comparatively poor at Ambedkar Park compared to other stations may be because of silty nature of substratum (Varshney, 2006). The untreated sewage is the highest contaminant of water in the underdeveloped countries like India. But sewage treatment plants are expensive, especially if primary, secondary and tertiary treatments are to be done. Many waterworks in India claim to remove all bacterial contamination by excessive chlorine treatment (Which itself is harmful to man), but it is observed that *E. coli* is the most resistant of all coliforms and other pathogens.

Dissolved organic matter in rivers, streams, lakes and ponds cause a steep rise in BOD with deoxygenating of water, killing of zooplankton, nekton and fish. Suspended organic matter is more hazardous to fish as it gathers round the each silt particle and this silt organic mass clogs the gills causing asphyxiation even though the DO in water may be above 5.00 mg l^{-1}.

Sugar factory in upstream near Gomti (prior entering the river in Lucknow) discharges the entire molasses and wastes into the Gomuti River every year where thousands of fish may be seen dead and floating downstream for miles as also on the banks. Similarly tannery waste near Gaga bridge, undo discharged by long channel into Ganga. Not only there is foul smell but the water discharged also contain high organic matter. The stretch of Ganga where it falls has become entirely lifeless and the pollution recovery zone appears after several miles downstream. Besides tannery also discharges tanning chemicals which are toxic and poisonous and highly hazardous to fish.

Table 1. Variation in Water Parameters of River Gomti at Lucknow during 2003- 2005

Parameters/Stations	Maa Chandrika Devi	Mohan Mekins	Ambedkar Park	Aquaduct
Temperature (°C)				
(a) Air	11.00-37.00 26.39	15.00-36.00 26.13	14.00-33.00 24.93	14.00-30.00 23.62
(b) Water	12.50-32.60 23.20	14.00-32.70 24.03	13.00-31.60 23.10	13.80-31.30 22.00
pH	7.50-8.40 7.90	7.20-8.20 7.20	7.20-8.00 7.60	6.90-7.90 7.30
DO_2 (mg l^{-1})	3.60-7.70 6.15	0.80-6.26 3.38	0.40-7.37 3.60	0.20-8.23 3.49
CO_2 (mg l^{-1})	Nil-20.00 7.80	Nil-30.00 8.18	Nil-36.00 8.02	Nil-26.00 7.40
Alkalinity (mg l^{-1})	100.00-360.00 236. 80	90.00-560.00 282.05	80.00-400.00 238.35	100.00-400.00 237.78
Hardness (mg l^{-1})	80.00-340.00 218.00	98.00-420.00 249.83	80.00-328.00 224.77	130.00-340.00 221.75
Phosphate (mg l^{-1})	0.05-13.00 2.40	0.60-8.00 2.72	0.69-3.00 2.55	0.52-11.20 3.31
Nitrite (mg l^{-1})	0.06-6.90 0.83	0.02-5.23 1.00	0.02-8.77 1.18	0.01-6.19 1.38
Nitrate (mg l^{-1})	3.60-74.40 22.70	5.82-74.70 30.60	6.56-69.68 26.08	6.80-70.20 23.45
COD (mg l^{-1})	3.60-203.00 36.05	6.40-59.20 21.34	3.20-219.00 37.35	8.00-187.20 34.70

Table 2. Variation in Soil Parameters of River Gomti in Lucknow during 2003- 2005

Parameters/Stations	Maa Chandrika Devi	Daliganj	Ambedkar Park	Aquaduct
pH	6.60-8.20 7.76	6.30-8.00 7.42	5.20-8.20 7.56	6.00-8.50 7.64
Ca CO_3 (%)	1.50-25.75 9.78	1.25-28.75 10.62	2.50-30.25 12.75	1.50-36.75 13.36
Water Retention Capacity (%)	24.00-46.60 38.03	29.00-57.00 38.80	17.00-40.00 32.74	29.00-59.00 39.04
Organic Carbon (%)	0.07-1.00 0.48	0.08-2.04 0.67	0.07-1.20 0.44	0.07-1.40 0.53

Maa Chandrika Devi

Protozoa was represented by only one genera i.e. *Polytoma*. The two genera *Cypris* and *Cyprinotus* belong to order Ostracoda. Order Rotifera has *Brachionus, keratella* and *Hexorthra*. The genera belong to the order Cladocera were *Daphnia, Moinodaphnia, Ceriodaphnia, Moina* and some unidentified forms. The genera *Cyclops, Mesocyclops* and *Diaptomus* represent the Copepoda. Diaptran insects were also reported. The worms were *Nais* and *Tubifex*. The *Hydra* polyp were observed rarely at this station.

The dominant genera commonly reported at this station were *Brachionus* (13.88%), *Daphnia* (5.55%), *Moinodaphnia* (6.94%), *Cyclops* (22.22%), *Mesocyclops* (11.11%), *Diaptomus* (8.33%) apart from worms (8.33%). Maxima for population density was reported in April, 04. Seasonally, higher density was observed during monsoon followed by pre-monsoon and post-monsoon.

Daliganj

The phyla Protozoa has only one genera *Euglena*. Ostracoda were represented by two genera *Cypris* and *Cyprinotus*. The genera belongs to Rotifera were *Filinia, Asplanchna, Conchilus, Testudinella, Brachionus, Trichocera, Monostyla* and *Hexorthra*. The cladocerans were *Daphnia, Moinodaphnia, Ceriodaphnia, Moina, Simocephalus, Macrothrix* and some unidentified forms. The genera *Cyclops* and *Diaptomus* belong to Copepoda. Worms were *Nais, Tubifex* and *Chironomus* larvae. Apart from all these *Hydra* was also found. Dominant genera reported at Daliganj were *Brachionus* (12.60%), *Hexorthra* (5.26%), *Daphnia* (4.21%), *Moinodaphnia* (17.89%), *Ceriodaphnia* (4.21%), *Cyclops* (20.00%) and worms (8.42%). Maxima for population density was observed in April, 04 and July, 04. Seasonally higher values for density were noticed during monsoon followed by pre-monsoon & post-monsoon.

Ambedkar Park

The phyla Protoza represented by three genera *Euglena, Vorticella* and *Polytoma*. The *Cypris* and *Cyprinotus* were the candidate genera belong to Ostracoda. The genera belongs to Rotifera were *Filinia, Asplanchna, Conchilus, Testudinella, Brachionus, Trichocera, Monostyla* and *Hexorthra*. The order Cladocera has *Daphnia, Moinodaphnia, Ceriodaphnia, Moina, Simocephalus* and unidentified forms. The Copepoda has *Cyclops, Mesocyclops Paracyclops* and *Diaptomus*. Diaptrans were also noticed apart from worms. (*Nais, Tubifex* and *Chironomus*) and *Hydra*. The dominant and common genera encountered at this station were *Brachionus* (4.97%), *Hexorthra* (3.73%), *Daphnia* (3.10%), *Moinodaphnia* (16.77%). *Ceriodaphnia* (18.63%), *Moina* (8.94%), *Cyclops* (11.18%), *Diaptomus* (4.35%), and worms (*Nais, Tubifex, Chironomus*(4.97%). Maxima for population density was noticed in July, 04 & Sept, 04. Seasonally high density was observed during monsoon followed by pre monsoon & post monsoon.

Aqua-duct

Phylum Protozoa has genera Euglena, Vorticella and Epistis. Ostracoda has only one genera Cypris. The rotiferan order has the genera Asplanchna, Brachionus, Keratella, Trichocera, Monostyla Pompholyx, Gastropodidae, Ascomorpha, & Sinatherina. Under cladocera the genera encountered were Daphnia, Moinodaphnia, Ceriodaphnia, Moina, Diaphanosoma, Simocephalus, Leydigia, Macrothrix & unidentified cladocerans. The copepoda has Cyclops, Mesocyclops. Paracyclops and Diaptomus. Diaptrans were also found apart from worms i.e. Nais, Tubifex and chironomus and Hydra. The dominant genera at this station were *Brachionus* (18.82%), *Ascomorpha* (3.53%), *Moinodaphnia* (21.18%), *Ceriodaphnia* (6.27%), *Moina* (4.71%), *Cyclops* (7.06%), & worms (7.65%). Maximum population density was reported in July, 04 followed by Oct. 04. Seasonally high value were noticed during monsoon followed by pre-monsoon & post-monsoon.

Because of their short life cycle plankters respond quickly to environmental changes and hence their standing crop and species composition are most likely to indicate the quality of water mass in which they are found. They strongly influence certain non biological aspects of water quality such as pH, colour, taste and odor and in a very practical sense they are a part of water quality. Certain taxa are often useful in determining the origin or recent history of a given water mass. Because of their transient nature and often patchy distribution, however, the utility of plankters as water quality indicators may be limited. Information on plankton as indicators is interpreted best in conjunction with concurrently collected, physico-chemical and other biological data. Among zooplankton copepods and the cladocerans were the most pre dominant groups in a sequence of rotifers, planktonic oligochaetes, diptra, ostracoda, protozoa and hydra at all the stations in general. Gurumayum *et al.* (2002) also reported dominance of cladocerans, rotifers and protozoans for the rivers of Meghalaya.

Seasonal variation in plankton abundance is a common phenomenon owing to change in climatic conditions thereby change in physico-chemical parameters. Mean maxima for zooplankton at all the four stations was observed during monsoon while minimum density was recorded during post-monsoon. Gurumayum *et al.* (2002) also reported poor zooplankton density but seasonally high population density was noticed during pre-monsoon. Higher zooplankton density during monsoon can be contributed to the rainfall which contribute unpolluted water. Decline in plankton density during post monsoon was also obserbed for rivers of Meghalaya (Gurumayum *et al.*, 2002), river Hoogly Chakraborty *et al.*, 1995) and river Damodar (Banarjee,1998). Carter, 1960 explained that in the tropics the amount of rain fall play a significant role in regulating the various seasonal biological rhythums.

Highest number of rotifer genera (11) were encountered at Daliganj while lower at Maa Chandrika Devi (5). The number of rotifer genera at Ambedkar Park and Aqua duct were 09 only. Cladoceran genera were highest (9) at Aqua duct followed by Daliganj (8), Maa Chandrika Devi (5) and Ambedkar Park (5). Copepode genera were 4 each at Ambedkar Park and Aquaduct while at Maa Chandrika Devi three and at Daliganj two only.

The population of zooplankton is comparatively low as compared to phytoplankton. Pathak *et al.* (2001) observed low population density for the three forerunners of river Brahmputra. Increasing number of rotifers next to copepodes and cladocerans indicate an aquatic regime with substantial load of organic matters (Baruah, 1993). Beside this protozoan also contributed in the community structure of zooplankton which too indicate organic load in the system. Relatively low abundance of zooplankton could be attributed to less availability of energy for their growth probably in a polluted environment. Higher concentration of nutrients especially NO_2 and NO_3 as well as COD were encountered during pre-monsoon may be due to precipitation and low influx of effluents in summer. Varshney (2006) also observed the similar condition for river Gomti in Lucknow. Phosphate values were high during monsoon probably due to runoff from agricultural fields. Higher values of nutrients coincide with high population density of phytoplankton and benthos during pre-monsoon (Varshney, 2006).

Relatively low abundance of plankton population and biomass in spite of high nutrient budget of the system was a function of great infestation of water quality by pollutants which indicate extreme state of deterioration of water quality leading to un-conducive ecological regime for fisheries development. The available energy in the system is largely lost and a little is available for luxuriant growth of pelagic organisms like plankton, an important component in aquatic grazing chain. The results are comparable with Gorai *et al.*, 2003 for the wetland.

Abundance of a particular group of organisms in a deteriorated and polluted environment may not culminate the energy at higher trophic level. Similarly the landing data procured from the State fisheries Department shows poor fish landing in Lucknow with declining trend. Among zooplankton copepods and the cladocerans were the most pre-dominant groups in a sequence of rotifers, planktonic oligochaetes, diaptra, ostracoda, protozoa and hydra at all the stations in general.

Mean maxima for zooplankton at all the four stations was observed during monsoon while minimum density was recorded during post-monsoon. Increasing number of rotifers next to copepodes and cladocerans indicate an aquatic regime with substantial load of organic matters.

References

APHA, (1998): Standard methods for the examination of water and waste water. American Public Health Association Washington D.C.

Baruah, A., 1993: Plankton variability of a tropical wetland, Kawar Lake, Begusarai (Bihar), *J. Freshwater Biol.*, 5 (1) 27-32

Banerjee, R. K., Mukhopadhyay, M. K., Pandit, P. K. Karamakar, H. C.,Mitra, K., Chatterjee, S. K., Das, B. B. and Sengupta, A., (1998): River Damodar and its environment, *Bull. No. 79, Central Inland Fish. Res.* Ins., Barrackpore. India

Chakraborty, N, Chakraborty, P. K., Vinci G. K. and Sugunan V. V., (2002): Spatiotemporal distribution pattern of certain plankton of river Hoogly, *J Inland Fish. Soc, India*, 27(1) 6-12

Gurumayum, S. D., Dainari, P., Goswami, B. S., Sarkar, A and Choudhury, M, (2002): Physico Chemical qualities of water and plankton of selected rivers in Meghalaya, *J Inland Fish. Soc, India*, 34 (2) 36-42

Gorai, B. K., Jha, B. C. and Pandey P. N. (2003): Plankton ecology of a wetland (Kaithkola), North Bihar, *ISFP News letter*, Vol 1&2 27-30.

Jal Nigam, 1993: Report on pollution abatment of river Gomti.

Pathak V., Choudhury, M., Laal, A., Bhattachariya, B. K., Sarkar, A. and. Mahavar L. R., (2001): Ecological status and fish production of Siang, Dibang and Lohit – the three fore number of river Brahmaputra, *J Inland Fish Soc India*, 33(1) 23-28.

Ward, H. B, and Wipple G.C; 1959: Fresh water biology (2nd Edn) John Wiley & Sons. Inc. New York. 1-1248

Needham, J.G. and Needham P.R; 1962: A guide to study of fresh water biology Holden Bay, San Francisco U.S.A., 1-107.

Edmondson, W.T. (Ed.), 1959: Freshwater Biology. New York, Wiley 1-1157

Varshney, P. K., (2006): Biodiversity and fisheries potential of river Gomti in relation to the prevailing environmental conditions in Lucknow, Project Report, CIFE, Lucknow Center, Lucknow 1-81

Table 3. Percentage Composition of Zooplankton of River Gomti at Lucknow (Maa Chandrika Devi)

Group/Months	Oct 03	Nov 03	Dec 03	Jan 04	Feb 04	Mar 04	Apr 04	May 04	Jun 04	Jul 04	Aug 04	Sep 04	Oct 04	Nov 04	Dec 04	Jan 05	Feb 05	Mar 05	Apr 05
PROTOZOA																			
Polytoma	--	--	--	--	--	--	12.50	--	--	--	--	--	--	--	--	--	--	--	--
OSTRACODA														--	--	--	--	--	--
Cypris	--	--	--	--	--	--	--	40.00	--	--	--	--	--	--	--	--	--	--	--
Cyprinotus	--	--	--	--	--	--	--	--	--	--	--	--	--	33.33	--	--	25.00	--	16.67
ROTIFERA														--	--	--	--	--	--
Brachionus	16.67	50.00	--	--	--	--	--	--	--	--	50.00	16.33	50.00	33.33	--	50.00	25.00	--	--
Keratella	--	--	--	--	--	--	12.5	--	--	--	--	--	--	--	--	--	--	--	--
Hexorthra	--	--	--	--	--	--	--	--	--	--	--	--	--	--	--	--	--	--	--
Scaridium			50.0					20.0											
Sinatherina																			
CLADOCERA														--	--	--	--	--	--
Daphnia	33.33	--	--	--	--	--	--	--	--	20.0	--	--	--	--	--	--	--	33.33	--
Moinodaphnia	--	50.00	--	--	50.00	--	--	--	--	--	--	16.33	--	33.33	--	--	--	33.33	
Ceriodaphnia	--	--	--	--	--	--	--	--	--	--	16.33	--	--	--	--	--	--	--	--
Moina	--	--	--	--	--	--	12.50	--	--	--	--	--	--	--	--	--	--	--	--
Alona												50.00							
COPEPODA														--	--	--	--	---	--
Cyclops	16.67	--	50.00	50.00	50.00	100.0	37.50	--	--	20.00	16.33	--	50.00	--	50.00	--	25.00	--	33.33
Mesocyclops	--	--	--	--	--	--	12.50	--	--	--	--	--	--	--	--	--	--	--	--
Diaptomus	16.67	--	--	50.00	--	--	--	--	--	20.00	--	16.33	--	--	50.00	50.00	25.00	--	16.67
DIPTERA	16.67	--	--	--	--	--	12.50	--	50.00	--	--	--	--	--	--	--	--	33.33	16.67
OLIGOCHAETES																			
Nais/Tubifex	--	--	--	--	--	--	--	20.00	50.00	20.00	16.33	--	--	--	--	--	--	--	16.67
Hydra	--	--	--	--	--	--	--	20.00	--	20.00	--	--	--	--	--	--	--	--	--

Table 4. Percentage Composition of Zooplankton of River Gomti at Lucknow (Daliganj)

Group/Months	Oct 03	Nov 03	Dec 03	Jan 04	Feb 04	Mar 04	Apr 04	May 04	Jun 04	Jul 04	Aug 04	Sep 04	Oct 04	Nov 04	Dec 04	Jan 05	Feb 05	Mar 05	Apr 05
PROTOZOA																			
Euglena	--	--	--	–	--	--	--	--	--	8.33	--	--	--	--	--	--	--	--	--
OSTRACODA																			
Cypris	--	--	--	–	--	--	9.09	--	--	--	--	--	--	20.00	--	--	--	--	--
Cyprinotus	--	--	--	–	--	20.00	--	--	--	--	--	--	--	--	--	--	--	--	--
ROTIFERA																			
Fitinio	--	--	--	–	--	--	--	--	--	--	--	--	--	20.00	--	--	--	--	--
Asplanchna	--	--	--	–	--	--	9.09	--	--	--	--	--	11.11	--	--	--	--	--	--
Conchilus	--	--	--	–	--	--	9.09	--	14.28	--	--	--	--	--	--	--	--	--	--
Testudinella	--	--	--	–	--	--	9.09	--	--	--	--	--	--	--	--	--	--	--	--
Brachionus	--	50.00	50.00	–	--	40.00	9.09	16.67	--	8.33	20.00	11.11	--	--	50.00	--	33.33	--	20.00
Trichocera	--	--	--	–	--	--	--	--	--	--	--	--	11.11	--	--	--	--	--	--
Monostyla	--	--	--	–	--	20.00	--	--	--	--	--	--	--	--	--	--	--	--	--
Hexorthra	--	--	--	50.00	--	--	--	16.67	--	--	--	--	--	--	--	--	--	--	--
Dicramphorus	--	--	--	–	--	--	--	--	--	--	--	--	--	--	--	50.00	--	--	--
Scaridium	--	--	--	–	--	--	--	--	--	8.33	--	--	--	--	--	--s	--	--	--
Gastropus		50.00	--	–	--	--	--	--	--	--	--	--	--	--	--	--	--	--	--
CLADOCERA																			
Daphnia	--	--	--	–	--	--	--	--	14.28	--	--	--	11.11	--	--	--	33.33	33.33	--
Moinodaphnia	33.33	--	--	–	--	--	27.27	--	14.28	25.00	--	44.44	22.22	--	50.00	--	--	33.33	20.00
Ceriodaphnia	--	--	--	–	--	--	9.09	--	14.28	--	20.00	--	11.11	--	--	--	--	--	--
Monia	--	--	--	–	--	--	--	--	--	--	--	--	11.11	--	--	--	--	--	--
Simocephalus	--	--	--	–	--	--	--	--	14.28	--	--	--	--	--	--	--	--	--	--
Macrothrix	--	--	--	–	--	--	--	--	--	--	--	--	11.11	--	--	--	--	--	--
Bosmina	--	--	--	–	--	--	--	--	--	8.33	--	--	--	--	--	--	--	--	--
Alona	--	--	--	–	--	--	--	--	--	--	--	33.33	--	--	--	--	--	--	--
Unideritified	--	--	--	–	--	--	--	16.67$	14.28#	--	--		--	--	--	--	--	--	--
COPEPODA																			
Cyclops	33.33		50.00	50.00	--	20.00	18.18	16.67	14.28	25.00	20.00	11.11	11.11	--	--	--	33.33	33.33	60.00
Diaptomus	33.33	--	--	–	--	--	--	--	--	8.33	20.00	--	--	--	--	--	--	--	--
OLIGOCHAETES																			
Nais/Tubifex/Chiro.	--	--	--	–	100.00	--	--	--	--	8.33	20.00	--	--	60.00	--	50.00	--	--	--
OTHER INV.																			
Hydra	--	--	--	–	--	--	--	33.33	--	--	--	--	--	--	--	--	--	--	--

Table 5. Percentage Composition of Zooplankton of River Gomti at Lucknow (Ambedkar Park)

Group/Months	Oct 03	Nov 03	Dec 03	Jan 04	Feb 04	Mar 04	Apr 04	May 04	Jun 04	Jul 04	Aug 04	Sept 04	Oct 04	Nov 04	Dec 04	Jan 05	Feb 05	Mar 05	Apr 2005
PROTOZOA																			
Euglena	--	--	--	--	--	--	50.00	--	--	--	--	--	--	--	--	--	--	--	--
Vorticella	--	--	--	--	--	--	25.00	--	--	--	--	--	--	--	--	--	--	--	--
Polytoma	--	--	--	--	--	--	--	--	--	--	--	--	--	--	--	--	--	--	14.76
OSTRACODA																			
Cypris	--	--	--	--	33.33	--	--	--	--	4.16	--	--	--	25.00	--	--	--	--	--
Cyprinotus	--	--	--	--	--	5.26	--	--	--	--	--	5.00	--	--	--	--	--	20.00	--
ROTIFERA																			
Fitinio	--	--	--	--	--	--	--	--	--	4.16	--	--	--	--	--	--	--	--	--
Asplanchna	--	--	--	--	--	--	--	--	--	--	12.50	--	8.33	--	--	--	--	--	--
Conchilus	--	--	--	--	--	--	--	--	--	--	--	--	--	25.00	--	--	--	--	--
Testudinella	--	--	--	--	--	--	--	--	--	4.16	--	--	--	--	--	--	--	--	--
Brachinous	--	--	--	--	--	5.26	--	11.11	--	4.16	12.50	5.00	8.33	--	16.66	16.66	16.66	-	--
Keratella	--	--	--	--	--	--	--	---	--	--	--	--	--	--	--	--	--	--	--
Trichocera	--	--	--	--	--	--	--	--	--	--	--	--	8.33	--	--	--	--	--	--
Monostyla	--	--	--	--	--	--	--	--	--	--	--	--	--	--	--	--	--	--	--
Hexorthra	--	--	--	--	--	--	--	--	--	4.16	--	--	--	--	--	--	--	--	--
Unidentified	--	--	33.33!	--	--	--	--	--	--	--	12.50#	--	8.33#	25.00*	--	--	--	20.00&	--
CLADOCERA																			
Daphnia	--	--	--	50.00	--	--	25.00	11.11	--	--	--	--	8.33	--	--	16.66	--	--	--
Moinodaphnia	--	50.00	--	--	33.33	15.78	--	22.22	40.00	45.76	--	15.00	25.00	--	--	--	--	20.00	--
Ceriodaphnia	--	--	--	--	--	57.86	--	11.11	--	12.48	25.00	60.00	8.33	--	--	--	--	--	--
Moina	--	--	--	--	--	--	--	11.11	--	--	--	--	8.33	--	--	--	50.00	--	52.36
Simocephalus	--	--	--	--	33.33	--	--	--	--	4.16	--	--	--	--	--	--	--	--	4.76
Unideritified	--	--	--	--	--	--	--	--	--	4.16$	12.50@	--	--	--	50.00ss	--	--	20.00	23.80
COPEPODA																			
Cyclops	50.00		33.33	50.00	--	10.52	--	--	20.00	8.32	--	10.00	8.33	--	16.66	16.66	16.66	--	14.28
Mesocyclops	--	--	--	--	--	--	--	--	--	--	--	--	--	--	--	--	--	--	--
Paracyclops	--	--	--	--	--	--	--	--	--	--	--	--	--	--	--	--	--	--	--
Diaptomus	50.00	--	33.33	--	--	--	--	--	20.00	4.16	12.50	--	--	--	16.66	--	--	20.00	--
DIPTERA	--	50.00	--	--	--	--	--	--	--	--	--	--	--	--	--	--	16.66	--	--
OLIGOCHAETEA	--	--	--	--	--	--													
Nais/Tubifex/Chiro.	--	--	--	--	--	5.26	--	11.11	20.00	--	12.50	5.00	8.33	25.00	--	50.00	--	--	--
OTHER INV.					--														
Hydra	--	--	--	--	--	--	--	11.11	--	--	--	--	--	--	--	--	--	--	--

$ Chidorus, # Acromorpha, @ Alona, * Ploesoma, & Syncheta Sinatherina

Table 6 Percentage Composition of Zooplankton of River Gomti at Lucknow (Aquaduct)

Groups/Months	Oct 03	Nov 03	Dec 03	Jan 04	Feb 04	Mar 04	Apr 04	May 04	Jun 04	Jul 04	Aug 04	Sept-04	Oct 04	Nov 04	Dec 04	Jan 05	Feb 05	Mar 05	Apr 05
PROTOZOA																			
Euglena	–	–	–	–	–	–	42.24	–	–	–	–	–	–	–	–	–	–	–	–
Vorticella	–	–	–	–	–	–	28.50	–	–	–	–	–	–	–	–	–	–	–	–
Epistlis	–	–	–	–	–	–	–	–	–	–	–	–	–	–	–	7.69	–	–	–
Paramaecium	–	–	–	–	–	–	–	–	–	–	–	–	–	–	–	–	–	25.00	–
OSTRACODA																			
Cypris	–	–	–	–	–	–	–	–	–	–	–	7.14	–	–	–	–	–	–	5.88
ROTIFERA																			
Asplanchna	–	–	–	–	–	–	–	–	–	–	–	–	9.5	–	–	–	–	–	–
Brachionus	33.33	50.00	–	50.00	–	27.27	–	20.00	–	8.00	7.69	28.57	57.12	20.00	50.00	–	33.33	25.00	
Keratella	–	–	–	–	–	–	–	–	–	–	7.69	–	–	–	–	–	–	–	–
Trichocra	–	–	–	–	–	–	–	–	–	–	23.02	–	–	–	–	–	–	–	5.88
Monostyla	–	–	–	–	–	–	–	–	–	–	–	7.17	–	–	–	–	–	–	–
Pompholyx	–	–	–	–	–	–	–	–	–	–	–	–	14.28	–	–	–	–	–	–
Gastropodidae	–	–	–	–	–	–	–	–	–	–	–	–	4.76	–	–	–	–	–	–
Ascomorpha																38.45			5.88
Sinatherina	–	–	–	–	–	–	–	–	–	–	–	–	–	40.0	–		–	–	
CLADOCER																			
Daphnia	–	–	–	–	–	–	14.28	–	–	8.00	–	–	4.76	–	–	–	–	–	–
Moinodaphnia	33.33	–	33.33	50.00	25.00	27.27	–	20.00	12.50	64.00	15.38	21.42	–	–	–	7.69	16.66	–	17.64

contd...

...contd.

Ceriodaphnia	--	--	--	--	--	27.27	--	10.00	25.00	4.00	7.69	21.42	9.52	--	--	--	--	--	--
Moina	--	--	--	--	--	--	--	10.00	--	--	--	--	--	--	--	--	--	--	41.20
Diaphanosoma	--	--	--	--	--	--	--	20.00	--	--	--	--	--	--	--	--	--	--	--
Simocephals	--	--	--	--	--	--	--	--	--	4.00	--	--	4.76	--	--	--	--	--	5.88
Leydigia	--	--	--	--	--	--	--	--	--	--	7.6	--	--	--	--	--	--	--	--
Macrothrix	--	--	--	--	--	9.09	--	--	--	--	--	7.14	--	--	--	--	--	--	--
Bosmina																	16.66		
Unideritified	--	--	--	--	--	--	--	--	12.50	4.00	--	--	--	20.00$	--	23.07&	--	--	--
COPEPODA																			
Cyclops	--	50.00	--	--	25.00	--	--	10.00	12.50	4.00	7.69	7.14	4.76	--	--	7.69	16.66	25.00	5.88
Mesocyclops	--	--	--	--	--	--	--	--	--	--	--	--	--	--	--	--	--	--	5.88
Paracyclops	--	--	--	--	--	--	--	10.00	--	--	--	--	--	--	--	--	--	--	--
Diaptomus	33.33	--	--	--	25.00	--	--	--	12.50	4.00	7.69	--	--	--	--	--	--	--	--
DIPTERA	--	--	--	--	--	--	--	--	12.50	--	--	--	--	--	50.00	--	--	--	--
OLIGOCHAET-EA																			
Nais/Tubifex/Chir-o.	--	--	66.67	--	25.00	9.09	--	--	12.50	--	15.38	--	--	20.00	--	23.07	16.66	25.00	
OTHER INV.																			5.88*
Hydra	--	--	--	--			14.2	--	--	--	--	--	--	--	--	--	--	--	--

& Plumatella, * Chaetogester

10

BIODIVERSITY OF CERTAIN WATER BODIES OF MADHYA PRADESH, INDIA

S.K. Kulshrestha

Abstract: Biodiversity of some lentic and lotic environments from central India has been described in order to provide base line data. The lentic water bodies include Upper, Lower and Mansarovar lakes of Bhopal, Gandhisagar reservoir and Undasa wetland. The lotic water bodies include rivers Betwa, Khan, Kshipra, Chambal and Narmada. Information on phytoplankton, zooplankton, periphyton, macrophytes, macrozoobenthos and fishes has been provided based on available studies.

Keywords: Phytoplankton, Zooplankton, Periphyton, Macrophytes, Macrozoobenthos, Fishes.

Introduction

Biodiversity is regarded as the variety of life and its many process. The number of species on earth described is estimated to 1.7 millions. However, the estimate of total number of species existing on earth may vary between 5 million to 12.5 million. Biodiversity includes countless millions of species, subspecies and local variants of species and the ecological processes and cycles that link organisms into population, communities, ecosystem and the entire biosphere.

Conservation of biodiversity is of critical importance as it may be of direct or indirect benefit to mankind besides looking into the threat of extinction and erosion of species. Wetlands are transitional areas between dry terrestrial and permanent aquatic ecosystems. These are highly productive ecosystems. The biodiversity of wetlands is of great importance. However, many lower groups of animals that occur in India have not received adequate attention, while many groups have not been reported. In this way, many species of aquatic plants have yet to be discovered.

The plateau of central India covers the northern part of Madhya Pradesh in the lower basin of Chambal river. It is formed by the Vindhyas and Deccan trap and bound by Bundi and Karauli

Ex. Professor of Zoology, Motilal Vigyan Mahavidyalaya, Bhopal hills on the western side. This area is marked by deep ravines of Chambal and its tributaries Kalisindh and Parwati rivers. The average rainfall is below 750 mm and the plateau undergoes a wide range of temperature fluctuations. The Bhoj Wetland of Bhopal is among 25 Ramsar sites identified in the year 2005. Besides man made and natural wetlands, river systems of Madhya Pradesh also form natural wetlands of the state. Out of 180 species of fish reported from Madhya Pradesh 40 species are under risk category. Information on aquatic organisms such as phytoplankton, zooplankton, periphyton, macrozoobenthos and aquatic insects is meagre. The present attempt is aimed to provide information about baseline data on some of the biodiversity of central Indian water bodies including Upper and Lower lakes and Mansarovar lakes of Bhopal, Gandhisagar reservoir, Undasa wetland, River Khan, River Kshipra, River Betwa, River Chambal and River Narmada.

Upper and Lower Lakes of Bhopal known as Bhoj Wetland have been identified as wetland of national importance by the Ministry of Environment and Forest, Government of India. The upper lake was formed by constructing an earthen dam in 11th century. Its catchment area is 361 sq.km. and submergence area at FTL 30.72 sq.km. It is mainly used for supply of water to Bhopal city. The lower lake was constructed in 1794. Its catchment area is 9.6sq.km. and submergence area at FTL 1.287 sq.km. The lower lake is used for washing, bathing and recreation.

Mansarovar Lake of Bhopal is a man made lake constructed recently on local drain and a tributary of river Kaliasot in the southern part of Bhopal city. Its catchment area is 8.29 sq.km. and submergence area 797 ha. At FTL.

Gandhisagar Reservoir was constructed on river Chambal in Mandsaur district of Madhya Pradesh. Its submergence area is 66,000 ha at FTL. It is mainly used for irrigation.

Undasa Wetland is located just outside the municipal limits of Ujjain city in Madhya Pradesh. The catchment area is 11.25 sq.miles and total water spread area of 5.12 sq.km.

River Khan originates from Nimoli tank located 16 kms. away from Indore. It meanders through Indore city and joins river Kshipra at Triveni Sangam 5 km south of Ujjain.

River Kshipra is a major tributary of the river Chambal and is source of water supply to Ujjain city. It originates near Mhow and traverses through Ujjain and joins river Chambal near Kalakheri.

River Chambal is a major tributary of river Yamuna. It originates near Mhow and meanders through the Malwa region of Madhya Pradesh. After traversing 312 kms in Madhya Pradesh, it enters Rajasthan, reenters M.P. and joins river Yamuna at 40 kms south west of Itawah in Uttar Pradesh.

River Betwa originates from Raisen district in MP and joins river Yamuna in Hamirpur district after meandering for 564 km.

River Narmada is the fifth largest west flowing river of India and one of the largest rivers of Madhya Pradesh. It originate from Amarkantak, traverses a length of 1312 km. and joins Arabian sea. Its 1077 km. stretch in Madhya Pradesh covers 87% of the Narmada basin.

Biodiversity

Aquatic organisms including plants and animals maintain their balance in natural condition. The plants include algal components of phytoplankton and periphyton and macrophytes of submerged or floating types. The animals are represented by zooplankton, benthic forms and nektons. Considerable variations in the number and population of organisms has been reported from water bodies of central India.

Phytoplankton

Phytoplanktonare chlorophyll bearing suspended microscopic organisms mainly consisting of algae. The algae are sensitive to their environmental conditions and different species of algae respond to different chemical or organic pollution. Algae are also used as indicators of pollution. Important studies on the algae and phytoplankton on Bhopal lakes include Sastry *et al.* (1970), Patil (1981), Kulshrestha (1989), Kulshrestha *et al.* (1989 a, b, 1991) and Misra *et al.* (2001). Kartha and Rao (2001) described phytoplankton from Gandhisagar reservoir. Shrivastava *et al.* (2003) described phytoplankton from Undasa Wetland. On rivers important studies include Adholia (1988) on Betwa river and Kulshrestha *et al.* (1987, 1989, 1991) on Khan, Kshipra and Chambal rivers. Most of these studies include seasonal variation in population density, group composition, species diversity, correlations with various physico-chemical characteristics and assessment of trophic status using various indices of water quality.

Sastry *et al.* (1970) reported 13 genera of Chlorophyceae, 10 genera of Bacillariophyceae, 3 genera of Euglenophyceae and a genera of Cyanophyceae from the Upper lake. Patil (1981) described 504 algal taxa of 141 genera in two lakes of Bhopal. Kulshrestha (1989) reported 93 species of phytoplankton including 50 species of Chlorophyceae, 18 species of Bacillariophyceae, 5 species of Euglenophyceae, and 20 species of Cyanophyceae in Lower lake and 85 taxa of phytoplankton from Mansarovar lake along with seasonal fluctuations in the community in relation to various physicochemical parameters and assessment of water quality using Nygaand index. Misra *et al.* (2001) assessed trophic status of Bhoj Wetland using Nyagaard index and reported 94 species of phytoplankton from upper lake and 41 species from lower lake. Kartha and Rao (1992) reported 41 species of phytoplankton from Gandhisagar reservoir while Shrivastava *et al.* (2003) reported 37 species from Undasa wetland.

In recent studies on rivers Adholia (1988) reported 85 species of Phytoplankton from river Betwa, Kulshrestha (1989), Kulshrestha *et al.* 1987, 1989 reported 49 taxa from river Khan, 44 taxa from river Kshipra, 45 taxa from river Chambal near Nagda and 64 taxa from river Chambal near Kota.

Zooplankton

Zooplanktons are important in fresh waters as a major trophic link between primary producers and higher consumer forms. Notable studies on the zooplankton of central India include Sastry *et al.* (1970) on Upper lake, Bhatnagar (1984), Singh (1984) and Kulshrestha (1989 a,b) on Lower lake and also on Mansarovar Lake. Kulshrestha *et al.* (1989 a,b,c, 1992) on river Kshipra, Khan and Chambal, Rao and Chaubey (1993) on Cladocua of Central India, Kartha and Rao (1992), Shukla *et al.* (2001) on Gandhisagar and Shrivastav and Rao (2003) on Undasa wetland.

Sastry *et al.* (1970) reported 14 genera of protozoans, 19 of Crustaceans, 9 of rotifers and 2 of miscellaneous invertebrates from Upper lake. Bhatnagar (1984) reported 14 genera of rotifiers, 5 of cladocerans and 3 of copepods from the Lower lake. Kulshrestha (1989) reported 56 taxa of Zooplankton from Lower Lake and 41 taxa from Mansarovar Lake belonging to five main groups *viz*. Rotifera, Cladocera, Copepoda, Ostracoda and Protozoa from Lower and Mansarovar lakes of Bhopal. Kartha and Rao (1992) Shukla *et al.* (2001) reported 60 species of Zooplankton including 13 of protozoa, 11 of Rotifera, 22 of Cladocera, 9 of Copepoda, 3 of Ostracoda and 2 of Coelenterata. Shrivastava *et al.* (2003) found 27 species of Zooplankton including 6 of Protozoa, 7 of Rotifera, 8 of Cladocera, 4 of Cepepoda & 2 of Ostracoda in Undasa wetland. Among rivers Kulshrestha (1989) found 42 species of Zooplankton in river Khan, 30 from river Kshipra, 32 from river Chambal near Nagda and 40 from river Chambal near Kota.

Periphytons

The periphytons include extremely heterogenous and complex association of organisms such as sessile micro-organisms growing on stones, sticks, aquatic macrophytes and other submerged surfaces. The abundance and composition of the periphyton community at a given location is governed by the water quality and, as such, they are useful as ecological indicators for detecting the degree of pollution. Indicator value of periphyton surpasses that of floating plankton since they are stationary and reflect the conditions existing therein. However, as compared to planktons, this group has received limited attention. In India, studies have been described from freshwater West Bengal and Himalayan Lakes. From central India, information is available on Bhopal lakes (Kulshrestha 1989; Kulshrestha and Johri 1991).

In Lower lake a total of 130 species periphytons were represented by attached algae, protozoa and rotifers. They included 43 species Chlorophycea, 23 species of Bacillariophyceae, 18 species of Cyanophyceae, 2 species of Euglenophyceae, 30 species of protozoa and 11 species of rotifers. In Mansarovar lake 100 species of periphyton were recorded including 35 species of Chlorophyceae, 21 species of Bacillariophyceae, 14 species of Cyanophyceae, 1 species of Euglenophyceae, 20 species of protozoans and 9 species of rotifers. Annual mean values, seasonal variations, percentage composition of different groups of periphyton at different sampling stations and relationship with various physicochemical parameters has been described in these studies.

Aquatic Macrophytes

Aquatic macrophytes include macroscopic vegetation such as macroalgae, ferns and angiosperms that grow in aquatic and wetland habitats. The macrophytes are sensitive indicators of water quality and condition of their habitats. They play an important role in production processes, nutrient dynamics and oxygen budget of water bodies. Sastry *et al.* (1970) described the occurrence and abundance of 15 species of emergent, 13 species of floating and 16 species of submerged macrophytes in relation to the water quality, fish production and depletion of nutrients in the Upper lake. A comprehensive account of macrophytes from Bhopal lakes has been given by Oommachan (1977), who reported 35 species. Bhatnagar (1984) reported 15 species of macrophytes in Lower lake while Kulshrestha (1989) reported 23 species from Lower lake and 16 species from Mansarovar lake. Shrivastava *et al.* (2003) reported 24 species of aquatic macrophytes from Undasa reservoir.

Macrozoobenthos

The macrozoobenthic organisms which occur in massive numbers on and in sediments play a significant role in the aquatic food chain leading to fish. These organisms include coelenterates, flatworms, nematodes, aquatic worms, ostracods, other crustaceans, mollascs and aquatic insects. These organisms have a tendency to concentrate in various niches of the aquatic environment. Studies on macrobenthic organisms from lakes of central Indian have been made by Oommachan and Belsare (1985, 1986) Saxena *et al.* (1988), Kulshrestha (1989), Kulshrestha *et al.* (1991), Kartha and Rao (1993) and Shrivastava *et al.* (2003). On rivers, studies on macroinvertebrates include Kulshrestha (1989), Kulshrestha *et al.* (1988, 1989), Khan and Kulshrestha (1993), Rao *et al.* (1999) and Shrivastava *et al.* (2001). In some of the studies, apart from qualitative aspects, quantitative aspects such as population density, groups percentage composition, bathymetric distribution, species diversity and the relationships between macroinvertebrates with water quality have also been attempted.

Fishes

Reservoir fish resources have received some attention recently. Bhatnagar (1984) reported 20 species from Lower lake while Kulshrestha (1989) reported 26 species from Lower lake and 17 species from Mansarovar lake. Rao *et al.* (1993) reported 41 fish species from Gandhisagar reservoir. Shrivastava *et al.* (2003) reported 18 species from Undasa wetland.

Dubey and Mehra (1959) reported 71 species of fish from Chambal river. Kulshrestha (1989) found 38 species of fish from Chambal, Kshipra and Khan rivers. Hora and Nair reported 40 species from Narmada river, while CIFRI surveys during 1958-66 found 77 species. Rao *et al.* (1991) observed 84 species out of which 22 are included in the list of endangered species. Fisheries resources of rivers have undergone drastic ecological changes due to human interference. It may be of interest to note that Talwar (1971) reported many species of fishes from India endangered or threatened as shown in the table given below.

Endangered and Threatened Fish Species in India (Talwar,1991)

Endangered	Family Cyprinidae	*Cyprinio semipolotus, Raiamas bola, Tor chelynoides*
	Family Cobitidae	*Enobarbichthys maculates*
Threatened	Family Cyprinidae	*Labeo fimbriatus, Labeo kontius, Labeo potail, Conoproktopterus curmuca, Puntius carnaticus, Neolissochilus hexagonolepis, Schizothoraichthys probates, Cirrhinus cirrhosus, Tor khudre, Tor putitora, Tor tor,*
Schizothorax richardsonii	Family Schilbeidae	*Silonia childreni*
	Family Pangasidae	*Pangasius pangasius*
	Family Sisoridae	*Bagarius bagarius*

Acknowledgements

I am grateful to all authors whose reference have been given in the text. I am thankful to Dr. S. N. Diwedi for inspiring me to write this article and to Dr. Pravin Tamot for valuable suggestions.

References

- Adholia, U.N. 1988. Phytoplankton of the river Betwa, Proc. Nat. Symp. Past, Present and Future of Bhopal Lakes , 45-48.
- Bhatnagar, G.P. 1984. Limnology of Lower Lake of Bhopal with reference to sewage pollution and eutrophication. Technical Report MAB Programme, Department of Environment, Govt. of India, New Delhi 1-77.
- Dubey, G.P. and Mehra, R.K. 1959. Proc. All India Congr. Zool. 1 (2) 647-665.
- Kartha, K.N. and Rao, K.S. 1993. A Study on the Piscifauna of Gandhisagar Reservoir with Particular Reference to the Abundance of commercial varieties of Fish. J. Fish. Tech. 30. 11-14.
- Khan, A.A. and Kulshrestha, S.K. 1993. Benthic Fauna in relation to Pollution A case study at River Chambal near Kota in Central India. Environment International 19. 597-610.
- Kulshrestha, S.K. 1989a. Limnological and Water Quality Status of the two Lakes in Bhopal. Technical Report to the Scientific commission for continuing studies on Effects of Bhopal Gas Leakage on Life Systems. 1-119.
- Kulshrestha, S.K. 1989b. Monitoring and surveillance of Residues of Organochlorine Pesticides and Heavy Metals in Fishes from selected Environmental components of Madhya Pradesh and Rajasthan. Technical Report to the Ministry of Environment and Forests, Govt. of India , New Delhi , 1-166.
- Kulshrestha S K, Adholia U N, Khan A A, Bhatnagar A and Saxena M 1987. Assessment of Water Quality of River Kshipra by algal analysis. In Environ. Pesticides and Toxicity. Eds. Dalela R.C. 8th Annual Session of AEB, 233-254.

- Kulshrestha S K, Adholia U N, Khan A A, Bhatnagar A, Saxena M and Baghail M. 1989a. Community structure of Planktons and Macrozoobenthos with reference to pollution of river Khan, India. Intern. J. Env. Studies 35, 83-96.
- Kulshrestha S K Tiwari A, George M.P. Saxena R, Johri M and Shrivastava M. 1989b. Studies on Rotifers of Mansarovar Reservoir of Bhopal with special reference to organic pollution, J. Hydrobiol 1. 43-47.
- Kulshrestha S K, Adholia U N, Khan A A, Bhatnagar A, Saxena M and Baghail M 1989c. Studies on pollution in river Kshipra. Zooplankton in relation to Water Quality. Int. J. Ecol. Environ. Sci. 15. 27-36.
- Kulshrestha S K, Adholia U N, Khan A A, Bhatnagar A, Saxena M and Baghail M 1989d. Community structure of Planktons and Macrozoobenthos in river Khan near Indore with reference to pollution. Intern. J. Environmental Studies 25. 83-96.
- Kulshrestha S K, Saxena R, George M P, Shrivastava M and Tiwari A 1989e. Phytoplankton of Eutrophic Mansarovar Reservoir of Bhopal. Int. J. Ecol. Environ. Sci. 15. 205-215.
- Kulshrestha S K, Shrivastava M, George M P, Saxena R, Tiwari A and Johri M 1991. Seasonal variations in Macrozoobenthic organisms of Mansarovar Reservoir, Bhopal. Proc. Nat. Acad. Sci. India 61 (B) 11. 153-162.
- Kulshrestha S K and Johri M 1991. Periphyton community of Lower Lake of Bhopal in relation to sewage pollution. In 'Aquatic Sciences in India' Indian Association for Limnology and Oceanography, 66-75.
- Kulshrestha S K, Adholia U N and Bhatnagar A 1992. Zooplankton community of river Chambal and its tributaries with special reference to industrial pollution. Intern. J. Environmental Studies 40. 207-216.
- Misra S M, Pani S, Bajpai A and Bajpai A K 2001. Assessment of Trophic status by using Nygaard Index with special reference to Bhoj Wetland Poll. Res. 20(2) 147-153.
- Oommachan L and Belsare D.K. 1985. Bathymetric distribution of mollusca in Lower lake of Bhopal. Proc. Nat. Sym. Pure and Appl. Limnology Bull Bot. Soc. Sagar 32. 109-113.
- Oommachan L and Belsare D K 1986. Bottom sediments and Bathymetric distribution of Oligochates in the Lower lake of Bhopal. J. Hydrobial 2(4). 57-62.
- Patil P, 1981. Algal Flora of Bhopal Lakes, Ph.D. Thesis Bhopal University, Bhopal.
- Rao K S, Chatterjee S N and Singh A K 1991. Studies on Preimpoundment Fishery Potential of Narmada Basin (Western Region) in the Context of Indira Sagar, Maheshwar, Omkareshwar and Sardar Sarovar Reservoirs. J. Inland Fish. Soc. India 23(1) 34-41.
- Rao K S, Shrivastava S, Chaube V 1999. Utility of Macroinvertebrates in Biological Water Quality in some Central Indian Rivers. In 'Inland Water Resources in India'. Ed. Rao. K S Discovery Publishing House. 223-246.
- Sastry C A Aboo K M, Bhatia H C and Rao A V 1970. Pollution of Upper Lake and its effects on Bhopal Water supply. Env. Hlth. 12(3), 218-238.
- Saxena A K, Tiwari R and Kulshrestha S K 1988. Population structure of certain Ephemeropterans in the Upper and Lower Lakes of Bhopal. Proc. Nat. Symp. Past. Present and Future of Bhopal Lakes. 49-54.

- Shrivastava S, Shukla S and Rao K S 2003. Biodiversity of Undasa Wetland Ujjain (India) with special reference to its conservation. J. Exp. Zool. India. 6(1) 125-135.
- Talwar, P. K. 1991. Animal resources of India, Protozoa to Mammalia. Zoological Survey of India. Pisces. 577 – 630.
- Tiwari R K, Saxena A K and Kulshrestha S K 1988. Evaluation of certain aquatic worms as indicators of water quality in Lower Lake of Bhopal. Proc. Nat. Symp. Past Present and Future of Bhopal Lakes. 67-76.

11

INLAND FISH PRODUCTION TRENDS, ISSUES AND SUSTAINABLE DEVELOPMENT

Dr. Salim Sultan

Abstract: India, particularly northern plains, has a unique case due to over dependence on monsoon and irrigation based agriculture. Land is an invaluable but fixed resource hence can not produce food crops beyond a certain limit. Adequate supply of protein food is an important aspect of food security. Perhaps the most significant impact of climate change is expected in respect to availability of water. Therefore restoring physical health of water bodies as well as increase in water spread area and its subsequent integrated use in fisheries may serve the purpose in food production, poverty alleviation, employment generation, rain water harvesting, biodiversity conservation and amelioration of environment up to a great extent. Integrated approach involving commitments for desired policies, economic institutions and social values should be pivotal to change the present trend of sequential events of this multi dimensional problem. Strategies for sustainable development must take into account the complex, reciprocal and ever changing relationship between production systems and the society. Synergistic use of natural resources coupled with proactive participation of all stakeholders have to contribute much with changed mind set with respect to exploitation and extraction of diminishing resources. Interventions for effective management are more demanding in these complex situations. The roadmap suggested for high inclusive and sustainable growth path by steering committee (Planning Commission, GoI), pertaining to inland fisheries, may be enumerated as demand driven diversification, input provisioning, institutional change in water governance, responsive technology and its dissemination through effective extension system, adequate credit, risk mitigation, remunerative prices supported by market reforms and participation of private investors. The *National Commission on Farmers* also emphasized the need for awareness of relevant issues on the part of decision makers, at the state and central level, and the political will to accord priority to these issues. Based on field observation author has assessed the major concerns, causes and possibilities to focus on high, inclusive and sustainable growth path in this sector among the major inland fish producer states of India.

Keywords: Agriculture, Sequentila events, Intervention

Deputy Director Fisheries, 7-Faizabad Road, Lucknow-UP, 226016; E-mail: salimsultan56@yahoo.in.

1. Introduction

The fisheries sector in India, besides nutritional security, provides employment to over 14 million people. Its contribution to national economy is significant, being 1.07 percent and 4.96 percent of the total and agricultural GDP respectively. The annual Indian fish production is around 7.00 million tonnes [2.9 marine + 4.1 inland]. Inland fish production is shared between aquaculture (87%) and capture fisheries (13%). Globally India ranks third in total fish production and second in aquaculture production. Production from marine sector is almost stagnant while inland sector is showing promising growth rate. Indian fish production has shown a phenomenal growth from a meager 0.75 [0.21-inland] million tonnes to 6.57 [0.361 (55%)-inland] *m t* in 2006-07. Aquaculture is fastest growing sub sector among agriculture and allied activities but existing resources continues to be under suboptimal utilization due to several impediments even after the passage of ten five year plans. India possesses latent capacity of about 9.0 million tonnes of fish annually. Planning Commission, GoI, has envisaged 6% annual growth rate for fisheries sector for 11th plan period.

Besides physical presence of aquatic resources other significant factors for the sustainability of inland fish production are hydrology, biology and management. Changes in climate will contribute to various aberrations in physical and biological systems which include shrinkage of glaciers, alterations in rainfall frequency and shift in distribution ranges of living organisms. The linkages of climate change impacts to productivity are dynamic, often interconnected, and context specific-reflecting geographic location; economic, social, and cultural characteristics. The developing countries are likely to suffer most from the negative impacts of these aberrations because of economic importance of climate sensitive sectors (e.g. agriculture and fisheries) and limited human, institutional and financial capacity to anticipate and respond to these changes. Interventions for effective management will be more demanding in these complex situations. With fish becoming a popular diet component, aquaculture should be promoted with a view of achieving nutritional security and generating further employment opportunities. Major concerns, causes and possibilities to focus on high, inclusive and sustainable growth among the major inland states of India have been assessed on the basis of field observation.

2. Production Scenario

Most people associate agriculture with food grains only but food grains account to only 26% of agriculture GDP, followed by the horticulture, 21; and livestock, 25%. Fisheries account for 4% and its share is rising. Between 1991 and 2003 food grain production rose only 30% while aquaculture output rose 102%. About one third population of India, living below poverty line, consumes protein deficient starchy food. This wide spread poverty induced protein deficient diet has doubled the need for enhancing protein rich affordable food products. In this context, use of aquatic resources for production purposes at par with terrestrial resources for fish production is imminent as fish is a valuable and easily accessible source of food, rich in protein. Like cereals, fish is also a staple diet for about 30% economically weaker Indians who cannot afford expensive nutritionally rich food items. Low grade fish generally serve their purpose. The present fish availability is only 25% of prescribed norms to 28% fish eating population of our country. With the declining trend in production of pulses, required protein demand cannot be fulfilled by land based agro or live stocks because of multiple limitations. Although aquaculture alone cannot fill the entire gap but it becomes more relevant and important in current protein production situation of our country because fish is economically and ecologically efficient converters as producing 1 kg beef takes 7 kg of grain, but 1 kg of fish needs only 1.6 – 2.0 kg grains.

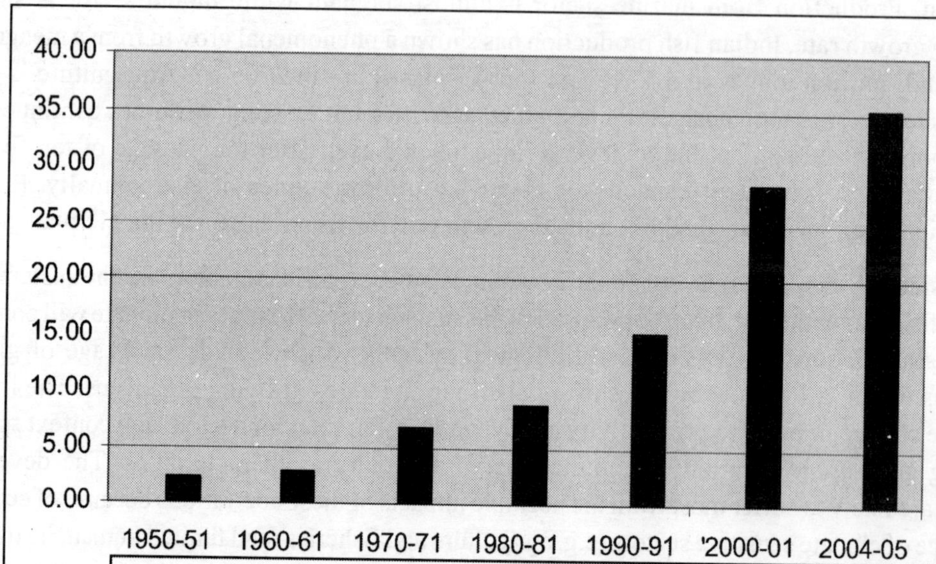

Source Dept of AH, Dairying & Fisheries, Ministry of Agriculture, New Delhi

The national inland fish production shows an increasing trend in all the states except Tamil Nadu in the recent past. The highest annual growth rate was observed in the state of Chattisgarh (156%) followed by Andhra Pradesh and Assam and the lowest in Maharashtra (1.5%). Figures show that farmed fish production is mostly in inland areas adjacent to the east coast and northern plains. West Bengal, Andhra Pradesh and Orissa occupy prominence both by resource utilization (87.55%) and production (93%). West coast states contribute only 12.45% area and 6.98% production. In addition to that even in northern plain best performing states are Punjab and Haryana where the farmers have better socio-economic status and investment capacity. Comparative Production trends (2003-04) for two different zones are as follows:

Biotic Status National Bureau of Fish Genetic Resources (ICAR) has validated the presence of 2183 fin fish species in India. NBFGR has evaluated 329 species and categorized extinct-1, extinct in wild-1, critically endangered-47, endangered-98, vulnerable-82, lower risk near threatened-67, lower risk least concern-3, lower risk conservation dependent-0 and data deficient-18 species (Lakra, 2006). Many species have not been collected since their original

S. No.	State	Pond area (ha)	Productivity (Kg/ha/yr)	Production (mt/yr)
1	West Bengal	2.76	4.2	9.88
2	Andhra Pradesh	5.17	6.5	6.8
3	Orissa	1.21	3.0	1.9
4	Uttar Pradesh	1.62	2.7	2.67
5	Bihar	0.95	2.5	2.66
6	Assam	0.23	1.4	1.81

description since long time (Sultan, 2006). As the species diversity is property at population level, while the functional diversity concept is more strongly related to the ecosystem stability and stresses, physical and chemical factors for determining population dynamics in the aquatic ecosystem. For optimal functioning of these production systems, apart from fish species, endurance of other biotic components is essential to maintain the integrity of these ecosystems. Carps contribute as much as 87% of the total aquaculture production therefore production enhancement can be achieved by only combining mainstay culture practices with strong traditional knowledge and by applying scientific strategies in management.

3. Eleventh Plan Expectations

Steering committee setup by GoI for formulation of 11th plan identified the following concerns for present dismissal position of agriculture and allied sectors deceleration in growth, degradation of natural resources, equity, efficiency and vulnerability. The road map suggested for high inclusive and sustainable growth path by the committee, pertaining to inland fisheries, may be enumerated as demand driven diversification, input provisioning, institutional change in water governance, responsive technology and its dissemination through effective extension system, adequate credit, risk mitigation, remunerative prices supported by market reforms and participation of private trade. The *National Commission on Farmers* also emphasized the need for awareness of relevant issues on the part of decision makers, at the state and central level, and the political will to accord priority to these issues. The working group recommended improvement in quality of fingerlings and other inputs with active involvement of private sector. This group proposed following targets for the four sub sectors are:

Mid Term Assessment 11th plan duration is 2007-08 to 2011-12; therefore, current year is appropriate to review the trends and future prospects. Facts mentioned in the following table indicate that states having comparatively more water area are still trailing to pickup the expected trajectory of growth. There is no correlation between resource potential and total production among various

Sub sector		Out put Share %	Proposed growth rate % per annum
Crops	Food grains	26	2.3
	Oil seeds	6	4.0
	Other Crops	14	3.0
Horticulture		21	5.0
Livestock		25	6.0
Fisheries		4	6.0
Forestry/logging		4	0.0

land locked states implying that unless these states having major potential, don't attain the required pace, our country cannot achieve the optimum production level. Main concerns are dependency on rainfall, under utilization of resources and resultant poor productivity. There is a need to rein force our efforts to enhance present pond productivity average of 2.20 t $ha^{-1}yr^{-1}$ to the level of 5 t $ha^{-1}yr^{-1}$ to justify the planned interventions in aquaculture. Only 40% of cultivable water area under aquaculture, therefore, reservoirs and floodplains still have tremendous scope for productivity enhancement. Under various Central Sector/Centrally Sponsored Schemes pond fishery get over emphasis leaving aside other two categories. The average national production level from small reservoirs is only 20-50 kg^{-1} ha^{-1} yr^{-1} which can be raised to 100-150 kg^{-1} ha^{-1} yr^{-1} or more easily.

The above identified problems of slow pace of growth, poor efficiency, and mode of exploitation and culture practices to provide much needed sustainability should be redressed individually through state specific action plans. Off late, National Fisheries Development Board has initiated the efforts to bring these water bodies on center stage; however, the results will take several years to come out of gestation. Present production status with respect to implying factors in major inland states is understandable from the following table. These issues are critical as supplementary water for pond, supply of fish seed, facilities to aquaculture at par with agriculture and technology dissemination are main concerns to be redressed for growth and sustainability of this sector.

4. Major Issues

Degradation of Natural Resource Base, Although water is critical for fish but it has non-consumptive use and acts only as substrate for growth as well as sustenance. While water is renewable resource, its availability at specific location and at different periods of the year is limited, being largely determined by the climatic, geographical and physical conditions. India receives an average annual rainfall of about 400 billion cubic meters (MoWR, 2003) but a major portion of this fresh water is wasted as surface runoff. Himalayan Rivers are important as country's water resource; therefore, creation of irrigation potential has been a major key to the agriculture development, production stability and food security. Fresh water ecosystems provide essential goods for livelihood such as food, shelter, and fuel; as well as services such as breaking down wastes and pollutants, purifying water, and maintaining soil fertility. Striking a balance between allocation of water for direct human use (domestic, agriculture, power generation and industry) and indirect human use (maintenance of ecosystem, goods and services) is necessary for productivity enhancement and sustainability. Habitat alterations due to impounding for water storage, canalization, mining activities in rivers, alteration of natural flow regimes, inundations and competition for water for various purposes has become a potential threat for fresh water. Loss of *Hilsa* and *Macro brachium* fishery in river Ganga due to Farakka barrage, *Tor tor* in Gandak due to Triveni barrage are now well established.

State	Area (000) Km^2	Water Resource (ha) & Productivity kg/ha/yr							
		Ponds (lac ha)		Reservoirs (lac ha)		F/ plains (lac ha)		Rivers (km)	
		Area	Prod/ty	Area	Prod/ty	Area	Prod/ty	Length	Prod/ty
UP	240	1.62	2900	1.61	4.68	1.33	43-357	31200	UM
Bihar	99.2	0.95	NA	0.966	0.54	0.05	40-200	3200	UM
Assam	78.43	0.31	3000	0.01	--	1.10	30	4820	UM
MP	308	0.60	1500	4.60	13.68	--	--	17088	2.55
Chat/garh	135	0.70	2485	0.83	20	0.05	NA	3573	UM
J/khand	79.78	0.30		0.73	2.5	0.15	--	34200	--
R/than	342.23	0.39	1200	2.70	24.89	--	--	5290	UM
Haryana	44.21	0.10	5200	0.0282	--	0.10		5000	UM
Punjab	50.36	0.07	6004	23000			50	17543	UM
India	3.27	23.81		20.31	20	7.98	17	191.024	--

State	Rainfall	Fish Production (m t/yr)		Fish Seed (million fry)		Status	Culture
	cm	Present	Potential	Production	Demand	Fish Ag.	Method
UP	100-120	3.49	6.50	1200	1500	No	Extensive
Bihar	120	260000	45000	350	600	Yes	//
Assam	100-200	180000	1000	222.8	700	--	//
MP	1370	61000	150000	455	1200	--	//
Chas/garh	1292	137753	--	249	600	yes	Semi
Jh/khand	160	NA	66100	180	110	no	Extensive
Rajsthan	10-100	22945	371639	845.6	450	--	Extensive
Haryana	10-100	50000	100000	300	300	Yes	Intensive
Punjab	10-100	86697	143538	150	246	Yes	Intensive
India	--	4.1	9000000	19000	NA	--	--

Deforestation has increased erosion and deposition of suspended matter resulting in habitat loss, destruction of spawning ground and species extermination. To deal with these aspects creation of database, effective linkages among various stakeholders are desired. In the present case, it is much more relevant because of wide distribution of water bodies in remote areas, fluctuation in water spread area and multiple uses of these critical resources.

Geochemical Factors. Each biological species possess distinct environmental niche and altered physicochemical situation disturbs the ecological fabrics. Ground water depletion increases percolation of water consequently reduces life span of water including flora and fauna. The major impacts of pollution and dynamiting are de-oxygenation, high BOD, rapid eutrophication and accumulation of harmful substances which causes fish kills, destruction of habitats and direct toxicity to organisms.

Land Use Pattern. Land use has profound effects on the condition of land, vegetation and water resources. It is determined by the physical attributes, specially climate, topography and soil, human attributes of an area which include social structures, cropping patterns, way of life and even the traditions, culture habits and temperament of inhabitants. Man is an effective agent who adversely affects renewable sources of his own livelihood through inappropriate agriculture practices unsuited to climate, slope and soil and cause extinction of biodiversity through hunting, fishing and disturbance of habitat. From land use pattern point of view water bodies are found in all types of land. Economic benefits from fish culture are more than crops besides ecological advantages. We would have to judiciously utilize various resources through adopting land use patterns, upgraded utilization levels, creation of infrastructure, and propagation of aquaculture in ponds, water governance and management of reservoirs, rivers, canals, and flood plains.

Livelihood Deceleration in growth influences more the poor due to direct dependence upon the use of natural resources for their livelihood. Socio-economically deprived fishers are away from main stream development because of their poor infrastructure risky jobs and uncertain income. Therefore resource degradation is likely to make livelihood more difficult for riparian population. Economically less beneficial ecosystems alienate the community to intensify efforts in future.

Institutional Failures Market, community and government are the main institutions responsible for the management of natural resources. Market failures occur when it does not achieve socially efficient allocation of resources. The role played by community level institutions has often been disregarded. Government failure arise when policy interventions are either not undertaken or create distortions that lead to socially sub optical outcomes. The lack of government initiatives to manage the vast resources has virtually created a situation of open access regime causing depletion of fish stocks in natural waters.

Biological Factors Preference for culture of some native species with apathy to other organisms, introduction of exotics natural habitats disturbing the ecological fabrics of most water bodies through competition or predation. Population of common carp, Thai mangur and tilapia constitute major yield in river Yamuna (Singh *et al.* 2008) a not good sign for biodiversity of our precious gangetic river system.

Risks, Uncertainties and Sustainability in Inland Fish Production. The growing shift in climate pattern coupled with recurring droughts and floods are already causing considerable losses to reservoir and wetland fisheries. The unscheduled events may reduce the congeniality of water bodies for fish production. There are numerous risks including physical and hydro-chemical properties of water, rate of evaporation and change in economic scenario which affect natural breeding of fishes, seed production and rearing of fish. Identification and Quantification of risks which can influence the aquaculture should be in place to reduce their impact.

Inadequate Extension Machinery Aquaculture extension services are probably weakest among all the agriculture and allied activities. State fisheries departments are insignificant, linkages between research institutes and farmers are negligible. ICAR institutes are mainly located in coastal areas and conduct high end researches with little concern to farmers' need particularly of inland states. Contrary to other sectors, never there was any specific scheme for such purpose in this sector.

5. The Way Forward

Inland fisheries requires improvement in existing systems to increase the adaptive capacity and resilience of people, institutions, systems and natural resources to cope with variability and its impact on their livelihood and economic opportunities. The development process must be able to assess the threats and corresponding measure to minimize the risks. Risks can be reduced in aquaculture through diversification, better marketing arrangements, insurance coverage and effective implementation of government programs etc.

National Fisheries Policy. Lack of broader policy framework at the center and comprehensive policy at state level, non availability of adequate and professionally skilled human resources, ineffective and redundant services delivery system and poor infrastructure have almost limited the scope of fisheries development in India. A policy is required to give over all direction and to achieve sustainable development on the pattern of other sectors. Inland aquaculture should be treated at par with agriculture.

National Water Use Policy. Ministry of water resource regulates use and distribution of water for social consumption, irrigation, power generation and all other purposes. Fisheries do not gets due place; therefore, during planning each activity should get recognition because any water management issue affects fishery directly.

Improved Governance Transparent and accountable policies regarding *property regimes* to ensure participation of target groups will increase the utilization of water bodies. Fish farmers' are usually resource poor hence empowerment is essential to strengthen their efforts because the poor often possess their own strategies.

Aquatic Resource Conservation Reversing the present trend and restoring physical health of water bodies through appropriate land use policies and strategies is a challenge of immense gravity. It presupposes the repairing of ecological damage to be tackled at the same time to enable transition to an ecologically adapted land use, the need for environmentally acceptable ways of addressing the issues in development, blending the best of modern sciences with traditional wisdom and the ecological prudence of local societies which is possible only by the application of science and technology, high measure of humanitarian solidarity and participatory efforts by all stakeholders (Sundaram, 2006). Agriculture carries a high responsibility for management of water resources in qualitative and quantitative terms for sustainability. Combined agrarian and aquarian reforms in the form of water usage and exploitation with a greater sense of responsibility towards endangered fauna may emolliate the situation.

Synergistic Use of Resources Agriculture, horticulture, animal husbandry and aquaculture, the main stay of human survival, consume water except the last one. Fishes have non-consumptive use of water; therefore, its availability sustains a large biomass of aquatic flora and fauna. The basic principle of integrated farming is synergic utilization of farm activities and conservation, including farm wastes. It minimizes the costs by effective utilization of all resources, which ensures optimum per unit area yield.

Institutional Reforms and Improved Quality of Growth. Growth is essential for poverty alleviation, but its quality matters most in order to benefit poor. Unpredictable risk and difficult choices needs responsive and accountable institutions. Aquaculture is vulnerable against all acts of God, therefore, to handle unforeseen situations effective mechanism for risk mitigation will increase the adaptability of fish farmers. Quality growth may be achieved through variability related economic planning, development of user friendly infrastructure, improvement in technology and provision of insurance coverage against the expected risks. Facility of insurance against risks in aquaculture is long standing demand of fish farmers to be finalized by concerned authorities.

Social intervention Public-Private and Community Partnership are unavoidable by virtue of their critical role in the management of water bodies. Empowerment of fisher community and especially all stakeholders through capacity building will facilitate the development of partnership with private as well as public investors. Religion and rivers are intertwined in India. Saving rivers could, literally, be a matter of faith. Awareness activities may be organized at special congregations like *Kumbh, Chhat* or various "*snans*". Civil movements by different organizations such as *Narmada Bachao Andolan, Barh Mukti Abhiyan, Ganga Mukti Andolan, Diara Mukti Andolan* and *Tehri dam* have came up for the cause. The Kali Bein river in Punjab, sacred for sikhs, flows again by the use of faith factor by Saint Balbir Singh. WWF had experimented with sadhus to educate the local populace about significance of dolphins in nearby villages of Narora (Pillay, 2006).

Creation of Data base and Surveillance of Population Dynamics. Among the aquatic organisms, most attention has been focused on fish as food because of their overwhelming economic importance, ignoring the vast portion undocumented so far. In the prevailing circumstances it is imperative to create a data base as well as review of ongoing activities both for different ecosystems and biotic diversity. There is a need of repeated surveys to know the emerging trends of various aspects due to highly dynamic nature of these ecosystems to facilitate formulation of management strategies.

Increased Investment, Institutional Credit and Subsidies. Since the beginning of government sponsored development schemes, agriculture is getting financial support through credit, HYV of seeds, fertilizers, power tariff, water and Minimum Support Price. While there is strong advocacy to reduce such subsidies on power and irrigation but in the case of fisheries, previously deprived sector, the facility is still to begin. The present allocation by GoI for 11th plan period is inadequate for successful implementation of various fisheries activities desired pace. Although a fishery is a state subject but funds for development come from GoI, therefore, enhanced allocation are needed for required growth. In depth analysis reveals that for inland sector will get a meager Rs 300 lac directly for development of culture operations.

S. No.	Name of the Scheme	Allocation (Rs lac)
A	Centrally Sponsored Schemes	
1	Development of Inland Fisheries & Aquaculture	300
2	Development of Marine Fisheries	350
3	National Scheme for welfare of Fishermen, Fisheries Training and Extension	180
	Total CSS	830
B	Central Sector Schemes	
1	Strengthening of Data base & Information Networking	25
2	Assistance to Fisheries Institutes	371
3	National Fisheries Development Board	1660
	Total CS	2056

Development of any sector is directly proportional to investment; hence, there is a strong case for enhanced investment, credit and financial assistance from government and private players to increase investment in aquaculture as earlier done for agriculture for creation of infrastructure and volume of trade.

Aquaculture Extension Strengthening of State Fisheries Department Vis-à-vis quantum of resource is compulsory to provide satisfactory extension services at end users. There is a need of at least one time demand based extension package to break the jinx and create awareness among rural masses about this sector. ICAR Institutions should be made responsible to serve farmers at their end in the inland states.

Development of Infrastructure. Like all other production systems, inputs are critical for aquaculture also. Presently availability and quality of inputs is inadequate, a main impediment in the growth of inland fisheries. Till now, inland states are not self sufficient in fingerling production, formulated fish feed production, post harvest facilities and capacity building of farmers. This situation will be more demanding with increase in area coverage and dissemination of technology. Present trends can be reversed only through target-oriented action plans to redress each segment of this sector to achieve stipulated growth pattern through diversification and intensification (Sultan, 2007).

SWOT analysis for assessing the potential of aquaculture in inland states

Strengths Availability of water resources, Availability of human resources, Availability of cheap live stock and agriculture wastes, Availability of aquaculture technologies.

Weakness Poor physical health of water bodies, ineffective policy for water resource management, weak extension machinery, Low investment capacity of fish farmers, inadequate dissemination of technical know-how, Poor infrastructure, non availability of fingerlings, feed and other inputs.

Opportunities Multiple uses of existing water resources, employment avenues for rural workforce, inclusive growth in rural areas, integrated fisheries for synergistic use of farm wastes, nutritional security for protein deficient population. Use of fish as ecological and economical converter.

Threats Unused water bodies will gradually become nonproductive due to silting, pollution and infestation. Inhospitable conditions will increase the extermination pace of aquatic biodiversity. Loss of opportunities in terms of potential and time.

Acknowledgements

Author is thankful to Director Fisheries, Uttar Pradesh for his valuable suggestions and encouragement during this work.

References

- Lakra, W. S. 2006. Conservation of Fishery Resources Diversity vis-à-vis India's Biological Diversity Act, 2002. Base paper Brain storming meeting at NBFGR, Lucknow. January 28, 2006.
- Ministry of Water Resources, 2003 Fresh Water Year 2003, Government of India, New Delhi.
- Pillay, V. 2006. Dolphin Home Gets a Name. Hindustan Times, Luck now. August 24, 2006.
- Sultan, S. 2006. Freshwater Ichthyo–biodiversity Changes and Challenges. Fish Introduction in India Status, Potential & Challenge. September 16-17, 2006. NBFGR, Lucknow.
- Sultan, S. 2006. Intensification and Diversification Strategies for Aquaculture Promotion in Uttar Pradesh. Fishing Chimes. 6 (1) 23-230
- Sundram, K.V. 2006. Sustainable management of land resources The challenges before the current millennium. National Seminar on sustainable Management of land resources through land care movement. January 12-13, 2006. Yojana Bhavan, Lucknow.
- Singh, A.K. Pathak. S. Sultan. A. Mishra. & W. S. Lakra. 2007. Spread of exotic fishes in river Yamuna. 8th Asian Fisheries Forum. November 20-23, 2007 Cochi India. BOD 020.

12

INFLUENCE OF DISSOLVED ORGANIC CARBON IN MODIFYING THE TOXIC EFFECT OF XENOBIOTICS ON PLASMA GLUCOSE, ALKALINE PHOPSPHATASE, LDH, TOTAL PROTEIN AND LIVER GLYCOGEN OF COMMON CARP

Subhendu Datta and Saurabh Sasmal[*]

Abstract: Raw cow dung, compost, F.Y.M., humus etc are commonly used in aquaculture practices can easily increase the level of organic carbon. Dissolved organic carbon, which is part of good aquaculture, should have some role in modifying the sub-lethal effect of xenobiotics. In the present study, it was observed that predetermined sublethal concentrations of metals (Cd, Hg and Pb) and synthetic pyrethroid (Cypermethrin) decreased the liver glycogen reserves and total serum total protein level, where as serum glucose, alkaline phosphatase and lactate dehydrogenase level of scale carp increased significantly in comparison to control group. When dissolved organic carbon (0.88 mg/L) was included with the test chemicals, values of all these biochemical parameters attained the level which was not significantly different from control level in all cases. The effect was significantly different from chemical (alone) treatment. Dissolved organic carbon when combined with test chemicals was able to restore almost the normal level of all biochemical parameters mainly due to their influence in reducing the chronic toxic effect of test chemicals on scale carp. It is concluded from this study that organic matter, have good potential to protect the fish from lethal and sub-lethal effects of inorganic (e.g. metals) and organic (pesticides) xenobiotics. Dissolved organic matter protects due to its high CEC, adsorbtion and complexation ability.

Keywords: Aquaculture practices, Biochemical, Chronic effect

Introduction

The increased use of heavy metals in industries and agriculture and their subsequent intrusion in indeterminate amounts into the environment has caused ecological and biological changes. The

Central Institute of Fisheries Education, Powarkheda Centre, Powarkheda, Dist. - Hosangabad, M.P., 461 110, India, Barkatullah University, Bhopal, M.P., India; E-mail subhdatta@gmail.com seriousness and longevity of the metal in the environment are compounded by the fact that they are non-degradable with significant oxidizing capacity and substantial affinity for nucleophiles in proteins and enzymes. Synthetic pyrethroid insecticides are extensively used in place of organochlorine, organophosphorus and carbamate insecticides to control various types of pests and increase agricultural production. Due to their multiple beneficiary qualities, farmers are attracted to use these compounds (low mammalian toxicity, high biodegradability etc.). Pesticides carried away by rains and floods to larger water bodies like ponds and rivers alter the physico-chemical properties of water. Owing to the excessive use of synthetic pyrethroids, the environment and water resources are being polluted, thus endergering aquatic life directly and humuan life indirectly. The lipophilicity of pyrethroids indicates that these chemicals will be absorbed by fish even from very low concentrations in water. These compounds have actually been found to be highly toxic to fish.

Physiological and biochemical parameters in fish blood and tissues could change when exposed to xenobiotics (heavy metals and pesticides) and that these parameters are extremely sensitive to these pollutants. It was also reported that different abiotic environmental parameters e.g. hardness, pH, alkalinity, dissolved organic carbon, calcium, carbonate and bicarbonate levels of water, soil colloidal sediments (clay and humus) and contact time with soil sediment can significantly reduce the acute toxicity of xenobiotics (Datta and Das, 2003; Datta *et al.*, 2002abcd; Datta *et al.*, 2003abc, Das *et al.*, 2005). Dissolved organic carbon (DOC) is an indication of organic fertilization level of pond. Raw cow dung, compost, F.Y.M., humus etc are commonly used in aquaculture practices can easily increase the level of organic carbon. So DOC which is a part of good aquaculture should have some role in modifying the sub-lethal effect of xenobiotics. Under that presumption the present work was undertaken to study the influence of dissolved organic carbon in modifying the effect of heavy metals (Cd, Pb, Hg) and pyrethroid pesticide (Cypermethrin) on some of the biochemical parameters (e.g. liver glycogen, serum glucose, LDH, alkaline phosphatase and total protein) of *Cyprinus carpio* communis.

Materials and Methods

Dissolved organic carbon stock solution was prepared from a mixture of malonic acid sodium salt, dipicolnic acid (pyridine 2,6-dicarboxylic acid) and sodium oxalate present in molar concentration of 1.11 µmol, 0.022 µmol and 0.44 µmol respectively (MacRae *et al.*, 1999 and Marr *et al.*, 1999). All essential chemicals were purchased from local market. The test animal, *Cyprinus carpio* communis (Scale carp) was selected for investigation because of its significance among carps in freshwater aquaculture and availability throughout the year. Advanced fingerlings (juveniles) of average length maximum 13.1 ± 0.72 cm and minimum 11.45 ± 0.35 cm and average weight maximum 30.8 ± 6.8 g and minimum 19.9 ± 2.6 g were selected. As test containers, 60 L plastic tubs were used. Simple aquarium aerator with two outlets was purchase from local market. Experiment was conducted in demineralised water. Acclimatization, conditioning and range finding and acute tests were carried out following standard method (Clesceri *et al.*, 1998). For this experiment fishes are acclimatized for 14 days.

For acute tests, after the conditioning, fishes were transferred to the experimental jars, which were numbered and arranged in completely randomized design. Six to ten toxicant concentrations and a control were maintained in each experiment. Three replications were given per treatment. In each replication 10 numbers of fish were placed randomly from the conditioning stock. Fishes were exposed under static conditions for 96 h. The fishes were not fed during the test. Mortality was recorded at every 12h interval. Concentration ranges of different test chemicals from which six to ten toxicant concentrations are chosen were; for Lead 5-30 ppm, for Hg (0.1-1.0 ppm); Cd (0.5-6 ppm), cypermethrin (8-80 ppb). Along with mortality, the behavioural changes of fish were also recorded. Maximum and minimum room temperatures during the experiment were also recorded with a centigrade thermometer. 96 h LC_{50} is calculated from the percentage mortality data, which was utilized for deciding the dose of sub-lethal tests for measuring the biochemical parameters.

Maximum and minimum room temperatures during the experiment were also recorded with a centigrade thermometer. Concentration of DOC was fixed at 0.88 mg/L for sub-lethal experiment (which is equivalent to 16 mg/L in natural conditions) from the result of acute tests taking the clue from Krishan (2005) and Sing (2005).

Experimental set up for sub-lethal tests

After 2 weeks of acclimatization, fishes were kept in the experimental tubs. Control treatment was without any chemical. Fishes were maintained in demineralized water only. Two treatments were given for each of the test chemicals. Out of which first treatment (Chemical Treatment) was set in predetermined concentration of test chemicals in demineralized water. Concentrations of chemicals were 3 ppm (96 h LC_{50} 3.7 ppm) for cadmium, 0.3 ppm (96 h LC_{50} 0.43 ppm) for mercury, 12ppm (96 h LC_{50} 19.9 ppm) for lead and 30 ppb (96 h LC_{50} 49.38 ppb) for cypermethrin. Second treatment was test chemicals with dissolved organic carbon (0.88 mg/L). 48 h contact time was allowed between the test chemicals and DOC. Fishes (advanced fingerlings) were released in the experimental tubs after this period (i.e. 2 days). 10 numbers of fishes were stocked in each tub which containing 40 L of demineralized water. Continuous aeration was given to maintain the dissolved oxygen level above 5 ppm. Length and weight of the fish were taken at the time of fish release into the tubs. Feeding regime was similar to that of acclimatization mentioned before. The loss of water through evaporation as made up at every alternate day to keep the concentrations constant. Experiments were continued for 14 days after which blood and liver samples were collected for biochemical analysis.

Chemical analysis

Parameters such as pH, dissolved oxygen, TDS and conductivity were measured at -2 days i.e. half an hour after the test chemicals were mixed with other chemicals (e.g. $CaCO_3$, DOC, bentonite) in the experimental tubs, which was 2 days before the release of fish in the test water medium, 0 day i.e. after the lapse of 48 h contact time and just before fishes were release in the experimental tubs, after 7 days from fish release in the experimental tubs and 14 day from fish release in the experimental tubs both in control and treatments. pH was measured by pH-meter (Model Water proof pH scan 1 tester, Eutech Instruments), TDS and conductivity were measured by electrical water analysis instruments (Model 161E of Environment and Scientific Instruments Co., Haryana, India), dissolved oxygen was measured by dissolved oxygen meter model Cyber scan 300/310 of Eutech Instruments Pvt. Ltd, Singapore. No water sampling was done. Analysis was done by dipping the electrical meter in the water kept in the tub for all treatments.

Blood collection

Blood samples were collected from five groups (each group consisted of 2 fish and the blood samples collected from the 2 fishes were kept in one vial) directly from the heart with the help of tuberculin or insulin syringe of 1 ml capacity. Immediately after collection of blood it was transferred in sterile glass vials of 15 ml capacity containing heparin (0.2 mg/ml blood) as anti-coagulant with gentle horizontal rotation. Blood samples were centrifuged at 4000 rpm for 15-20 minutes to separate the plasma from the red blood cells. Plasma was separated from the erythrocytes (red blood cells) within 60 minutes of blood collection. Supernatants were collected in 5 ml sterile storage vials. Then plasma was stored at +2°C to +8°C temperature until it was analyzed for biochemical parameters. Biochemical parameters were estimated within 7 days but glucose in plasma was determined with in 24 h. Blood samples were collected in the evening of 14th day.

Estimation of Liver Glycogen (G)

Liver glycogen was estimated by anthrone method (and observed the O.D (absorbance) at ThermoSpectronic UV1, Merck Ltd., at at 620 nm (Seitfer *et al.*, 1950; Carrol *et al.*, 1956).

Estimation of plasma biochemical parameters

Glucose (Gl)

Glucose in plasma was determined by using MERCK photometric determination kit (Merckotest®) based on Glucose oxidase – peroxidase method with the help of Thermo Spectronic UV1, Merck Ltd, at 500 nm absorbance and the value was expressed as mg/dl (Barham and Trinder, 1972).

Total protein (TP)

Plasma total protein was determined by using MERCK photometric determination kit (Ecoline®) with the help of spectrophotometer (Model ThermoSpectronic UV1, Merck Ltd.), at 540 nm absorbance and the value was expressed as g/dl (Doetsch and Gadsden, 1975).

Alkaline phosphatase (ALK)

Alkaline phosphatase in plasma was determined by using MERCK photometric determination kit (Ecoline®). Alkaline phosphatase enzyme of serum was measured with the help of with the help of ThermoSpectronic UV1, Merck Ltd, at 405 nm absorbance and the value was expressed as U/l (Seligman *et al.* 1951).

Lactate dehydrogenases (LDH)

In plasma, the lactate dehydrogenases was determined by using MERCK photometric determination kit (Ecoline®) with the help of ThermoSpectronic UV1, Merck Ltd, at 340 nm, 334 nm or 365 nm absorbance and the value was expressed as U/l (Morgenstern *et al.* 1966).

Data analysis

Acute toxicity (96h LC_{50}) of heavy metals lead (Pb), cadmium (Cd), mercury (Hg) and pesticide (Cypermethrin) in demineralised water on scale carp was determined by Trimmed Spearman-Karber method. All the data of biochemical parameters were statistically analyzed in IBM-PC using MS Excel 2003 and a software, statistical package (SPSS) for windows (version 10.0, SPSS Inc., Chicago, IL).

Result

Chemical parameters of demineralised water were as follows pH – 6.8, calcium – 2 ± 0.2 ppm, total hardness – 20 ± 3.5 ppm, carbonate – 0 ppm, bicarbonate – 8 ± 2.0 ppm, dissolved oxygen – 6.5 ± 0.4 ppm, T.D.S. – 7.9 ± 1.2 ppm, conductivity – 12 ± 2.8 μmhos/cm.

Different chemical parameters e.g. pH, dissolved oxygen, T.D.S. and conductivity in control and treatments were measured and trend of these chemical parameters over 14 days period (Table 1 and 2). It was observed that water pH slightly lower in DOC treated treatments with respect to control. In Chemical (alone) treatments pH didn't change with respect to control. Duration of exposure time had no influence on pH. Dissolved oxygen was found to decrease, T.D.S. and conductivity was found to increase with the progress of time in all treatments. There was no definite trend for dissolved oxygen, where as TDS and conductivity were found to increase when treatments were compared with control.

Average room temperature during this experimental period was Max. 31.9 ± 1.9 Min. 25 ± 0.8°C.

Influence of DOC on the effect of Cd, Pb, Hg and Cypermethrin on biochemical parameters of *Cyprinus carpio* var *communis* (Scale carp)

In the control experiment, levels of different biochemical parameters were

Glycogen 39.51 mg/g, Plasma glucose 43.75 mg/dl, Total protein 3.9 g/dl, Alkaline phosphatase, 56.36 U/l, Lactate dehydrogenase 55.8 U/l.

Table 1 Chemical parameters of water in control and chemical treatments

Treatments	Sampling days	pH	Dissolved Oxygen (mg/l)	T.D.S. (ppm)	Conductivity (mhos/cm)
Control	-2 days	6.6	6.1	30.1	0.064
	0 day	6.6	5.84	65.4	0.121
	7 day	6.6	5.72	135	0.228
	14 day	6.6	5.61	201.4	0.315
Lead	-2 days	6.6	6	92.1	0.107
	0 day	6.6	5.77	132	0.201
	7 day	6.6	5.74	204	0.344
	14 day	6.6	5.76	315.4	0.489
Mercury	-2 days	6.6	5.98	47.1	0.093
	0 day	6.6	5.74	76.9	0.116
	7 day	6.6	5.70	125	0.178
	14 day	6.6	5.70	198	0.241
Cadmium	-2 days	6.6	6.1	75.1	0.112
	0 day	6.6	5.35	89.1	0.148
	7 day	6.6	5.10	145	0.274
	14 day	After 8 days from fish release in the experimental tubs, all fishes died.			
Cypermethrin	-2 days	6.6	6.1	48.1	0.121
	0 day	6.6	5.75	79.2	0.141
	7 day	6.6	5.71	148	0.201
	14 day	6.6	5.66	275	0.293

Table 2. Chemical parameters of water in different chemical and dissolved organic carbon treatments

Treatments	Sampling days	pH	Dissolved Oxygen(mg/l)	T.D.S.(ppm)	Conductivity (mhos/cm)
Lead and DOC	-2 days	6.6	6.10	99.1	0.135
	0 day	6.5	5.80	114	0.173
	7 day	6.5	5.74	209	0.331
	14 day	6.5	5.70	313	0.477
Mercury and DOC	-2 days	6.6	6.01	73.4	0.148
	0 day	6.5	5.76	107.8	0.192
	7 day	6.5	5.67	187	0.275
	14 day	6.5	5.64	302	0.401
Cadmium and DOC	-2 days	6.6	6.0	98	0.141
	0 day	6.5	5.75	121.5	0.172
	7 day	6.5	5.65	157	0.232
	14 day	6.5	5.62	196.2	0.325
Cypermethrin and DOC	-2 days	6.6	6.11	63.1	0.132
	0 day	6.5	5.75	91.4	0.159
	7 day	6.5	5.51	168	0.248
	14 day	6.5	5.50	294	0.355

In Chemical treatments

Glycogen levels in Cd-treatments 8.83 mg/g, Hg-treatment 18.1 mg/g, Pb-treatment 29.39 mg/g and in Cypermethrin-treatment it was 23.72 mg/g.

Plasma glucose levels in Cd-treatments 198.47 mg/dl, Pb-treatment 82.12 mg/dl, Hg-treatment 125.22 mg/dl and in Cypermethrin treatment 104.17 mg/dl.

Total protein level was 1.9 g/dl in Cadmium treatment, 3.02 g/dl in the Lead treatment, 2.62 g/dl in the Mercury treatment and 2.82 g/dl in the Cypermethrin treatment.

Alkaline phosphatase level was 105.38 U/l in Cadmium treatment, 86.5 U/l in Lead treatment, 82.4 U/l in Cypermethrin treatment and 88.49 U/l in Mercury treatment.

Lactate dehydrogenase level was 80.0 U/l in Cadmium treatment, 69.3 U/l in Lead treatment, 72.5 U/l in Cypermethrin treatment and 71.3 U/l in Mercury treatment.

In Chemical and DOC treatments

When dissolved organic carbon was mixed with chemicals levels of different biochemical parameters were found to come closer to the values control treatment (Fig 1-4).

Glycogen levels were 37.23, 35.18, 36.85 and 36.01 mg/g for for Pb, Cd, Hg and Cypermethrin.

Plasma glucose levels were 52.25, 60.19, 52.37 and 57.20 mg/dl for Pb, Cd, Hg and Cypermethrin.

Total protein levels were 3.63, 3.03, 3.35 and 3.41 g/dl for Pb, Cd, Hg and Cypermethrin.

Alkaline phosphatase levels were 62.42, 68.48, 69.58 and 65.54 U/l for Pb, Cd, Hg and Cypermethrin.

LDH levels were 61.22, 64.51, 65.18, 66.6 U/l for Pb, Cd, Hg and Cypermethrin.

Standard error of values all these treatments were given in parenthesis in Table 3.

Discussion

Chemical parameters

Lower pH in DOC treatments than control was observed as the solution of dissolved organic carbon was prepared by mixing three synthetic organic acids (i.e. oxalic, malonic and pyridine dicarboxylic acid). Dissolved oxygen was found to decrease, T.D.S. and conductivity was found to increase

Table 3. Values of different biochemical parameters with and without the influence of dissolved organic carbon.

Biochemical parameters	*Control*	*Pb*	*Pb +DOC*	*Cd*	*Cd +DOC*	*Hg*	*Hg +DOC*	*Cyp*	Cyp +DOC
Glycogen	39.51 (0.43)	29.39 (1.45)	37.23 (1.39)	8.83 (1.24)	35.18 (1.61)	18.1 (0.96)	36.85 (0.78)	23.72 (1.35)	36.01 (1.15)
Glucose	43.75 (1.66)	29.39 (1.45)	52.25 (1.65)	198.47 (7.93)	60.19 (2.44)	125.2 (4.96)	52.37 (3.68)	104.17 (2.80)	57.20 (2.50)
TP	3.92 (0.22)	82.12 (2.24)	3.63 (0.11)	1.95 (0.20)	3.03 (0.11)	2.62 (0.13)	3.35 (0.16)	2.82 (0.19)	3.41 (0.18)
ALK-P	56.36 (2.28)	86.47 (1.70)	62.42 (1.30)	105.38 (1.98)	68.48 (2.04)	88.49 (1.96)	69.58 (2.31)	82.44 (1.60)	65.54 (2.02)
LDH	55.77 (3.12)	69.28 (2.27)	61.22 (2.28)	79.97 (2.89)	64.51 (2.89)	71.30 (2.87)	65.18 (3.12)	72.57 (2.51)	66.60 (2.90)

Figures in parentheses indicate standard error

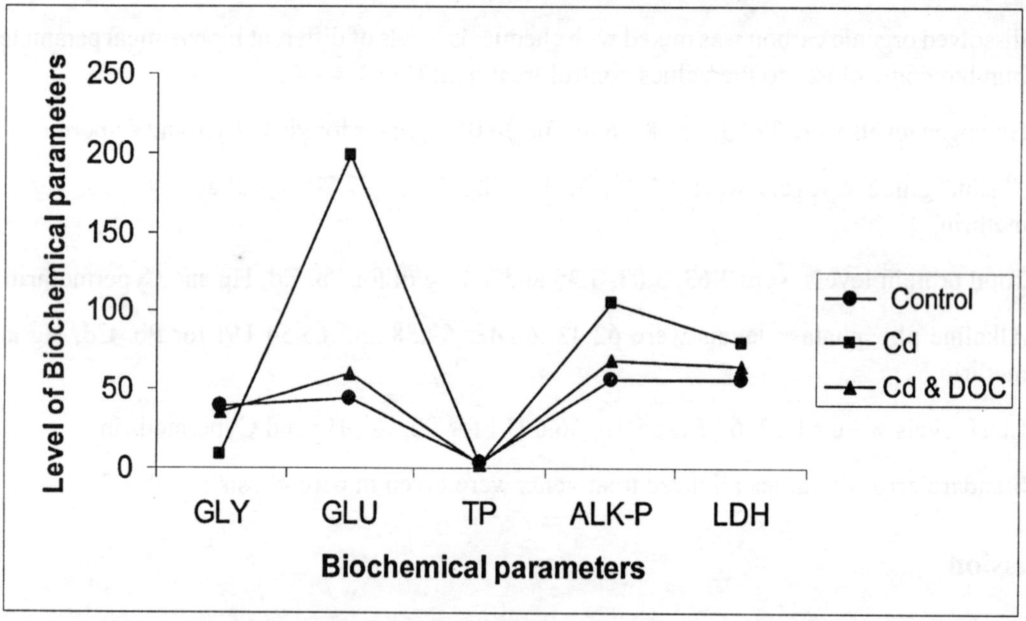

Fig 1. Influence of DOC on the effect of Cadmium on biochemical parameters of Scale carp

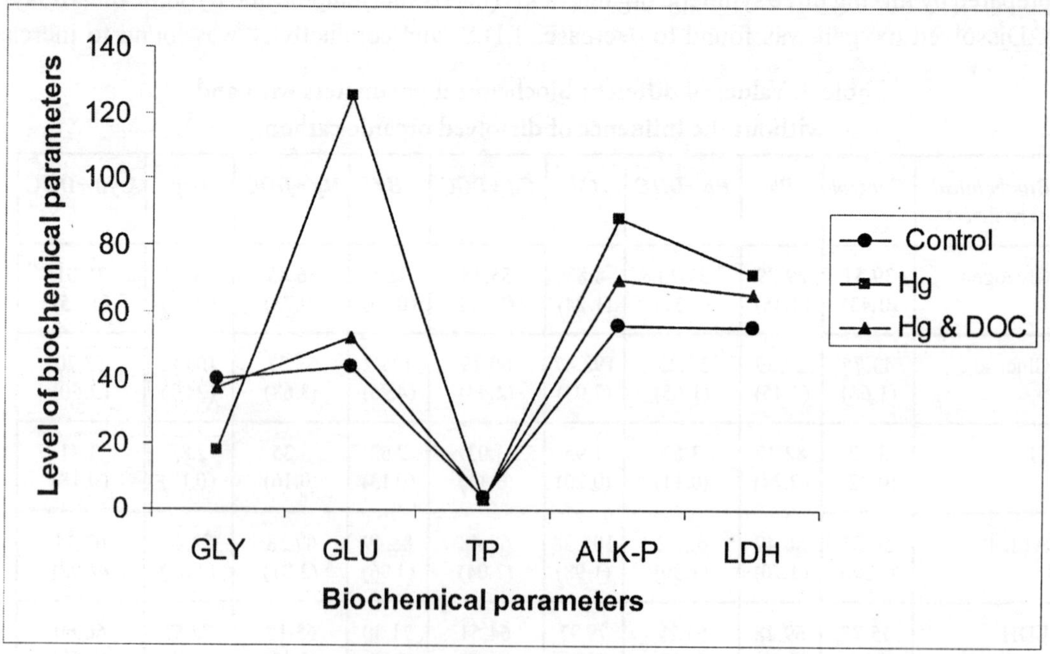

Fig 2. Influence of DOC on the toxicity of Mercury on the biochemical parameters of Scale carp

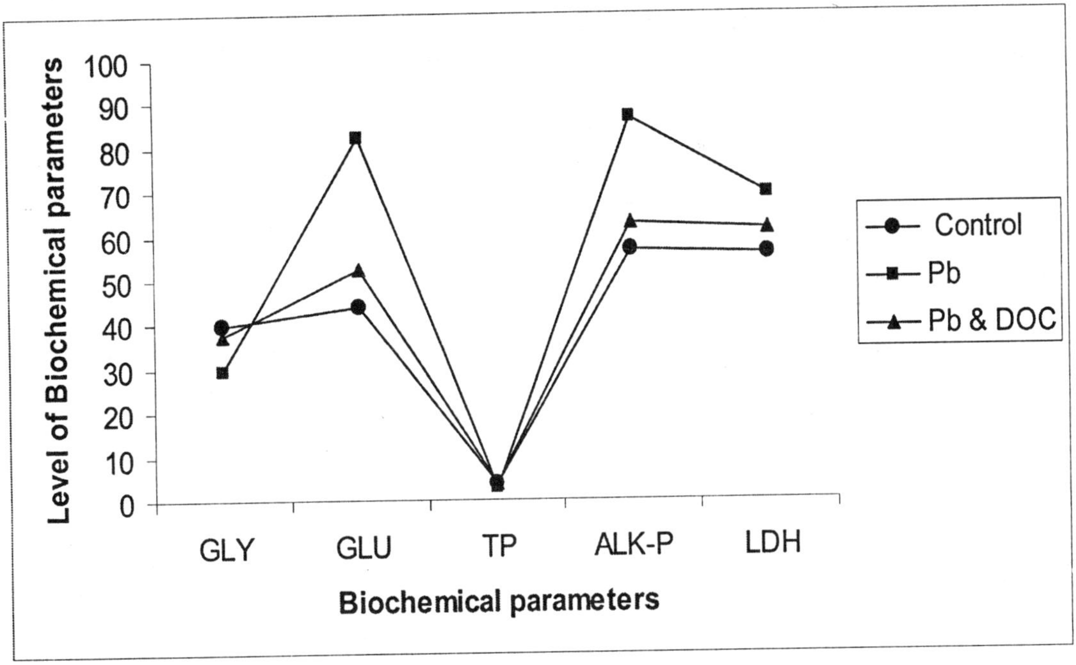

Fig 3. Effect of DOC on the toxicity of Lead in the Biochemical parameters of Scale Carp

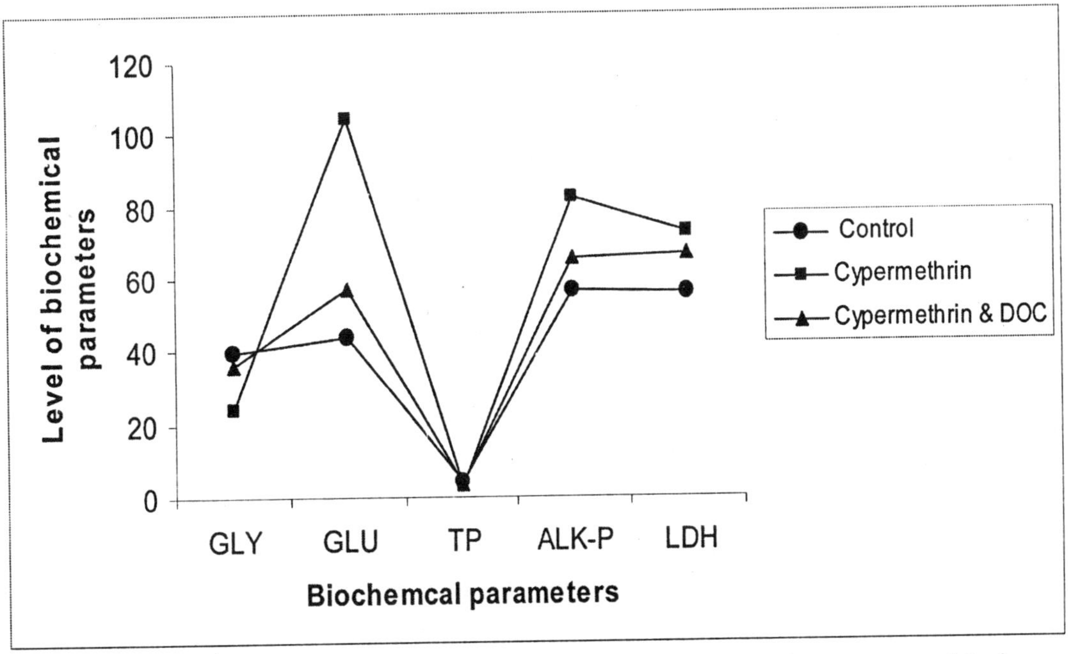

Fig 4. Effect of DOC on the toxicity of Cypermethrin on biochemical parameters of Scale carp

with the progress of time in all treatments; this mainly due accumulation of unfed materials, fecal matter etc. When treatments are compared with control there was no definite trend for dissolved oxygen, which may be due to continuous aeration. TDS and conductivity were found to be slightly higher in treatments with respect to control. This may be due to the chemical itself and the stress exerted by the chemical on fish resulting in higher physiological activities, which in turn responsible for lower food intake and higher excretion.

Biochemical parameters

It is known that physiological and biochemical parameters in fish blood and tissues could change when exposed to xenobiotics (heavy metals and pesticides) as these parameters are extremely sensitive (Philip *et al.*, 1995; Das and Mukherjee, 2003; Reddy and Yellamma, 1991). Thus, it was argued that several biochemical parameters in fish blood and tissues could be used as an indicator of xenobiotics toxicity (Toguyeni *et al.*, 1997). It is very clear from the figures 1 to 4 that increased values of all biochemical parameters due to the stress induced by toxicants were almost normalized with in the experimental period.

Liver glycogen and serum glucose. The results of the present study showed that predetermined concentrations of heavy metal (Cd, Hg and Pb) and synthetic pyrethroid (Cypermethrin) decreased the liver glycogen reserves and increase serum glucose level of *C. carpio* after 14 day exposure significantly in comparison to control groups. When dissolved organic carbon (0.88 mg/L) was included with the test chemicals glycogen content and glucose level attained almost normal level in all cases. Effect of DOC was significantly different from chemical (alone) treatment.

The decrease found in the liver glycogen reserves of *C. carpio* in this study might be the result of Cd, Hg and Pb stimulating the activities of enzymes that work in glycogenolysis. Carbohydrates are stored as glycogen in fish tissue and organs like the muscle and liver in order to supply the energy needs when there are hypoxic conditions, intensive stocking and lack of food (Wendelaar Bonga, 1997). Some investigations also showed that heavy metals could decrease the glycogen reserves in fish (Levesque *et al.*, 2002) by affecting the activities of enzymes that play a role in the carbohydrate metabolism.

Serum glucose levels in fish were affected by many stress factors, including heavy metals (Canly,1995) and pyrethroides (Das and Mukherjee, 2003). The increase in serum glucose levels in *O. mykiss* and *H. fossilis* exposed to Cd and *Channa punctatus* and *C. carpio* exposed to Hg were linked to the decrease in glycogen reserves in the muscle and liver tissues (Richard *et al.*, 1998; Canly, 1996). The same holds good for this experiments also. However DOC was able to restore the almost normal level of liver glycogen reserves and serum glucose when combined with test chemicals mainly due to their influence in reducing the toxic effect of these chemicals.

Serum total protein. It was observed that total protein level in serum decreased significantly under the influence of test chemicals (Pb, Cd, Hg and cypermethrin) when compared with control in all cases. But when dissolved organic carbon was included with test chemicals serum total protein level increased significantly and reached almost to control level. Liver is the exclusive site of synthesis of albumin (the most abundant of the plasma protein) and all blood coagulation proteins (except factor VIII). Synthesis of coagulation protein tends to be diminished in liver disease. The proteins in the plasma decrease rapidly in severe, acute liver injury (Diana, 1982). In the present study decrease of total serum protein content might be because of liver injury due to xenobiotics toxicity, excessive energy metabolism due to stress. dissolved organic carbon (0.88 mg/L) when combined with test chemicals were able to maintain almost normal level of serum protein mainly due to reduced chronic toxic effect of test chemicals on scale carp.

Alkaline phosphatase. Serum alkaline phosphatase level of fish increased significantly in all chemical treatments as compared to control treatments. This is may be because of liver and/or hepatic necrosis due to the exposure of fish into chronic toxic level of xenobiotics. Under stress liver, intestine and kidney can generate higher level of alkaline phosphatases.

Serum alkaline phosphatases activity may be elevated in both acute and chronic liver diseases and in the regenerative process that occurred in the liver following injury. Modest increases in serum alkaline phosphatases do occur in hepatic necrosis (Kaneko *et al.*, 2000).

In DOC treated tests of Cd, Pb, Hg and cypermethrin alkaline phosphatases level decreased significantly in comparison to only chemical treatments. DOC was able reduced the sub-lethal toxicity of all four test chemicals and thereby decreased the alkaline phosphatase level significantly.

Serum LDH. The serum lactate dehydrogenase levels of fish increased significantly in chemical treatment than control. Serum LDH activity is tissue non specific, however, necrosis of muscle, liver and hemolysis are the major cause of elevations.

In cases, LDH is the most sensitive indicator of liver cell damage. Hyperglycemia and hyperlactemia occurs simultaneously (Philip *et al.*, 1995) in many cases where impaired carbohydrate metabolism was observed due to external environmental stress.

The serum LDH level decreased to normal level in Chemical-DOC treatments. Adverse effects of xenobiotics on serum LDH was completely neutralized by $CaCO_3$, DOC and bentonite.

It was observed that effect of DOC treatments was highest in case of cadmium. This is because in chemical control treatment of cadmium (at 3 ppm Cd), all fishes died on 8th day of the experiment (for Hg, Pb and cypermethrin no fish died in chemical control treatment). But in Cd-DOC treatments no fishes were died for the entire period of experiment, also affected biochemical parameters reached almost to the normal level.

Dissolved organic carbon was able reduce the toxic effect of organic (cypermethrin) and inorganic (Cd, Pb, Hg) due to its complexation ability. Complexation of metals by soluble inorganic and organic ligands in water bodies often lowers free metal ion activity and thereby bioavailability and toxicity. Dissolved organic substances are of basic importance for biding of metals into complexes. In this experiment lead, mercury and cadmium underwent complexation reaction with dissolved organic matter, which facilitated metal removal by sweeping organically bound metals into particulate phases via colloid aggregation. Cypermethrin due to its hydrophobic nature was absorbed on dissolved organic carbon which reduced its bioavailability in water medium and thereby toxicity to fish.

Conclusion

It is concluded from this study that organic matter either raw or partially decomposed (e.g. cow dung, humus, FYM) which are part of good aquaculture, have good potential to protect the fish from lethal and sublethal effects of inorganic (e.g. metals) and organic (pesticides) xenobiotics. Organic matter protects due to its high CEC, adsorbtion and complexation ability. Water ecosystems which has balanced quantity of organic matter can lower probability to be affected by pollutants than an ecosystem which don't have organic matter in optimum level. Therefore, fishes in an aquatic ecosystem can be affected faster by a pollutant if it doesn't contain organic matter in optimum level.

Acknowledgement

Authors are grateful to the Director, Central Institute of Fisheries Education, Mumbai, India for providing necessary facilities for the experiment.

References

- Arnold M. Seligman, Howard H. Chauncey, Marvin M. Nachlas, Leon H. Manheimer, and Herbert A. Ravin. 1951. The colorimetric determination of phosphatases in human serum. *J. Biol. Chem.* 190 (1) 7-15.
- Canly, M. 1995. Effects of mercury, chromium and nickel on some blood parameters in the carp Cyprinus carpio. Tr. J. of Zoology 19305-311.
- Canly, M. 1996. Effects of mercury, chromium and nickel on glycogen reserves and protein levels in tissues of *Cyprinus carpio*. Tr. J. of Zoology. 20161-168.
- Carroll, N. V., R. W. Longly and J. H. Roe. 1956. The determination of glycogen in liver and muscle by use of anthrone reagent. J. Biol. Chem. 220583-593.
- Clesceri, L.S., Greenberg, A.E and Eaton, A.D. (Eds.) 1998. Standard methods for the examination of water and wastewater (20th edition). American Public Health Association (APHA), American Public Works Association and Water Pollution Control Federation. Washington. New York.
- Das, B.K. and Mukherjee, S.C. 2003. Toxicity of cypermethrin in *Labeo rohita* fingerlings biochemical, enzymatic and haematological consequences. Comp. Biochem. Physiol. C. Toxicol. Pharmacol. 134(1)109-121.
- Das, S.C.S., Datta, S., Pal, A.K. and Das, R.C. 2005. Effect of air temperature on acute toxicity of arsenic (As) and mercury (Hg) to *Cyprinus carpio* var communis. Environ. Ecol. 23(2)311-314.

- Datta, M. and Kaviraj, A. (2003). Ascorbic acid supplementation of diet for reduction of deltamethrin induced stress in freshwater catfish Clarias gariepinus. Chemosphere. 53(8)883-891.
- Datta, S. and Das R.C. 2003. Influence of some abiotic environmental factors on acute toxicity of inorganic lead (Pb^{2+}) to *Cyprinus carpio* var communis (Linn.) and *Catla catla* (Ham.) in simulated toxic aquatic environment. Toxicol. Environ. Chem. (U.K.). 85(4-6)203-219.
- Datta, S., Das, S.C.S. and Das, R.C. 2002d. Influence of soil sediment factors and contact time with complexing agents of sediment on the toxicity of inorganic mercury (Hg^{2+}) to scale carp in aerobic aquatic environment. In XXI Annual Conference of Society of Toxicology, India held at Regional Occupational Health Centre (Eastern), ICMR, Salt Lake City, Kolkata, on 3-5th Dec, 2002. Abstract No. FC-21.
- Datta, S., Das, S.C.S. and Pal, A.K. 2002b. Effect of contact time with soil sediment on the acute toxicity on mercury to common carp. In Sixth Indian Fisheries Forum's Abstract. Asian Fisheries Society (Indian Branch). Central Institute of Fisheries Education, Mumbai, 17-20 Dec., 2002. Abstract No. AE-51.p.132.
- Datta, S., Masala, S.H. and Das, R.C. 2002c. Influence of some abiotic factors on the acute toxicity of cadmium to *Cyprinus carpio*. In Sixth Indian Fisheries Forum's Abstract. Asian Fisheries Society (Indian Branch). Central Institute of Fisheries Education, Mumbai, 17-20 Dec., 2002. Abstract No. AE-56.p.137.
- Datta, S., Masala, S.H. and Das, R.C. 2003b. Influence of some abiotic factors on the acute toxicity of Cadmium to *Cyprinus carpio*. J. Indian Fish Assoc. 3041-52.
- Datta, S., Rongsenkumzuk and Das, R.C. 2004a. Impact of calcium, carbonate and bicarbonate on the toxicity of inorganic lead to freshwater fish in simulated toxic aquatic environment. In Abstracts of All India Seminar on "Sustainable Aquaculture for Augmentation of Export with Special Reference to Environment, Engineering and Value Addition". p.67. Organised by Agricultural Engineering Division. The Institute of Engineers (India), West Bengal State Centre, held on 3-4 Sept, 2004 at Kolkata.
- Datta, S., Shah, P.S. and Das, R.C. 2002a. Soil sediment and hardness of water reduce acute toxicity of deltamethrin to Scale carp. Pestic. Res. J. 14(2)327-336.
- Datta, S., Shah, S.I.G. and Das, R.C. 2004d. The significance of calcium, carbonate and bicarbonate ions on the toxicity of cadmium to a freshwater teleost. In Abstracts of All India Seminar on "Sustainable Aquaculture for Augmentation of Export with Special Reference to Environment, Engineering and Value Addition". p.70. Organised by Agricultural Engineering Division. The Institute of Engineers (India), West Bengal State Centre, held on 3-4 Sept, 2004 at Kolkata.
- Datta, S., Singh, H. and Das, R.C. 2003a. Influence of soil sediment factors on the acute toxicity of inorganic mercury to Catla catla. Environ. Ecol. 21(3)542-551.
- Datta, S., Singh, K.D. and Das, R.C. 2004b. Influence of some ions (Ca^{2+}, CO_3^{2-}, HCO_3^-) on the toxicity of incrganic mercury to scale carp. In Abstracts of All India Seminar on "Sustainable Aquaculture for Augmentation of Export with Special Reference to Environment, Engineering and Value Addition". p.68. Organised by Agricultural Engineering Division. The Institute of Engineers (India), West Bengal State Centre, held on 3-4 Sept, 2004 at Kolkata.
- Datta, S., Zargar, B.A. and Das, R.C. 2004c. Effect of changing salt concentration on the toxicity Cypermethrin to *Cyprinus carpio*. In Abstracts of All India Seminar on "Sustainable Aquaculture for Augmentation of Export with Special Reference to Environment, Engineering and Value Addition".

p.69. Organised by Agricultural Engineering Division. The Institute of Engineers (India), West Bengal State Centre, held on 3-4 Sept, 2004 at Kolkata.

- Denise Barham and P. Trinder.1972. An improved colour reagent for the determination of blood glucose by the oxidase system. Analyst, 97, 142-145.
- Diana, J.S. 1982. An experimental analysis of the metabolic rate and food utilization of northern pike. Comp. Biochem. Physiol. 71395-398.
- Kaneko, J. J., Harvey, J. W. and Bruss, M. L. (Eds.) 2000. Clinical Biochemistry of Domestic Animals. Fifth edition. Acadmic Press. p. 932.
- Karl Doetsch and Richard H. Gadsden, 1975. Determination of Urinary Total Protein by Use of Gel Filtration and a Modified Biuret Method, Clinical Chemistry 21 (8) 778-781.
- Levesque, H.M., Moon, T.W., Campbell, P.G.C. and Hontela, A. 2002. Seasonal variation in carbohydrate and lipid metabolism of yellow perch (Perca flavescens) chronically exposed to metals in the field. Aquat. Toxicol. 60257-267.
- MacRae, R.K., Maest, A.S. and Meyer, J.S. 1999. Selection of a organic acid analogue of dissolved organic matter for use of toxicity testing. Can. J. Fish. Aquat. Sci., 56(8)1484-1493.
- Marr, J.C.A.L., Cacela, J.A., Hansen, J.A., Meyer, J.S. and Bergman, H.L. 1999. Bioavailability and acut toxicity of copper to rainbow trout (Oncorhynchus mykiss) in the presence of organic acids simulating natural dissolved organic carbon. Can. J. Fish. Aquat. Sci., 56(8)1471-1483.
- Philip, G.H., Reddy, P.M. and Sridevi, G. 1995. Cypermethrin induced in vivo alterations in the carbohydrate metabolism of freshwater fish, *Labeo rohita*. Ecotoxicol. Environ Saf. 31(2)173-181.
- Reddy, A.T. and Yellamma, K. 1991. Perturbations in carbohydrate metabolism during cypermethrin toxicity in fish, Tilapia mossambica. Biochem. Int, 23(4)633-641.
- Richard, A.C., Daniel, C., Anderson, P. and Hontela, A. 1998. Effects of sub chronic exposure to cadmium chloride on endocrine and metabolic functions in rainbow trout Oncorhynchus mykiss. Arch. Environ. Contam. Toxicol. 34(4)377-381.
- Seitfer, S., Dayton, S., Novic, B. and Muntwyler, E., 1950, The estimation of glycogen with the anthrone reagent. Arch. Biochem, 25 191-200.
- Stanley Morgenstern, Richard Flor, Gerald Kessler and Bernard Klein. 1966. The Automated Determination of NAD-Coupled Enzymes II. Serum Lactate Dehydrogenase. Clinical Chemistry 12 274-281.
- Toguyeni, A., Fauconneau, B., Boujard, T., Fostier, A., Kuhn, E.R., Mol, K.A. and Baroiller, J.F. 1997. Feeding behaviour and food utilisation in tilapia, Oreochromis niloticus, effect of sex ratio and relationship with the endocrine status. Physiol. Behav., 62273-279.
- Wendelaar-Bonga, S.E. 1997. The stress response in fish. Physiol. Rev. 77, 591-625.

13

GEOGRAPHICAL INFORMATION SYSTEMS (GIS) AS A TOOL FOR MANAGING INLAND WATER RESOURCES FOR SUSTAINABLE FISHERIES

*A.S.Ninawe and S. S. Ghatge**

Abstract: Maharashtra state is blessed with vast inland water recourses, which are spatially scattered over the state. Maharashtra state added 5.9 lakh tonnes to the India's fish production, of which 1.4 lakh tonnes is produced from the inland waters (2006-07). The Vidarbha region of Maharashtra with it's available reservoir potential area of around 70,000ha contributes more than 80% of the total inland fish production of the state. The average productivity of the reservoirs in Vidarbha region is low as compared with the national productivity of the small reservoirs (i.e. 50 kg /ha/ year). Planning for management, development and conservation of the inland water resources in a sustainable way is thus a vital issue, which requires reliable and up-to-date information on the extent and distribution of resources over time and space. Geographic Information System (GIS) and Remote sensing (RS) have emerged as powerful tools in providing such reliable spatial information on various natural resources, and in organizing, analyzing and displaying spatially explicit data. This paper deals with the some possible aspects of use of GIS in managing the inland water resources in the context of Vidarbha region of Maharashtra state.

Keywords: Inland water resources, Remote sensing

Introduction

Fisheries sector plays an important role in the Indian economy. It contributes to the national income, exports, food and nutritional security and in employment generation. This sector is also a principal source of livelihood for a large section of economically underprivileged population of the country, especially in the inland areas.

Vice-Chancellor, Maharashtra Animal and Fisheries Sciences University, Nagpur (Maharashtra). *Assistant Professor (Fish Biology), College of Fishery Science, Nagpur (Maharashtra).

The state of Maharashtra located in the north centre of Peninsular India, with a command of the Arabian Sea on the west coast, has been blessed with vast inland water resourses; fresh water tanks and ponds of 0.59 lakh ha, reservoirs of 2.79 lakh ha, brackish water area of 0.19 lakh ha, 16000 km length of rivers and canals; the total inland water bodies area is approximately 3.48 lakh ha. Maharashtra state added 5.9 lakh tonnes (2006-07) to the India's fish production, of which 1.4 lakh tonnes is produced from the inland waters which are spatially scattered over the state.

The Vidarbha region of Maharashtra which is situated in between latitude 17° 57′N to 21° 46′N and longitude 75°57′E to 80°59′E and comprises eleven districts *viz*; Nagpur, Amaravati, Akola, Yavatmal, Buldhana, Chandrapur, Wardha, Gadchiroli, Bhandara, Gondia, and Washim. The region with it's available reservoir potential area of around 70,000ha, contributes more than 80% of the total inland fish production of the state i.e. 80,259 tonnes in 2006-07.

Out of nine agro climatic zones of Maharashtra, three zones viz., assured rainfall zone, moderate to moderately high rainfall zone and high rainfall zone are in Vidarbha. By virtue of its geographical situation in monsoon belt, parts of Vidharbha region are endowed with fairly good rainfall and consequently extensive water bodies suitable for aquaculture, especially in the districts of Gondia (Lake District), Bhandara and Nagpur. The inland aquacultural water resources in the form of ponds and tanks have been distributed over almost all the districts of Vidharbha. The average productivity of the reservoirs in Vidarbha region is low as compared with the national productivity of the small reservoirs (i.e. 50 kg /ha/ year).

Improper and unplanned exploitation of any resource leads to decrease in its potentiality and collapse of the resource. Hence, there should be proper plan to exploit natural resources. For proper planning it is necessary to know resource profile and associated infrastructure to suggest management measures for their sustainable use.

As pressures on fisheries resources continue to increase, it has become evident that the data needed to make informed management decisions are either lacking or are inaccessible. Gathering this needed information through monitoring and research is an important step towards better informed management; however, simply gathering this information will not solve the problems associated with managing that information and making it readily available. Unless advanced information management technologies are instituted in resource management agencies, effective utilization of the information to better manage our resources will not occur. Fishery is a State subject in the Indian context and as such the primary responsibility for development rests with the State Governments.

Planning for management, development and conservation of these inland water resources in a sustainable way is thus a vital issue, which requires reliable and up-to-date information on the extent and distribution of resources over time and space. Geographic Information System (GIS) and Remote sensing (RS) have emerged as powerful tools in providing such reliable spatial information on various natural resources, and in organizing, analyzing and displaying spatially explicit data. GIS is a powerful tool capable of organizing, analyzing and displaying spatially explicit data, yet it is not being extensively used in fisheries science.

Prospects of GIS in managing the inland water resources

An application of GIS for resource management is limited only by the imagination of how to combine the different datasets. GIS offers advantage over conventional approaches because of the speed and accuracy with which it handles the large spatial datasets in a user-friendly manner. The GIS integrates various thematic information derived from satellite data with other collateral data such as socio-economic and infrastructural data which are significant in arriving at the status of the particular area and decision making. Such an approach helps in initiating a sustainable management of the fisheries resources.

The slow emergence of GIS in fisheries is due to socio-economic constraints such as the level of education or financial resources with which to purchase technology. But in present situation, Institutions like National Remote Sensing Centre, Hyderabad and Indian Institute of Remote Sensing, Dehradun are willingly providing indigenous trained man power and the available technology is also becoming cheaper with emergence of open source GIS software. Meaden (2000) indicated GIS is a promising technology for fisheries science and management. Nowadays, it escapes no one that the spatial dimension has to be taken into account when it comes to designing a fisheries management plan. The different socioeconomic components and the distribution of exploitable resources vary greatly from one area to another. This is a constraint to any attempt to apply a generalized set of management methods set on a national scale and tells us how essential it is to consider local specificities.

Several authors like Kieser *et al.* (1995) enumerated the advantages and disadvantages of using GIS in fisheries science while demonstrating the strength of using GIS in visualization in studies of fisheries acoustics scatter data, especially when these datasets are combined with various environmental parameters. GIS is being increasingly employed in aquaculture. An overview of GIS methodology appropriate for aquaculture is provided by Meaden and Kapetsky (1991) and a review of applications was made by Kapetsky and Travaglia (1995). Most applications of GIS in aquaculture are for site selection over rather limited geographic areas, but strategic assessments of aquaculture potential have not been made for larger areas.

A continuous monitoring and analysis project using GIS has been initiated by the Institute for Fisheries Research, School of Natural Resources and Environment, University of Michigan, for assessment of fisheries resources from the Great Lakes, *Viz* Lake Huron, Lake Erie, Lake Michigan, Lake Ontario and Lake Superior. The results are exemplary which give clear idea regarding the resources from these water bodies on real-time basis. Similarly, application of GIS for major reservoirs of Vidharbha and elsewhere in India is possible for their sound management and increased production.

Salam *et al.* (2003) demonstrated the usefulness of RS, Geographical Positioning System (GPS) and GIS techniques to estimate the area of land available for different types of aquaculture development in Khulna region, southwestern Bangladesh. Murthy *et al.* (1992) prepared an inventory map of Uttar Pradesh on type, number of water bodies and seasonal variation in water spread along with turbidity and presence or absence of aquatic vegetation. He suggested, that on basis of suitable criteria, selection of potential aquaculture sites is primarily possible by RS inputs. Narayan *et al.* (1983) surveyed Idukki district of Kerala and Bundelkhand region of Uttar Pradesh using multi-band aerial photographic survey (Feb 1978 to Feb 1979) and Landsat imagery. They generated various thematic maps and statistical data on various resources and concluded that information obtained by their survey was extremely useful for resource managers and decision makers. On similar lines base line resource data on natural and infrastructural facilities can be prepared for sustainable utilization of available resources.

Rautela *et al.* (2002) assessed the negative impact of construction of large dams using Remote Sensing and GIS tools. It is estimated that the Tehri reservoir in the upper catchments of Ganga river would directly affect 2687 ha of agricultural land and another 3347 ha around the reservoir rim would be rendered unfit for cultivation. Such type of studies in Vidharbha would help to know possible impacts of damming on agriculture as well as the fish biodiversity, especially in case of semelparous, catadromus migratory species like *Anguilla bengalensis* which is available in the region.

Wang *et al.* (2004) demonstrated the use of satellite images to monitor the changes in water quality and found a very high correlation between the satellite-based estimates and in situ measurements. Thimmaiah (2007). Using SOI toposheet No. 48 O/10 (1976) and IRS P6 LISS IV Mx (2005) studied the shoreline changes of Bhadra reservoir in Karnataka from 1976 to 2005. The area change was calculated using ArcGIS 9.1 software. He observed reduction in reservoir area of about 2586ha. Also the Maximum Sustainable Yield (MSY) of 114 tonnes/ year was worked out on basis of Morpho Edapic Index (MEI).

With GIS constant monitoring of shoreline change, yearly available water spread can be studied by employing minimum labor and time. Potential yield of a water body can be predicated from primary production, Morpho-Edaphic Index, and Maximum Sustainable Yield. It is possible to work out the MSY of reservoirs and decide the stocking density with the help of MEI. According to the Welcomme's formula adopted by Government of India (Sugunan, 1997), Morpho Edapic Index (MEI) is an essential factor in determining the stocking densities for reservoirs. GIS and RS with help o f the available utilities are nowadays becoming important tools in assessment of MEI.

Conclusion

Baseline information on the extent of reservoir, rivers and canals, ponds and tanks, fish hatcheries, transportation routes, can be generated using the satellite images. On the basis of capability of GIS tools a continuous monitoring of the changes in water quality, shoreline changes, Water spread, prediction of fish yields and estimating stocking rates for reservoirs is possible. Potential sites for aquaculture, stock enhancement, cage and pen culture can be easily worked out using RS and GIS. To conclude these tools will possibly help to augment the average productivity of the reservoirs in Vidarbha region and require initiatives at local and regional level.

References

- Kapetsky, J.M., & Travaglia, C. 1995. Geographical information systems and remote sensing An overview of their present and potential applications in aquaculture. pp. 187–208, in Nambiar, K., & Singh, T. (eds). Aquaculture towards the 21st Century. INFOFISH, Kuala Lumpur, Malaysia.
- Kieser, R., Langford, G. and Cooke, K. 1995. The use of Geographic Information Systems in the acquisition and analysis of fisheries acoustic data. In Prcoceedings of the ICES International symposium on Fisheries and plankton Acoustics. June1995, Aberdeen Scotland.
- Meaden, G. J. and Kapetsky, J. M. 1991. Geographic Information System and Remote sensing in Inland Fisheries and Aquaculture. FAO Fisheries Technical Paper, No. 318, Rome, Itlay 262p.
- Murthy, T.V.R., Garg, J.K., Palria, S. and Singh, T.S., 1992. Application of remote sensing for Inland Aquaculture site selection in Uttar Pradesh. In Proc. Of National Symp. On Rem. Sen. For Sustainable Development, pp.213-217.
- Meaden, G. J. 2000. GIS in Fisheries Management. Geocoast, 1(1), pp.82-101.
- Narayanan, L. R. A. and Sahai, B. 1983. Integrated resources survey and applications. Sadhana, 6(4) Dec. pp. 287-295.
- Rautela, P., Rakshit, R., Jha, V. K., Gupta, R. J., Munshi, A., 2002. GIS and remote sensing-based study of the reservoir-induced land-use/ land-cover changes in the catchment of Tehri dam in Garhwal Himalaya, Uttaranchal (India). Current Science., 88(3)308-311.
- Sugunan, V. V., 1997. Guidelines for management of small reservoirs in India. Fishing Chimes, 1723-27.
- Salam, M. A., Lindsay, G. R., Beveridge, C. M. M., 2003. A comparison of development opportunities for crab and shrimp aquaculture in southwestern Bangladesh, using GIS modeling. Aquaculture, 220477-494.
- Thimmaiah G. N. 2007. Appraisal of Bhadra reservoir fiheries using GIS. M.F.Sc. Thesis. Central Institute of Fisheries Education, Mumbai 94p.
- Wang, Y., Xia, H., Fu, F. and Sheng, G., 2004. Water Quality Change in Reservoirs of Shenzhen, China Detection Using LANDSAT/TM Data. Science of the Total Environment., 328(1-3) 195-206.

14

ECOLOGY AND FISH DIVERSITY OF THE RIVERS IN TRIPURA, INDIA

Devashish Kar, Anjam Hussain Barbhuiya, Bubul Das

Abstract: The province of Tripura is flanked by both hills and plains having four principal rivers flowing through its terrains, viz., (a) River Monu (b) River Khowai; (c) River Gomati and, (d) River Feni. All these rivers flow into Bangladesh independently. In addition to the above, there are other little smaller rivers, viz., River Howrah, River Deo, River Muhuri, Juri, Kakri etc. Of these, the rivers Monu and Gomati originate from the Longtarai Hills while the river Khowai originates from the Atharamura Hill ranges. Individually, the number of species recorded in each of the above-mentioned rivers were 10 species in the River Monu, 11 species in River Feni, 14 species in River Muhuri, 15 species in River Dhalai, 15 species in River Deo, 7 species in River Juri and 5 species in River Kakri. Results of studies on habitat parameters, physico-chemical characteristics of water and fish diversity have been discussed in the chapter.

Keywords: Fish diversity, Tripura rivers, North-East India, Hotspot

Introduction

Biodiversity has immense importance in the context of recent advances made in Science and Technology and this provides a wide spread need for conservation of natural ecosystem in aqua regime. Further, Biodiversity is essential for stabilization of ecosystem, protection of overall environmental quality and for understanding the intrinsic worth of all species on the earth (Gadgil and Kar, 2000). Fishes make-up most of the abundant classes of vertebrates, both in terms of number of species and number of individuals. They exhibit enormous diversity of size, shape, number and biology; as well as, in the habitats in which they live. 21,723 species of fish under 4044 genera, 445 families and 50 orders have been recorded in the world (Jayaram, 2010). Jayaram (2010) mentioned 2500 species of fishes in the Indian region of which 930 are freshwater inhabitants. Talwar and Jhingran (1991) estimated 2546 species of fish belonging 969 genera, 245 families and

Department of Life Science, Assam (Central) University, Silchar-11, India; E-mail: *devashishkar@yahoo.com*

40 orders. These represent 11.72 % of species, 23.96 % of genera and 57 % of families and 80 % of the global fishes.

The North-Eastern region of India (25°55′12.09" to 27°42′21.28"N and 89°38′22.44" to 97°04′9.67"E) contains a large number of diverse types of water bodies which include both lotic and lentic systems. It comprises of seven provinces and is one of the Hotspots of Biodiversity. These provinces with their unique topography, diverse physiographic features and varied water quality patterns are a lucrative field for ichthyological studies with innumerable hill streams, rivers, and wetlands. Nevertheless, the region is fed by two major drainage systems, viz., the Brahmaputra River system (forming the Brahmaputra valley) and the Barak River system (forming the Barak valley).The drainages in Tripura mostly enter Bangladesh independently (Kar, 2003).

The Province of Tripura ((23°49′51.18" to 23°53′ 57.42"N and 92°12′ 50.43" to 92°24′ 34.76"E) is flanked by both hills and plains with four principal rivers, viz., Monu, Khowai, Gomati and Feni. In addition to the above, there are other smaller rivers, viz., Muhuri, Juri, Kakri, Dhalai, Deo, Haora, etc. of these, the rivers Monu, Khowai, Gomati and Feni have been studied in great detail by Kar (Kar, 2003, 2004, 2005 a,b,c, 2007) and Kar *et al.* (2002 a,b,c). However, not much works have been done on the fishes of the rivers Deo, Juri, Kakri and Muhuri. As such, a preliminary study has been undertaken to record the panorama of Fish diversity along with water characteristics and fish catching devices in the rivers Juri, Kakri, Deo, Monu, Dhalai, Feni and Muhuri rivers in Tripura.

The River Juri originates from Mauglarai Hills and the River Kakri originates from Baugsul Hill in Damchhara and flows into Bangladesh independently. The River Deo originates from the Zampui Hill ranges, flows northwards through the valley between Zampui and Shakhan ranges, locally called Kanchan valley with Dharmanagar Sub-division. Later, the River Deo turns westward cutting through the Shakhan and meets the River Monu in mid-valley between the Shakhan and Longtarai ranges in Kailashahar Sub-division (Kar, 2005 a, b).

The river Monu originates from the Longtarai Hills in the Shakhan Tang range and it takes course through the valley between Shakhan and Longtarai ranges to the northern direction. After meeting the river Deo, it runs further northward and passes through the Su-divisional Headquarters at Kailashar and enters Bangladesh. The River Monu is *c* 93 km long with catchment area of *c* 2228 km^2. The River Dhalai is said to have originated from the Longtarai hill ranges and flowing the Ambasa region, it enters Bangladesh (Kar, 2005 c).

The river Feni is said to have originated from the escarp of the Chittagong Hill tracts and flows down beside the village Shilachhari along the Indo-Bangladesh border. Finally, it enters Bangladesh through a village called Amlighat (Kar, 2007).

The River Muhuri originates from the Kailashi hills and flows down by the side of Shantirbazar and Belonia Sub-division and finally enters Bangladesh as reported by Kar (2007).

Compared to the average annual fish production of 6-7 kg per ha from the Indian water bodies, the overall fish production from the water bodies of Tripura is not very less, as reported by Kar (2007). However, contrary to the demand, the present fish production rate can not very much fulfil the needs of the consumers. Loss of habitat, unsustainable harvest and fish disease (notably Epizootic Ulcerative Syndrome or EUS) are some of the reasons for the depleting trend of fish yield. Under this background, a preliminary study has been undertaken to record the present panorama of fish diversity in certain rivers across the length and breadth of the Province of Tripura (2007).

There have been extensive studies on the Freshwater (FW) fishes of India, notably, Hora (1921), Menon (1999), Jayaram (2010), Talwar and Jhingran (1991).

However, such works from the NE region of India are scanty and restricted mainly to the works of Dey (1973), Sen (2000), Ghosh and Lipton (1982), Sen (1985), Dey and Kar (1990), Kar and Dey (2002), Kar *et al.* (2002 a, b, c; 2003, 2004), Kar and Sen (2007), Kar (1990, 2003, 2004, 2005 a, b, c, 2007). But most of them are concerned with taxonomy (Biodiversity), biology and aquaculture.

Material and Methods

General survey of the Fish Biodiversity was done using standard procedures of Armontrout (1990). Also, NBFGR Manual (2000) was consulted for studying the habitat parameters Headwater to downstream studies were based on River Continuum Concept as advocated by Vannote, *et al.* (1980).

Fish samples were collected through experimental fishing using caste nets (dia. 3.7 m and 1.0 m), gill nets (vertical height 1.0 m- 1.5 m; length 100 m -150 m), drag nets (vertical height 2.0 m), triangular scoop nets (vertical height 1.0 m) and a variety of traps. Camouflaging technique was also used to catch the fishes. Fishes have been preserved at first in concentrated formaldehyde in the field itself and then in 40 % formalin. Fishes have been identified after standard literature of Menon, (1999); Talwar and Jhingran, (1991); Jayaram (2010). Yield statistics were extrapolated after Dey and Kar (1990), and Kar (1990) from daily catch statistics recorded at the landing stations following FAO (1974). The physico-chemical characteristics of water have been analysed after the methods in APHA (1995) while the different aspects of fish catching devices were studied after Kar and Dey (1991, 1993, 1996) and Kar (2007).

Result and Discussion

The results of the present study pertaining to aspects, like habitat inventory of the studied rivers, their physico-chemical characteristics and Fish Biodiversity are presented in Tables 1,2,3 respectively.

Table 1 Habitat Inventory of the Surveyed Rivers in Tripura

Sl. No.	Name of the Rivers	Place	Date	Valley Segment	Breadth of River (approx.) m/km	Depth of the River (approx.) m	Microhabitat Type	Substratum
1	Juri	Dharmanagar	14.10.07	Alluvial	55.0 m	3.0 m	Run	Mainly silt
2	Kakri	Dharmanagar	25-10-07	Alluvial	41.0 m	2.9 m	Run	Mainly silt
3	Deo	Pechartal	8-11-07	Alluvial	96.0 m	4.2 m	Run	Sandy
4	Monu	Kailasahar	24-10-07	Alluvial	83.0 m	4.8 m	Run, some turbulence	Mainly Sand and Silt
5	Dhalai	Ambasa	23-10-07	Colluvial	47.0 m	3.7 m	Run with some turbulence	Mainly Sand and Silt with some gravels
6	Feni	Sabrum	30-10-07	Colluvial	85.0 m	3.9 m	Run-Sheet with some turbulence	Mainly Sand and Silt
7	Muhuri	Belonia	31-10-07	Colluvial	72.0 m	2.9 m	Mainly Run	Mainly Sand

Table 2. Physico-chemical Characteristics of the Surveyed Rivers in Tripura

Sl. No	Name of water body	Temp. of Water (°c)	Turbidit-y (NTU)	pH	D.O.(mg/-l)	FCO_2(mg/l)	T.A.(mg/l	Conductivity (imho/cm)
1	River Deo	27.0	160	6.74	5.9	14	104	159
2	River Manu	28.5	212	7.1	7	8	110	163
3	River Dhalai	26.8	70	6.6	4.9	12.5	103.5	157
4	River Feni	27.2	90	6.63	3.4	10	63	100
5	River Muhuri	28.1	344	6.58	3.6	9	42	69

Table 3. Ichthyo diversity in the Surveyed Water bodies in Tripura

Abvr.- Ma Manu, Fe Feni, Mu Muhuri, Dh Dhalai, De Deo, Ju Juri, Ka Kakri. "+"= Present and "–"= Absent

Sl.No.	Name of Fish	Ma	Fe	Mu	Dh	De	Ju	Ka
1	*Notopterus notopterus* (Pallas)	+	-	-	+	-	-	-
2	*Gudusia chapra* (Ham–Buch)	+	-	-	-	-	-	-
3	*Salmostoma bacaila* (Ham–Buch)	+	-	+	+	-	-	-

4	*Aspidopar*ia morar (Ham–Buch)	-	-	-	-	+	-	-
5	*Barilius bakeri Day*	-	-	-	-	+	-	-
6	*Barilius barna* (Ham–Buch)	-	-	-	+	-	-	-
7	*Esomus danricus* (Ham–Buch)	-	+	-	-	-	+	-
8	*Amblypharyn*godon mola (Ham –Buch)	-	+	-	-	+	-	+
9	*Puntius* chola (Ham–Buch)	-	-	-	-	+	+	-
10	*Puntius vittatus Day*	+	+	+	+	-	-	-
11	*Puntius puntio* (Ham–Buch)	-	-	+	+	+	-	-
12	*Puntius sarana orphoides* (Valenciennes)	-	-	-	+	-	-	-
13	*Cirrhinus mrigala* (Ham–Buch)	-	-	-	-	+	+	-
14	*Catla catla* (Ham–Buch)	-	-	-	-	-	+	+
15	*Labeo caeruleus* (Day)	-	-	+	-	-	-	-
16	*Labeo boga* (Ham–Buch)	-	+	+	+	-	-	-
17	*Labeo gonius* (Ham–Buch)	-	+	+	-	-	-	-
18	*Labeo rohita* (Ham–Buch)	-	+	-	-	+	+	-
19	*Crossochilus latius latius* (Ham –Buch)	+	-	-	-	+	-	-
20	*Psilorhynchus suction* (Ham–Buch)	-	-	-	+	-	-	-
21	*Botia Dario* (Ham–Buch)	+	-	-	-	-	-	-
22	*Botia rostrata Gunther*	-	-	-	-	+	-	-
23	*Lepidicephalus guntea* (Ham–Buch)	-	+	-	-	+	-	-
24	*Mystus tengara* (Ham–Buch)	-	+	-	-	-	-	-
25	*Mystus bleekeri* (Day)	+	-	+	-	-	-	-
26	*Eutropiichthys murius ((*Ham–Buch)	+	-	-	-	-	-	-
27	*Pangasius pangasius (Ham –Buch)*	-	-	-	-	-	+	-
28	*Amblyceps mangois (Ham –Buch)*	-	-	-	-	+	-	-
29	*Bagarius bagarius (Ham –Buch)*	-	-	-	-	+	-	-

30	*Gagata gagata* (Ham–Buch)	-	-	+	-	+	-	-
31	*Glyptothorax telchitta* (Ham–Buch)	-	-	-	+	-	-	-
32	*Clarius batrachus* (Linn.)	-	-	-	-	-	-	+
33	*Heteropneustes fossilis* (Bloch)	-	-	+	-	-	-	-
34	*Xenentodon cancila* (Ham–Buch)	-	-	+	-	-	-	-
35	*Macrognathus aral* (Bloch & Schneider)	-	-	+	+	-	-	-
36	*Macrognathus pancalus* (Ham–Buch)	-	+	-	+	-	-	-
37	*Mastacembelus armatus* (Lacepede)	-	-	-	+	+	-	-
38	*Chanda nama* (Ham–Buch)	-	+	+	-	-	+	-
39	*Parambassis ranga* (Ham–Buch)	+	-	-	-	-	-	-
40	*Nandus nandus* (Ham–Buch)	-	-	-	+	-	-	-
41	*Glossogobius giuris* (Ham–Buch)	-	-	+	-	+	-	-
42	*Anabas testudineus* (Bloch)	+	-	-	-	-	-	+
43	*Colisa fasciatus* (Schneider)	-	+	-	+	-	-	-
44	*Channa punctatus* (Bloch)	-	-	+	+	-	-	+

Ichthyological diversity in aquatic ecosystem refers to variety of fish species. Depending on context and scale, ichthyodiversity can refer to alleles and genotypes within a piscian population, co-species or life forms within a fish community. Further, biodiversity is essential for stabilization of ecosystem, protection of environmental quality, for understanding intrinsic worth of all species on the earth (Jhingran, 1991).

Studies revealed that, there is variation in valley segment in having mostly alluvial valley segment in the plains and colluvial in the upstream rheophilic region. Colluvial valley segment gives a favourable habitat for colonization by fishes as reported by Kar (2007). A variety of fish covers, notably turbidity, deep pools, overhanging vegetation, bottom-free big boulders, etc., were recorded as fish covers which protect the fishes from predation and hunting. Most of the rivers surveyed are mesoriparian. Riparian vegetation are scanty and consists mainly of evergreen trees with some amount of shrubs and herbs. The water bodies are subjected to human interference in the form of release of domestic and agricultural wastes as well as sometimes Tea garden refuges. The riparian land use pattern consists of human habitation, cultivation, etc (Table 1). Temperature, turbidity, pH, Dissolved oxygen (DO), Free carbon di-oxide (FCO_2), Total Alkalinity (TA), and Conductivity were the parameters studied with regard to physico-chemical characteristics of water (Table 2).

The results revealed variations among the studied parameters. DO portrays higher values (7.0 mg/lit) in River Manu. Lower DO values (3.4 mg/lit) were recorded in River Feni. Turbidty shows highest value (344.0 NTU) in River Muhuri and lowest (70.0 NTU) in Dhalai. River Manu showed maximum pH (7.1) while river Muhuri showed minimum value of pH (6.58). Further, River Deo showed maximum value of FCO^{-}_{2} (14.0 mg/lit) and River Manu showed the lowest value (8.0 mg/l). The River Manu displayed highest value of TA (110 mg/lit) while River Muhuri portrayed the lowest value (42.0 mg/lit). Further, highest value of conductivity was exhibited by River Manu (163.0 micromhos/cm) and lowest value of conductivity was portrayed by River Muhuri (69.0 micromhos/cm) (Table 2).

Similar results were obtained by other workers like Dey (1973); Kar (2007). 44 ichthyospecies belonging to 33 genera, 19 Families and 7 orders have been recorded from the surveyed rivers in Tripura. Of these, 30 species (81.82%) belonged to the Primary FW group while the remaining (18.18%) species belonged to Peripheral class. Kar (1990, 2003, 2007), Kar and Dey (2002), Kar *et al.*, (1996, 2002 a, b, c; 2005, 2006). Kar and Sen (2007); studied the Fish diversity along with habitat inventory parameters in a number of rivers and wetlands in Barak drainage in Assam and in the provinces of Mizoram and Tripura. In the present study, *Salmostoma bacaila* and *Labeo boga* were recorded in most of the rivers indicating their abundance. On the other hand, *Eutropiichthys murius, Amblyceps mangois* and *Labeo caeruleus,* etc., were recorded only from few of the surveyed rivers, indicating their rarity (Table 3).

Acknowledgement

The authors gratefully acknowledges the financial assistance received from the Ministry of Environment and Forests, Government of India and wish to express their gratitude to the local people for their help in field studies.

References

- APHA. (1995). Standard Methods for the Examination of Water and Wastewater, pp. xxxix + 1193 American Public Health Association (USA).
- Armontrout N.B. (1990). Aquatic Inventory Bureau of Land Management, Eugene District (USA) , pp. 32.
- Dey S.C. (1973). Studies on the Distribution and Taxonomy of the Ichthyofauna of the hill streams of Kamrup-Khasi-Garo Regions of Assam with special reference to the Functional morphology of some rheophillic Fishes. D.Sc. Thesis, xi + 299, University of Calcutta (India).
- Dey S.C. and Kar, D (1990). Fish Yield trend in Sone, a tectonic lake of Assam. Matsya 15-16 39-43.
- FAO. (1974). Design of Fisheries Statistical Surveys Inland Waters. Food and Agricultural Organisation. Fisheries Tech. Pap., 133 1-122.
- Gadgil, M. and Kar, D. (2000). Conserving India's Biodiversity Let People Speak pp.583-589. In Setting Biodiversity Conservation, pp. 707, World Wide Fund (WWF) for Nature-India (New Delhi).

- Ghosh, S.K. and Lipton, A.P. (1982). Ichthyofauna of the NEH Region with special reference to their economic importance. ICAR Research Complex, NEH Region, Shillong, Spl. Bull. 1 119-126.
- Hora, S.L. (1921). Fish and Fisheries of Manipur with some observations on those of the Naga Hills. Rec. Indian Mus., 22 165-214.
- Jayaram, KC (2010). The Freshwater Fishes of the Indian Region, xxiv + 616, Narendra Publishing House (Delhi).
- Jhingran, V.G. (1991). Fish and Fisheries of India, pp. xxiii + 727, Hindustan Publishing Corporation (New Delhi).
- Kar, D. (1990). Limnology and Fisheries of Lake Sone in the Cachar district of Assam (India). Matsya, 15-16 209-213.
- Kar, D. (2003). Fishes of Barak drainage, Mizoram and Tripura pp. 203-211. In Environment, Pollution and Management (Eds.) Kumar, A.; Bohra, C.; and, Singh, L.K., pp. xii + 604, APH Publishing Corporation (New Delhi).
- Kar, D. (2004). A Glimpse into the Fish Bioresources of North-East India with a note on their Management, Conservation and Biotechnological Potential. Proc. (Invited Lecture) DBT-sponsored National Symposium on Biodiversity Conservation and Sustainable Utilisation of Environmental Resources Tripura University, 9 p.
- Kar, D. (2005 a). Fish Genetic Resources and Habitat Diversity of the Barak drainage pp. 68-76. In Aquatic Ecosystems, Conservation, Restoration and Management (Eds.) Ramachandra, T.V.; Ahalya, N.; and, Rajsekara Murthy, C., pp. xiii + 396, Capital Publishing Company (Bangalore).
- Kar, D. (2005 b). Fish Fauna of River Barak, of Mizoram and of Tripura with a note on Conservation. J. Freshwater Biol., 16.
- Kar, D. (2005 c). Fish Diversity in the Major Rivers in Southern Assam, Mizoram and Tripura Proc. 2nd International Symposium on GIS and Spatial Analyses in Fisheries and Aquatic Sciences, 2-6 Sep 2002, University of Sussex, Brighton (UK), (Eds.), Vol.2, Nishida, T.; Kailola, P.J.; and, Hollingworth, C.E., pp.679-691, Fisheries and Aquatic GIS Research Group, Kawagoe, Saitama (Japan).
- Kar, D. (2007). Fundamentals of Limnology and Aquaculture Biotechnology, pp. xiv + 609 Daya Publishing House (New Delhi).
- Kar, D.; Laskar, B.A.; Mandal, M.; Lalsiamliana; and, Nath, D (2002 a). Fish Genetic Diversity and Habitat parameters in Barak drainage, Mizoram and Tripura. Indian J. Environment and Ecoplanning, 6 (3) 473-480.
- Kar, D.; B.A. Laskar; D. Nath; M. Mandal.; and, Lalsiamliana. 2002 b. Tor progenius (McClelland) under threat in River Jatinga, Assam. Science and Culture, 68 (7-8) 211.
- Kar, D.; Laskar, B.A.; and, Nath, D (2002 c). Tor sp. (Mahseer fish) in river Mat in Mizoram. Aquacult, 3 (2) 229-234.
- Kar, D. and Dey, S.C. (2002). On the occurrence of Advanced fry of Hilsa (Tenualosa) ilisha (Hamilton-Buchanan) in Chatla Haor Seasonal wetland of Assam. Proc. Zool. Soc. Calcutta, 55 (2) 15-19.
- Kar, D.; Dey, S.C. Dey; and, Datta, N.C. (2003). Welfare Biology in the New Millennium, pp. xx + 97, Allied Publishers Pvt. Ltd. (Bangalore).
- Kar, D.; Roy, A.; and, Dey, S.C. (2004). An overview of Fish Genetic Diversity of North-East India. In Proc. National Workshop on Rational use of Water Resources for Aquaculture (Eds.) Garg., S.K. and Jain, K.L., 18-19 March 2004, CCS Haryana Agricultural University, 1 164-171.

- Kar, D and Sen, N. (2007). Systematic List and Distribution of Fish Biodiversity in Mizoram, Tripura and Barak drainage in North-East India. ZOOs' Print Journal, 22 (3)2599-2607.
- Kar, D. and Dey, S.C. (1991). Gill nets in Lake Sone of Assam with their Economics and Impact on Fishery. J. Applied Zool. Res., 2 (2) 76-79.
- Kar, D. and Dey, S.C. (1993). Variegated Encircling gears in Lake Sone of Assam. J. Appl.Zool. Res., 4 (2) 171-175.
- Kar, D. and Dey, S.C. (1996). Scooping Gears of Lake Sone in Assam. J. Applied Zool. Res., 7(1) 65-68.
- Kar, D.; Dey, S.C.; Kar, S. Michael, R.G.; and, Gadgil, M. (1996). Ichthyoecology, Management and Conservation Fish Resources of Lake Sone in Assam (India). Tiger Paper (FAO, UN), XXIII (3) 27-32.
- Kar, D.; Nagarathna, A,V.; Ramachandra, T.V.; and, Dey, S.C. (2005). Fish Diversity and Conservation aspects in an aquatic ecosystem in North-East India. ZOOS' Print Journal. 21 (7) 2308-2315.
- Kar, D; Laskar, B.A.; and, Nath, D. (2006). Length-weight relationship of Neolissochilus hexagonolepis (McClelland) and Garra lissorhynchus (McClelland) from River Jatinga in Cachar district of Assam. Indian J. Fish, 52.(4) 495-496.
- Menon, A.G.K. (1999). Checklist Freshwater Fishes of India. Occasional Paper No. 175, pp. xviii + 366, Zoological Survey of India (Calcutta).
- NBFGR (2000). Manual on Habitat Inventory. National Bureau of Fish Genetic Resources (ICAR) (Lucknow), pp. 27.
- Sen, N. (2000). Occurrence, Distribution and Status of Diversified Fish Fauna of North-East India pp.31-48. In Fish Diversity of North-East India, (Eds.) Ponniah, A.G. and Sarkar,U.K, pp. 228, National Bureau of Fish Genetic Resources, ICAR (Lucknow).
- Sen, T.K. (1985). The Fish Fauna of Assam and the neighbouring North-Eastern States of India. Records of Zoological Survey of India. Occasional Paper No. 64 1-216.
- Talwar, P.K. and Jhingran, A.G. (1991). Inland Fishes of India and Adjacent Countries, Vol. I & II, pp. 1158, Oxford and IBH Co., Pvt. Ltd. (New Delhi).
- Vannote, R.L.; Minshall, G.W. Cummins, K.W.; Sedell, R.; and, Cushing, C.E. (1980). The River Continuum Concept. Canadian Journal of Fisheries and Aquatic Sciences, 37 130-137.

15

DIVERSITY OF PHYTOPLANKTON AT SASTHAMCOTTAH LAKE, KOLLAM, KERALA, INDIA

Ruchira Chaudhary and Rachana S. Pillai

Abstract: The biological constituents of water include planktons and there is always an inter-relation between diversity of phytoplankton and the physico-chemical parameters of water. The study shows the biodiversity of phytoplankton of Sasthamcottah Lake of Kollam district of Kerala. Biodiversity of phytoplankton of four stations of Sasthamcottah Lake was intensively studied for a period of one year i.e. between 2007 and 2008. The physico-chemical parameters of water studied during this period are pH, Total Hardness, Alkalinity, Phosphate, Nitrate, Total Nitrogen, B.O.D., C.O.D., Electrical Conductivity and Turbidity. The analysis of physico-chemical parameters as well as the counting and identification of phytoplankton were done using the standard methods. The mean number of species per stations ranged between 8 to 12 did not vary significantly. Shannon Weaver diversity index H indicated significant differences among the four stations. The Simpson's Index of Dominance indicated the dominance of species of Chlorophyceae family in the four stations. The spatial Index of similarity was determined between the four sites. The variation in the physico-chemical parameters of the lake at the four sites accounted for the existence of species. The present study will help in understanding the biodiversity of the lake as well as the factors affecting it and also in its conservation and management.

Keywords: Phytoplankton, Alkalinity, Electrical Conductivity, Dominance, Chlorophyceae

Introduction

Wetlands constitute an important and integral part of the ecosystem. According to Waller wetland is an eco-system of lowland areas where water is commonly accumulated to induce characteristic water loving and water tolerant vegetation and anaerobic humic soils. The flora is unique and diverse and is the energy fixing base for complex food chains that result in diversified wildlife

Department of Zoology, Govt. Motilal Vigyan Mahavidyalaya, Bhopal, (M.P.), India; E-Mail: ruchira_2006@yahoo.co.in resources. Biological diversity is a term used to describe the number, variety, and variability of organisms in a particular habitat or ecosystem. Presence of phytoplankton in a certain area serve as an indication of condition in the environment (Superales and Zafaralla, 2008). They constitute the principal primary producers. They inhabit even inhospitable and prohibitive environment and serve as important markers in elucidating any shift in the ambient environment. So intricate is their association with the water quality that a continuous study of the biochemical processes such as nutrient uptake and cycling, algal growth in water of various qualities, zooplankton dynamics and dissolved oxygen cycling may provide an insight for conservation, biomonitoring and desirable aquatic life. The knowledge of presence or absence of certain indicator species is favoured in assessing the degree of stress and disturbance on the ecosystem. The role of phytoplankton in managing bioenergetics of lakes their secretions of enzymes in purification of polluted aquatic habitats has been known for a long time (Shastree, 1993). In India several workers have attempted to study the phytoplankton in different ecosystem (Nair and Mishra, 1988; Kaushik *et al.*, 1991; Daganonkar and Saksena, 1992; Sarkar and Chowdhury, 1999, Rao *et al.*,1992; Leeper and Taylor, 1995; Hosetti *et al.*, 1995; Shivappa *et al.*, 2000) but looking to their varied nature, investigation on water bodies are still meagre.

Study Area

The lake considered for investigation during present work is Sasthamcottah lake (Fig. 2) which is also designated as a 'Ramsar Site'. This lake is situated in the Kollam district of Kerala (India) covering an area of 373 hectares. The lake lies between 0'-5' N latitude and 76°35' to 76°40' E longitude in the South west coast of India. Sasthamcottah Lake is situated at a height of 33m above MSL. The lake has a capacity to hold 22,390 million litres of water and is the main source of drinking water for half a million people of Kollam district of Kerala.The four sampling sites are Station 1-lake site near D.B. College, Station 2-Kadavu, Station 3-Punnakad, Station 4-Ookenmukku (Fig.-1).The present study is an attempt to study the biodiversity of phytoplankton of the lake and the physico-chemical factors affecting them.

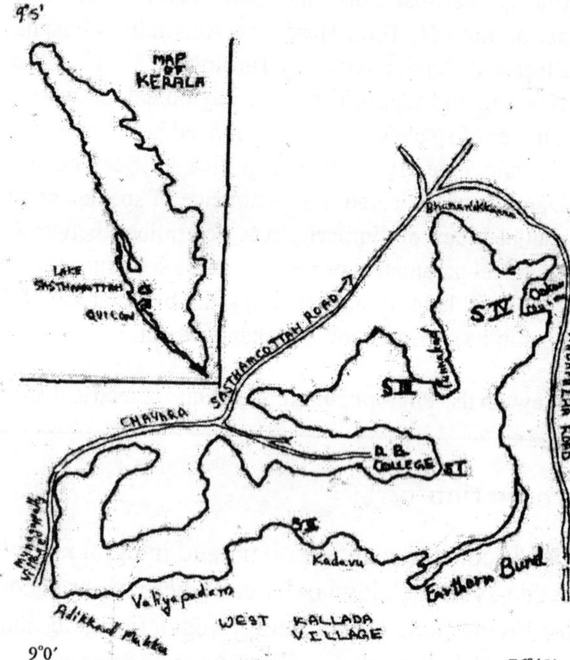

Fig. 1 Map of Sasthamcottah lake, Kerala, India indicating study sites

Fig. 2 Photograph of Sasthamcottah Lake, Kerala

Materials and Methods

In the present study for the physico-chemical analysis of water the samples were collected in clean polyethylene containers for one year from June 2007 to May 2008 regularly in three parts at an interval of 30 days from four sampling sites. The physico-chemical parameters like pH, hardness, total alkalinity, B.O.D., C.O.D., turbidity, electrical conductivity, total nitrogen, phosphate and nitrate were analyzed in the laboratory as per Standard Methods for the examination of Water and Waste Water (APHA, 1988) and as per Workbook on Limnology (Adoni, 1985). For phytoplankton identification the water samples were concentrated and preserved in Lugol's solution. Counting was done using Plankton Counting Chamber. The plankton were systematically identified using microscope and keys by APHA, Fritsch and Workbook on Limnology (Adoni,1985).

The species composition was analyzed using the simple percentage and ecological statistics e.g. species richness, species diversity, dominance, similarity etc. The species compositions were compared using the species diversity value in each site. Species richness was arrived at by simply enumerating the varying genera identified. For species diversity analysis Shannon Weaver Index of Diversity (H') was used as per the formula

$$H' = -\Sigma \{\pi \ln \pi\}$$

Where pi = importance probability for each species

H' = Diversity index

The Simpson's Index of Dominance was used in calculating the species dominance

$$C = \Sigma \pi^2$$

Where C is the Index of Dominance

The index of similarity (IS) between sites in terms of species composition was determined using the formula

$$IS = 100 \times \frac{2C}{A + B}$$

Where

IS= Index of Similarity

C= Number of common species to both sites (compared sites)

A= Number of species in site 1

B= Number of species in site 2 respectively

Results and Discussion

Water Quality

The physico-chemical properties of the four sampling stations are shown in Table-1. Variation was observed in the monsoon and summer months in the physico-chemical parameters.

Composition of Phytoplankton

The phytoplankton found on studying the Sasthamcottah Lake consisted of mainly three families Chlorophyceae (63.21%), Bacillariophyceae (31.57%) and Dinophyceae (5.22%) (Fig.3). Chlorophyceae consists of Cosmarium, Pediastrum, Eudorina, Coelastrum, Desmidium, Micrasterias, Spirogyra, Closterium. The species identified in Bacillariophyceae mainly consists of Navicula, Cyclotella and Diatoma. Ceratium was the only species identified in Dinophyceae. The presence of these different varities of phytoplankton can be attributed to the physico-chemical parameters of the lake (Table-2).

Comparison of the sampling sites

The four sampling sites were compared based on species richness, species diversity and species dominance.

Species Richness. Species richness reflects the number of different species present at a particular site. At station 1 i.e. lake site near D. B. College three families of phytoplankton were present with a total of twelve species. At station 2 i.e. Kadavu also had a total of total of twelve species belonging to three families. At station 3 i.e. Punnakkad it had a total of eleven species belonging to groups Chlorophyceae and Bacillariophyceae. Station 4 i.e. Ookenmukku had eight species belonging to

Table 1. Mean physico-chemical parameters of Sasthamcottah Lake in 2007-08

	pH	T Hardness	Nitrate	B.O.D.	E.C.	TN2
Months						
June	7.49	65.83	0.064	4	59.38	4.1
July	7.52	67.8	0.058	1.3	67.25	4.45
Aug	7.52	70.2	0.054	1.14	66.05	3.48
Sept	7.39	67.44	0.052	1.28	62.18	4.1
Oct	7.32	63.75	0.043	1.48	57.9	3.9
Nov	7.27	60.76	0.05	1.8	55.13	4.31
Dec	7.2	58.5	0.06	2	54.18	4.75
Jan	7.14	55.13	0.066	2.29	50.6	4.33
Feb	7.17	54.31	0.12	2.28	52	4.54
Mar	7.21	59.1	0.12	2.52	53.15	4.51
Apr	7.29	65	0.13	2.3	54.88	4.81
May	7.39	63.11	0.125	2.59	55.28	5.46

Table 2. physico-chemical parameters of Sasthamcottah Lake in 2007-08

PARAMETERS	STATION S1	STATION S2	STATION S3	STATION S4
pH	7.348	7.333	7.277	7.34
Total Hardness	62.89	64.15	59.81	62.28
Nitrate	0.079	0.08	0.081	0.077
B.O.D.	2.331	1.897	1.529	1.71
C.O.D.	8.138	7.984	8.029	7.73
Phosphate	0.104	0.112	0.099	0.159
E.C.	56.38	57.95	59.47	55.52
Alkalinity	80.7	83.98	92.92	76.9
Total Nitrogen	4.203	4.288	4.504	4.546
Turbidity	0.726	0.834	0.757	0.699

Chlorophyceae family alone. The occurrence of species richness at the stations 1 to 3 can be attributed to the various physico-chemical parameters like alkalinity, phosphate, nitrate, B.O.D., C.O.D., electrical conductivity and total nitrogen which showed a positive co-relation with the phytoplankton. The species found in Station 4 belonged to only Chlorophyceae family as the values of some physico-chemical parameters were comparatively higher like total hardness, pH etc.

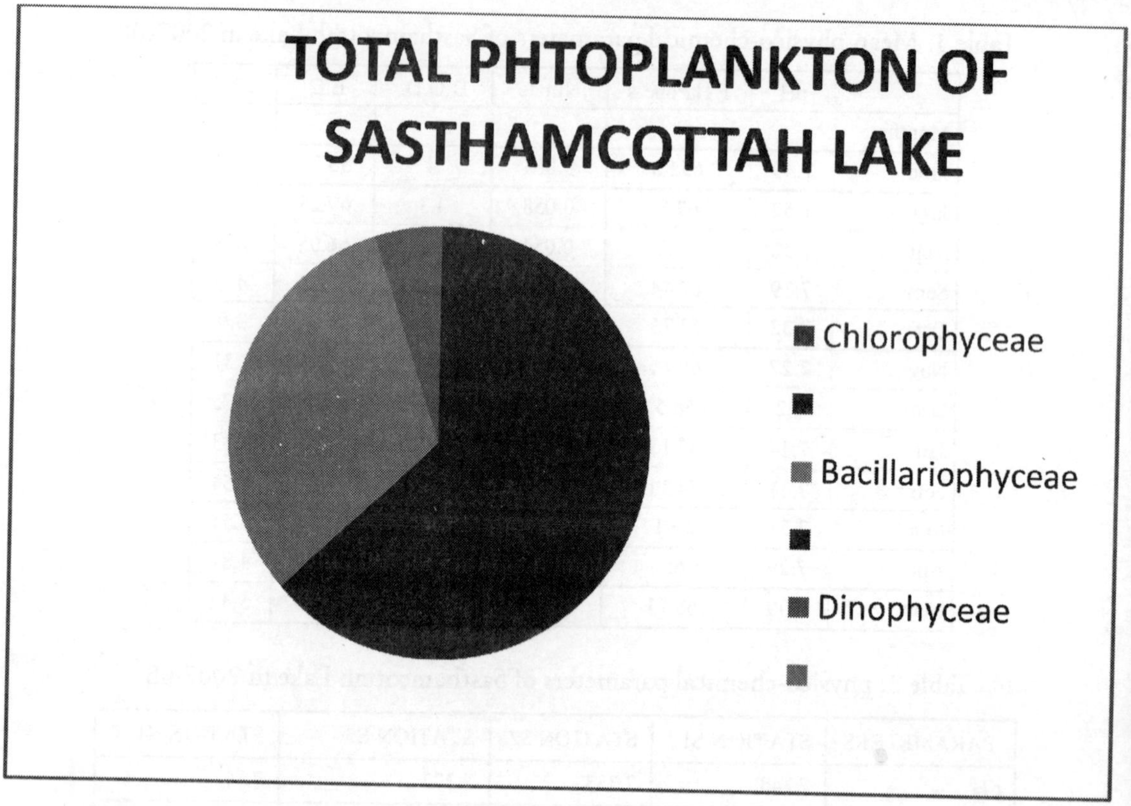

Fig. 3 Percentage Composition of Phytoplankton of Sasthamcottah Lake, Kerala

Species Diversity. The species diversity was determined using Shannon-Weaver's diversity index (H)'. At station 1 the diversity value was 1.33, at station 2 the diversity value was 1.31, at station 3 the diversity value was 1.43 and station 4 the value was 0.85. Stations 1 to 3 have diversity values greater than 1 indicating greater diversity and station 4 has diversity value less than 1 indicating less diversity (Fig.4). High diversity indicates that the niche space, habitat and food sources are adequate for the survival of many species (Babour, 1996). In many situations, diverse populations make for an increased stability which means the communities are less affected. Diversity of plankton is fairly dependent on quality of water and climatic factors. Various physico-chemical and biological characteristics must be simultaneously taken into consideration for understanding the fluctuations of plankton population (Davies, 1955). Shivappa *et al.* (2000) reported eleven species of algae from Savalanga tank, Shimoga belonging to diatom (7), green algae (2) and blue green algae (2). Daganonkar and Saxena (1972) have reported 25 species of algae from temple tank at Gwalior. Temple tank was dominated by green and blue green algae indicating pollution and eutrophication of the water body (Hosetti *et al.*, 1995).

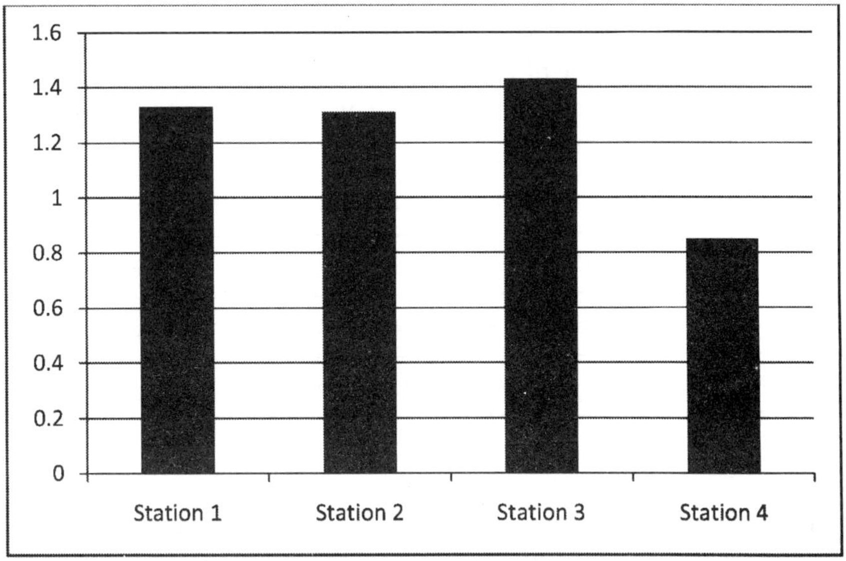

Fig. 4 Species diversities of the four study sites of Sasthamcottah Lake, Kerala

Species Dominance. In station 1 the dominant species were of Bacillariophyceae which had a dominance value of 0.348. At station 2 also the dominant species were of Bacillariophyceae having value of 0.348. At station 2 also the dominant species were of Bacillariophyceae having value of 0.349. At station 3 also the dominant species were of Bacillariophyceae family with values of 0.339. At station 4 the dominant species were of Chlorophyceae family having value 0.13 as there were number of other species of other families (Table 2). Bacillariophyceae was the most dominant family in 3 stations and occurred throughout the period of collection. The density of Bacillariophyceae population was found to be closely associated with pH. It is observed that the acidic waters do not support an abundance of Bacillariophyceae while in alkaline waters with pH above 7 their density is more. In the present study the average pH ranged around 7.5. Goel *et al.* have stated that Bacillariophyceae occurs in all types of waters. Navicula and Cyclotella indicate pollution in the present water body that occurred commonly. Chlorophyceae and Dinophyceae were more during summer season. The phosphate showed lower values during July to December but there was a definite increase in phosphate concentration during January to May. From the present observation it is difficult to point out any single factor which is responsible for the fluctuation in abundance of plankton.

Similarity Index The similarity index was determined based on the genera of phytoplankton present in a pair of stations. At the generic level a high similarity of 100% was observed between station 1 and station 2 i.e. lake site near D.B. College and at Kadavu. Station 1 and 4 and staion 2 and 4 were biologically disparate. The results imply that different degrees of similarity are revealed when genera are utilized for the similarity index. The physico-chemical parameters and the seasonal fluctuations have come into play in the biological similarity of places.

During the study period all the four stations had high alkalinity, total hardness, high conductivity, total alkalinity and pH which act as the factors for the diversity of phytoplankton. High nitrate have made the lake one of the most productive as indicated by studies of Unni *et al.* (2002). Gopal *et al.* (1988) also reported differences between dominance of diatoms i.e. Bacillariophyceae during winter months.

Table 3 Analysis of Similarity between Phytoplankton in 4 sampling sites of Sasthamcottah Lake

Taxon	Pair 1	IS	Pair 2	IS	Pair 3	IS	Pair 4	IS	Pair 5	IS	Pair 6	IS
Level	S1 & S2	%	S1 & S3	%	S1 & S4	%	S2 & S3	%	S2 & S4	%	S3 & S4	%
Genera	12 &12	100	12 & 11	95.65	12 & 8	80	12 & 11	95.65	12 & 8	80	11 & 8	84.21

Conclusion

lanktonic study is a very useful tool for the assessment of water quality in any type of water body and also contributes to an understanding of the basic nature and general economy of the lake. The present study ensures that the phytoplankton diversity of the four stations of Sasthamcottah lake were mainly dependent upon nutrient level and the other physico-chemical factors of the water body. Species richness was of group Chlorophyceae in all stations. Station 3 had the largest species diversity. In dominance Bacillariophyceae was the dominant group and Chlorophyceae was the sub-dominant group. This type of large diversity indicates that the lake has become polluted and the process of eutrophication has set in.

References

- Adoni A.D., (1985). Workbook on Limnology, Indian MAB Committee, GOI.
- APHA, (1988). Standard Methods for the Examination of Water and Waste Water, 19th Edition, American Public Health Association, New Delhi.
- Babour, M.T., (1996). Measuring the health of aquatic ecosystems using biological assessment techniques A Natural Perspective in Effects of watershed Development and Management on aquatic Ecosystems. Proceedings of an Engg. Conf. ASCE American Soc. Of Civil Engineers, New York, New York 18-33.
- Dagaonkar, A and Saksena, D.N., (1992). Physico-chemical and biological characterization of Temple Tank, Kaila Sagar, Gwalior. J. Hydrobiology 8 11-19.
- Davies, E.C. (1955). The Marine and Freshwater Plankton. Michigan State Univ. Press, East lassing U.S.A.
- Fritsch, F.E. (1975). The Structure and Reproduction of Algae. The Cambridge Univ. Press, Enston Road, London N.W.
- Gopal *et al.* (1988). Comparative study of species composition, density and species diversity of the phytoplankton in a non-polluted and a sewage receiving freshwater reservoir. Arch. Hydrobiol. Supple. 79 291-323.

- Goel, P.K., Kulkarni A, Y. , Katavkar, S.D. and Trivedy R.K. (1993). Studies on diurnal variation in some physico-chemical characteristics and phytoplankton of freshwater polluted pond. Indian J. of Environ. Prot. 12 (97) 503-508.
- Hosetti, B.B., Shivaprasad, R.S. and Natraja, S. (1995). Limnology of a eutrophic pond with reference to physico-chemical factor at Navile, Shimoga, In Recent Researches in Aquatic Ecosystems. (Eds. Ashutosh, G. And Agarwal, N.K.) Daya Publishing House, Delhi 35 99-105.
- Kaushik, S., Agarkar, M.S. and Saksena, D. N. (1991b). Water quality and periodicity of phytoplanktonic algae in Chambal tal, Gwalior, Madhya Pradesh. Bionature, 11(2) 87-94.
- Leeper, D.A. and Taylor, B.E. (1995). Plankton composition, abundance and dynamics in a severely stressed cooling reservoir. Journal of Plankton Research, 17(4) 821-843.
- M.W. Waller. (1982). Fresh Water Marshes, Univ. Minnesota Press Minniapolis, U.S.A.
- Nair, R.M.S., Seth, O.N. and Mishra, K.D. (1988). Limnological studies on a village Pond, Imalia (Vidisha) II. Phytoplanktons. Proc. Nat. Symp. Past, Present and Future of Bhopal Lakes 203-208.
- Pant, M.C., Sharma,A.P. and Chaturvedi, O.P. (1983). Phytoplankton population and Diel variation in sub-tropical lake. J. Environment Biol. 4(1) 15-23.
- Rao, K.S., Thilak, A., and Shukla, A.N. (1992). Observation on Phytoplankton Community trends in GSR with Relation to its Eutrophication. J. Hydrobiol. VIII (1) 51-62.
- Sarkar, S.K. and Chowdhury, P.B. (1999). Role of Some Environmental Factors on the Fluctuation of Plankton in a Lentic Pond at Calcutta. In Limnological Research in India (Ed. S.R. Mishra). Daya Publishing House, Delhi 35 108-132.
- Shastree, N.K., Pathak, S., Islam, M.S. (1993). Phytoplankton as biological indicators in lentic hydrosphere. Adv. In Limnol. (ed.) H.R. Singh. 113-132.
- Shivappa, D., Ravi, G.S. and Hosetti, B. B. (2000). Ecological characteristics of water from Savalanga tank, Shimoga. Indian J. Environ. And Ecoplan. 3(1) 137-141.
- Superales, B. Jerry and Zafaralla, T. Macrina. (2008). Distribution of benthic macro invertebrates at Maubog Bay of moro Gulf, western Mindanoa, (Philippines). J. Of Env. Res. and Develop. 303-313 Vol. 2, No. 3.

16

AGE AND GROWTH OF INDIAN MAJOR CARP (*CATLA CATLA*, HAM. 1822) IN AASAN POND OF SOUTHERN RAJASTHAN, INDIA

N.C. Ujjania, M.P.S. Kohli and L.L. Sharma***

Abstract: In the present chapter the growth performance of Indian major carp *Catla catla*, was studied using key scales in small water body namely Aasan pond (75 ha) located in Banswara district of southern Rajasthan on West Coast River System of India. The annual rings or annuli were observed to estimate selected growth parameters. In the scale samples maximum +6 annuli were found and use to assess the growth data. The growth parameter for *C. catla* viz. Index of species average size (Øh) 11.018 cm and Index of weight growth (ØCw) 83.864 g were recorded. Growth rate estimation on the basis of scale studies revealed that growth of carps in small water body like Aasan pond of southern Rajasthan could be divided into two phases the first, phase of immaturity, which lasts up to two years and second, phase of sexual maturity.

Keywords: Indian major carps, Growth, Scales, Rajasthan, Maturity, Catla

Introduction

Catla catla belongs to Cyprinididae family and highly commercially importante species for freshwater bodies in India. This species inhabits in freshwater rivers, reservoirs, lakes, ponds and beels. This is the fastest growing species among carps (Mitra, 1942). Jhingran (1952) provided a comparative account of growth of *C. catla*, *L. rohita* and *C. mrigala*. Chogunovas (1963) has given a descriptive account of age and growth in fishes. Scales are frequently used to determine the age of fish (Van, 1957; De and Dutta, 1990; Johal and Kingra, 1992; Singh *et al.* 1998), Seshappa (1999) reviewed the literature on age determination of Indian fishes using scales and other hard

Associate Professor, Department of Aquatic Biology, VNSGU, Surat - 395007 (Gujarat),
*Central Institute of Fisheries Education, Fisheries University Road, Versova, Mumbai - 61 (India)
**College of Fisheries, MPUAT (Udaipur) - 01 (Rajasthan) India, E-mail ncujjania@hotmail.com

parts. Ricker (1979) considered the scale method as the most reliable method compared to others. Menon (1953) has initially reviewed the literature on age and growth study of fishes from tropical and subtropical waters. Jhingran (1957, 1959 & 1969), Kamal (1969), Natarajan *et al.* (1976), Jhingran and Khan (1979), Khan (2000), Madan Mohan (2006), Licandio *et al.* (2006), Sarkar *et al.* (2006) and Prashant *et al.* (2008) estimated the growth rates of fishes using scales. Johal and Tandon (1992) interpreted that riverine populations of *C. catla* have a better rate of growth than the population of reservoirs using the scales. Prakash and Gupta (1986) studied the growth rate of *C. catla*, *L. rohita* and *C. mrigala* of Govindgarh Lake by means of scales and described the growth rate in these species is rapid for the first five years of life after which growth increment is very little. The growth rates were found to be high during August-October and February-April and low during May-June.

Age and growth studies are most important aspect of fishery biology. Growth rate of a fish can be enumerated from the estimations of age such studies can give information on stock composition, age and maturity, life span, mortality, growth, production etc. so, this study is highly significance for the management and conservation of fish population in natural water bodies. Johal and Tandon (1983) studied age and growth of *C. catla* and *C. mrigala* from Lake Sukhna, Chandigarh with the use of scale and found that the growth rate of these fishes was lowest as compared to the populations of other localities. On the basis of scale studies they interpreted that *C. catla* up to 3+ years of life increases rapidly in length but at 4+ years and onwards there is an appreciable increase in weight as compared to length. They also found that the condition factor increases with increase in length and weight of fish. The slow growth of fish in Sukhna Lake was assigned to oligotrophic nutrient status of lake. Singh and Sharma (1995) studied the age and growth of a Himalayan teleost Schizothorex richrdsonii from the Garhwal hills using scales. They concluded that the growth rate of fish was not uniform throughout year or its life span. It may show alternative fast and slow rates of growth depending upon favorable or adverse ecological conditions. Singh *et al.* (1998) have studied the key scale of *L. rohita* for finding information on age, growth and harvestable size from Lake Jaisamand. Jain (2000) has studied the biology and fisheries of *C. catla*, *L. rohita* and *C. mrigala* of Siliser reservoir.

Materials and Methods

Collection of samples

Samples of the commercially important Indian major carp *C. catla* (Ham.) were collected from the selected water body during the fishing year 2000-2001(January, 2000 to June 2001). Scale samples (5-10) were collected from 50 fish of different size groups of fish from above lateral line and below the dorsal fin from the each water body randomly over the period of entire fishing year. Intact scales were preserved in paper envelops with data on Total Length, Standard Length, Body Weight and date of collection of individual fish noted on the envelope.

Analysis of samples

Study of scales

Preserved scales were dipped in 1 % KOH solution for 5 minutes and washed with tap water 2-3 times with gentle rubbing by fingertips to remove mucus and other material. The scales measurements were taken for radius of scale and radius of all the annuli from the focal point and were studied in dry mountings using VEB microfilm reader with the magnification of 10.5 'X' at the Department of Zoology, Punjab University, Chandigarh.

A graph was plotted between total length and scale radius. The length of the fish at the time of formation of annuli could be estimated using following relationship given by Bagenal and Tesch (1978).

$$Ln = a + \frac{Sn\,(L - a)}{S}$$

Where

Ln = Length of fish when the annulus 'n' was formed.

L = Length of fish when scale sample was obtained.

Sn = Radius of annulus 'n'

S = Total scale radius

a = Correction factor (intersecting point between T L and Scale Radius)

Relative values based on scale studies

The growth parameters such as growth characteristics (C_{th}), specific linear growth (C_l), growth constant (C_{lt}), specific rate of weight increase (C_w), index of species average size (Øh) and index of population weight growth intensity (ØC_w) were also estimated. The following formulae have been used (Balon, 1971 and Johal and Tandon, 1983)

$$W = aLb$$

$$\text{Log } W = \text{Log } a + b \text{ Log } L$$

$$C_{th} = \frac{\text{Log } L_n - \text{Log } L_{n-1}}{0.4343}$$

$$C_{lt} = \frac{\text{Log } L_n - \text{Log } L_{n-1}}{0.4343} \times \frac{t_2 + t_2}{2}$$

$$C_l = \frac{L_n - L_{n-1}}{L_{n-1}} \times 100$$

$$C_w = \frac{W_n - W_{n-1}}{W_{n-1}} \times 100$$

$$\Phi h = \frac{\sum h = 1}{nj + a} \quad h = nj + a \text{ (Balon,1971)}$$

$$\Phi C = \frac{\sum Cw = 1}{nj + a} \quad Cw = nj + a \text{ (Balon,1971)}$$

Where

L_n, L_{n-1} = total length of fish at ultimate and penultimate age

W_n, W_{n-1} = Weight of fish at ultimate and penultimate age

j = juveniles

a = adult

h = absolute increase in length

t_1, t_2 = time intervals between ultimate and penultimate age

Results and Discussion

An age and growth rate of *C.catla* from AP was estimated on the basis of scale radius (Plates 1). Correlation among total length of fish and scale radius was also determined. A highly significant correlation ($P< 0.01$) was observed between total length of fish and scale radius (Table 1). In *C. catla* initial growth rate during first two years was observed constant in case of AP (Table 2 and Table 3). Specific rate of weight increase (C_w) was high in earlier years of life, which decreased, in later years. Specific rate of linear growth (C_l) was observed to decrease with increasing age. Growth constant average ($C_{lt(Av)}$) was more at first age year, which reduced, in subsequent years of age. Annual length increment (h) did not show regular pattern as it was observed in case of annual weight (w) increase. Index of species average size (h) 11.018 was determined in Aasan pond.

Growth rates estimation of *C. catla* from this small water body suggests of population variability. Though all the growth related parameters in the present study were observed to follow standard trends true for Indian major carp a high degree of correlation was found between total body length and lateral scale radius. Johal and Tandon (1992) also reported a high degree of correlation between these two variables for *C. catla* of Gobind Sagar and Harike reservoir.

Gradual decrease in the values of the specific rate of linear growth and the specific rate of weight increase with the increase in age suggests that growth rate of carps is more at early stage of life and gradually decreases as the age advances. Johal and Tandon (1992) reported three growth phases for *C. catla* on the basis of observed growth constants. In the present study only two growth phases were observed for all the three species from these water bodies. According to Chugunovas (1963) majority of fish populations show only two growth phases of life corresponding to sexual immaturity and sexual maturity. Third phase corresponds to old age. It is difficult to generate information on third phase as there is regular fishing in most of the water bodies and fish is harvested before it reaches this stage. Singh *et al.* (1998) observed that with increase in age there is decrease in the specific rate of linear growth, specific rate of weight increase and annual length increment. In the present study an increasing trend in growth was observed for *C. catla* during round the year.

Similarly, a high value of annual length increment (h) was found in *C. catla* of AP at fourth and fifth year. Tandon and Johal (1983) also observed the phenomenon of growth compensation in *C. catla* from Gobind Sagar Lake in the age class four. Prakash and Gupta (1986) in their study on the comparative growth rates of *C. catla, L. rohita* and *C. mrigala* of Govindgarh Lake by means of scales reported that growth rates in these species were rapid for the first year after which growth increment was very little. The growth increment was more for *C. catla* at fifth years It was observed that this species have been found to compensate for slow growth rate at early years. In AP lower growth of *C. catla* may be due to low level of plankton biomass.

Table 1. Regression of scale radius on total body length of *Catla catla*, from Aasan pond

n	Mean L ± SD	Mean Wt ± SD	a ± SE	b ± SE	r^2 *
50	62.560 ± 12.052	10.720 ± 2.287	0.141 ± 1.017	0.169 ± 0.016	0.911

* All the values of r^2 are significant P< 0.01

Table 2. Growth rate of *C.catla*, of Aassan pond

Parameters	Year class					
	l_1	l_2	l_3	l_4	l_5	l_6
L	24.232	33.423	41.874	51.349	59.607	66.108
h	24.232	9.191	8.451	9.475	8.258	6.501
Øh	11.018					
C_l	37.929		25.284	22.627	16.082	10.906
C_{th}	7.798		7.534	8.541	7.657	6.170
C_{lt}	0.482		0.338	0.305	0.223	0.155
$C_{lt(av)}$	0.482			0.255		
W	234.057	596.256	1149.917	2115.702	3251.411	4384.007
w	234.057	362.199	553.661	965.785	1135.709	1132.596
Cw	154.748		92.856	83.204	53.68	34.833
ØCw	83.864					

L Back calculated length in cm.

C_{th} Growth characteristics

w Annual weight increase in gm.

h Index of species average size

C_{lt} (Av) Growth constant average

C_l Specific rate of linear growth

W Calculated weight in gm.

h Annual length increment in cm.

C_{lt} Growth constant

C_w Specific rate of weight increase.

C_w Index of weight growth intensity

Table 3. Annual Growth of *Catla catla* from Aasan pond

Age Group	No. of specimens	Total length at the time of capture (cm)		Average Back calculated length (cm)					
				l_1	l_2	l_3	l_4	l_5	l_6
1+	4	Min	39.00	29.72					
		Max	40.00	31.34					
		Mean	39.50	30.53					
2+	6	Min	46.50	25.51	34.91				
		Max	49.00	26.70	39.70				
		Mean	47.50	26.122	37.01				
3+	9	Min	51.00	22.12	32.16	39.70			
		Max	57.50	26.91	37.29	46.57			
		Mean	54.00	24.22	34.63	42.78			
4+	15	Min	62.00	23.42	34.70	44.71	54.68		
		Max	66.00	25.57	34.73	46.44	58.18		
		Mean	64.00	24.49	34.71	45.57	56.42		
5+	11	Min	68.00	20.18	29.36	40.94	48.65	60.23	
		Max	73.00	27.60	36.75	45.70	57.26	66.10	
		Mean	70.07	23.72	33.30	43.21	54.10	62.36	
6+	5	Min	71.50	18.75	26.31	39.09	43.76	54.81	64.00
		Max	75.00	23.55	33.06	42.80	49.05	57.96	68.31
		Mean	73.50	21.78	30.61	39.09	47.14	56.85	66.10
Total	50	Min	39.00	18.75	26.31	39.09	43.76	54.81	44.00
		Max	75.00	31.34	39.70	46.57	58.18	66.10	68.31
		Mean	62.56	24.23	33.42	41.87	51.34	59.60	66.10

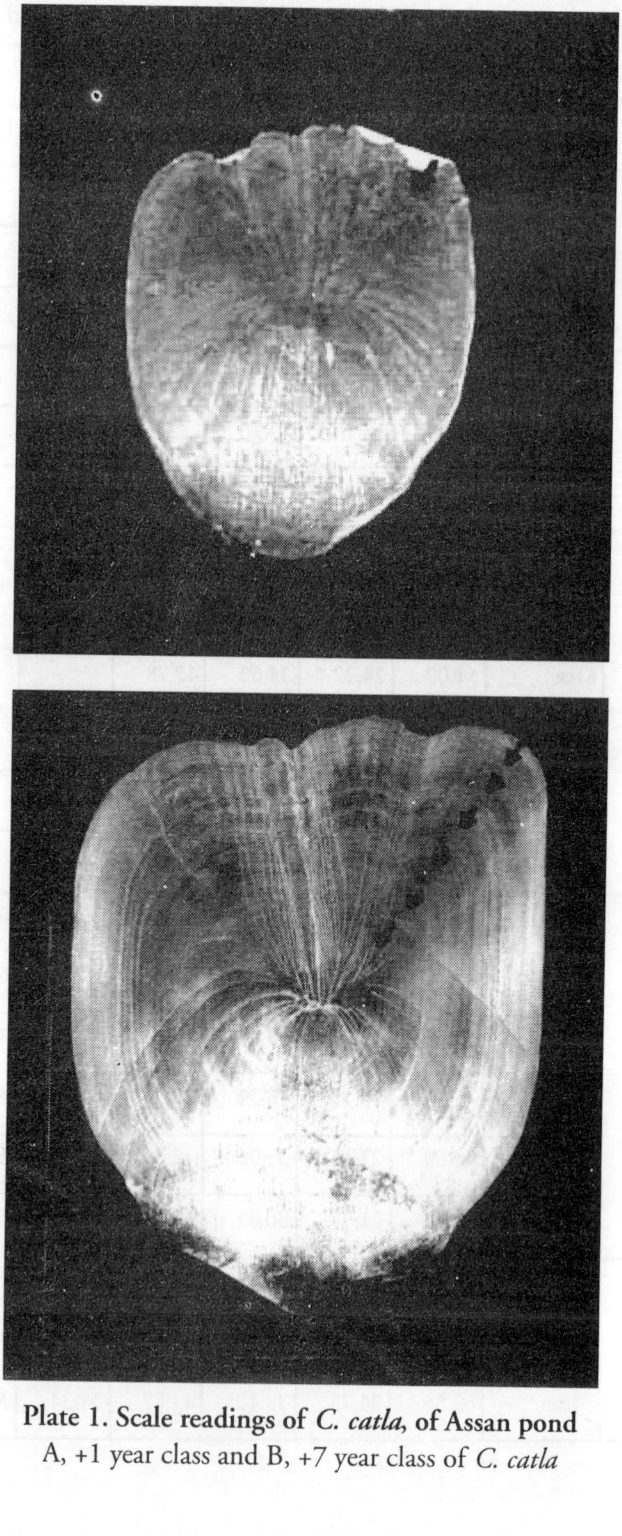

A

B

Plate 1. Scale readings of *C. catla*, of Assan pond

A, +1 year class and B, +7 year class of *C. catla*

Acknowledgement

The authors are grateful to the Director, CIFE, Mumbai, Dr. M.S. Johal, Professor, Zoology Department, Panjab University, Chandigarh, and General Manager of Rajasthan Tribal Area Development Cooperatives Federation, Udaipur for providing needful help to prepare this manuscript.

References

- Bagenal, T. B. and Tesch, F. W. 1978. Age and growth. In methods for assessment of fish production in freshwaters, 101-136. IBP Hand book No. 3 (Eds. T.B. Bagenal), 3rd edition, Blackwell Scientific Publication Ltd, Oxford.
- Balon, E. K. 1971. A short methodological outline for production survey of freshwater fish population (with example from Lake Kariba). UNDP/FAO, Central Fisheries Research Institute, Zambia 1-18.
- Chugnovas, N. I. 1963. Handbook for the study of age and growth of fishes. The National Science Foundation, Washington. 132 p (English Trans).
- De, D. K. and Dutta, N. C. 1990. Age, growth, length-weight relationship and relative condition in *Hilsa, Tenualosa ilisha* (Ham.) from Hooghly Estuarine System. Indian J. Fish. 37(3) 199-209.
- Jain, A. K. 2000. Studies on catla-rohu hybrid in relation to its parents *Catla catla* and *Labeo rohita.* Ph. D. thesis, South Gujarat University, Gujarat.
- Jhingran, V. G. 1952. General length-weight relationship of three major carps of India. Inst. Sci. India (B), 18(5) 449 - 460.
- Jhingran, V. G. 1957. Age determination of Indian major carp, *Cirrhinus mrigala* (Ham.) by means of scales. Nature, London 179 468-469.
- Jhingran, V. G. 1959. Studies on the age and growth of Cirrhinus mrigala (Ham.) from the river Ganga. Proc. Nat. Inst. Sci. India 25B (3) 107-137.
- Jhingran, V. G. 1969. Derivation of average length of different age groups in the fishes. J. Fish. Res. Bd. Can. 26(1) 3073-3076.
- Jhingran, V. G. and Khan, H. A. 1979. Synopsis of biological data on the mrigal, *Cirrhinus mrigala* (Hamilton,1822). FAO Fish. Syn. 70.
- Johal, M. S. and Kingra, J. S. 1992. Growth parameters of *Catla catla* (Ham.). Him. J. Env. Zool. Vol. 6 1-6.
- Johal, M. S. and Tandon, K. K. 1983. Age, growth and length-weight relationship of *Catla catla* and *Cirrhinus mrigala* (Pisces) from Sukhna Lake Chandigarh (India). Vestn. Cesk. Spol. Zool. 47 87-98.
- Johal, M. S. and Tandon, K. K. 1987. Age and growth of *Cirrhinus mrigala* (Pisces Cyprinidae). Vestn. Cesk. Spol. Zool. 51 252-280.
- Johal, M. S. and Tandon, K. K. 1992. Age and growth of the carp *Catla catla* (Hamilton, 1822) from the northern India. Fish. Res. 14 83-90.
- Kamal, M. Y. 1969. Studies on the age and growth of *Cirrhinus mrigala* (Ham.) from commercial catches at Allahabad. Proc. Nat. Acad. Sci. India 35B (1) 72-92.
- Khan, H. A. and Jhingran, V. G. 1975. Synopsis of biological data on the *Rohu, Labeo rohita* (Hamilton, 1822). FAO Fish. Synopsis.

- Khan, M. A. 2000. On age and growth determination of certain catfishes from Rehand reservoir. Rec. Zool. Surv. India 98 (Pt I) 123-130
- Khandker, N. A. and Haque, K. A. 1970. Determination of age of carp *Labeo rohita* by scale method. The Dacca Univ. Stud. Vol. XVIII 15-20.
- Licandeo R. R., Barrientos C.A., Maria Teresa Gonzalez. 2006. Age, Growth rates, sex changes and feeding habits of notothenioid fish Eleginops maclovinus from the central southern Chileon coast. Envirn. Biol. Fish. 77 51-61
- Madan Mohan. 2006. Age and growth of snow trout Schizothorex richardsonii from Kumao hills. J. Zool. 26 (2) 163-168
- Menon, M. D. 1953. The determination of age and growth of fishes of tropical and subtropical waters. J. Bombay Nat. Hist. Soc. 51(3) 623-635.
- Mitra, G. N. 1942. Rate of growth in the first year of life of *Labeo rohita* and *Catla catla* in the different districts of Orissa. Proc. Sci. Congr. 29(3)159.
- Natarajan, A. V., Desai, V. R. and Mishra, D. N. 1976. On the natural occurrence of inter-generic *Catla-Rohu* Hybrid in Rihand (Uttar Pradesh) with an account of its potential role in reservoir fisheries development in India. J. Inland Fish. Soc. India 8 83-90.
- Prakash, Shree and Gupta, R. A. 1986. Studies on the comparative growth rates of three major carps of the Govindgarh Lake. Ind. J. Fish. 33(1) 45-53.
- Prashant K. Deepak, Uttam, K. Sarkar, Raje S. Negi, Samir K. Paul. 2008. Age and Growth profile of Indian Major Carp *Catla catla* from rivers of Northern India. Acta. Zoologica Sinica 54 (1) 136-143
- Ricker, W. E. 1979. Growth rates and models. In Fish Physiology, Vol. VIII (Eds. W.S. Hora, D. R. Randal and J. R. Brett). Academic Press NY, 677-743.
- Sarkar U.K., Deepak P.K., Negi R.S. 2006. Age structure of Indian carp *Labeo rohita* (Hamilton Buchanan) from different wild population. Environment and Ecology 24 (4) 803-808
- Seshappa, G. 1999. Recent studies on age determination of Indian fishes using scales, otoliths and other hard parts. Indian J. Fish. 46(1) 1-11.
- Singh, Deepak and Sharma, R. C. 1995. Age and growth of a Himalayan teleost Schizothorax richardsonii (Gray) from the Garhwal Hills (India). Fish. Res. 24 321-329.
- Singh, Shalendra 1994. Some aspect of fishery and biology of *Labeo rohita* (Hamilton) from Lake Jaisamand, Udaipur Rajasthan. M.Sc. Thesis Rajasthan Agricultural University, Bikaner.
- Singh, Shalendra, Sharma, L. L. and Saini V. P. 1998. Age, growth and harvestable size of *Labeo rohita* (Ham.) from the lake Jaisamand, Rajasthan, India. Indian J. Fish. 45(2) 169-175.
- Singh, Shalendra, Sharma, L. L. and Saini V. P. 1998. Age, growth and harvestable size of *Labeo rohita* (Ham.) from the lake Jaisamand, Rajasthan, India. Indian J. Fish. 45(2) 169-175.
- Tandon, K. K. and Johal, M. S. 1983. Occurrence of the phenomenon of growth compensation in Indian major carps. Indian J. fish. Vol. 30(1) 180 182.
- Van, Oosten. 1957. The skin and scales. In The physiology of fishes Vol I (Ed. Brown, M.E.), NY. Academic press 207-244.

17

SOCIO-TECHNO-ECONOMIC APPRAISAL FOR SUSTAINABLE INTEGRATED DEVELOPMENT OF KOLLERU LAKE-A WETLAND OF INTERNATIONAL IMPORTANCE

*Ajit Kumar Roy and G.S.Saha**

Abstract: Kolleru Lake, located between the deltas of the Krishna and Godavari rivers, in Krishna and West Godavari districts of Andhra Pradesh is the largest natural freshwater lake in India. Rich in flora and fauna, it attracts migratory birds from Australia, Siberia, Egypt and Philippines between the months of October and March. During this season, the lake is visited by an estimated 2,00,000 birds. Recognizing the rich bio-diversity it was listed as a wetland of International importance in the Ramsar Convention in 1971. Presently known as "Carp Pocket of India" Kolleru Lake is the hub of aquaculture activities, and fish produced in this area is being supplied to almost all the eastern and northern states of the country. Keeping in view huge potential of aquaculture development in Kolleru Lake, a survey involving a sample of 221 practicing aquaculturists in Kolleru Lake area was conducted to collect data on various socio and techno economic aspects. The investigation has brought to light many important findings like impact of natural productivity, holding size, education, experience, locally available poultry manure & rice bran, feeding practices, primary occupation etc. on sustainable production of commercially important fish species. There is a chain of forward and backward linkage activities associated with production of fish like icing, packaging, storage, processing, marketing and transportation etc. which have not only made significant contribution to production, employment and income generation but also revitalized the socio-economic condition of the surrounding areas. Identified the impact of socio-economic and farm specific management variables on productivity through econometric analysis. Reported the alternative livelihood initiative taken by Govt. of A.P. in post demolition scenario with lively illustrations that may have far reaching consequences in restoration of both the wetland as well as aquaculture. Findings may be useful for a number of reasons and will serve as an eye opener for researchers, planners, and social scientists for future policy planning for all round integrated development of similar wetlands of India.

Keywords: Flora and Fauna, Transportation

Ex. Principal Scientist & Head Social Science Section & Co-coordinator, Bioinformatics of CIFA (ICAR), Bhubaneswar
*Social Science Section, Central Institute of Freshwater Aquaculture, Po Kausalyaganga, Bhubaneswar - 751 002, Orissa; E-mail akroy1946@yahoo.co.in

Introduction

Kolleru is one of the biggest shallow fresh water lakes in Asia which is located between the deltas of Krishna & Godavari River in the state of Andhra Pradesh, India. The lake serves as a natural flood balancing reservoir for the two rivers and has sustained the rich native flora and fauna as well as the people living around it.

It used to be an ideal habitat to hundreds of species of birds including migratory birds, many species of fishes, fresh water turtles and amphibians etc. and world's one of the very few breeding sites of Grey Pelican. The largest fresh water lake in Asia, Kolleru Lake is a virtual paradise for nature lovers. Rich in flora and fauna, it attracts migratory birds from Australia, Siberia, Egypt and Philippines between the months of October and March. During this season, the lake is visited by an estimated 2,00,000 birds, which include Open Bill Storks, Painted Storks, Glossy Ibises, White Ibises, Teals, Pintails, Shovellers, Red-Crested Pochards, Blackwinged Stilts, Avocets, Common Redshanks, Wigeons, Gadwalls and Cormorants, among Others. It is known the world over for the famous Kolleru Bird Sanctuary. This 673 sq.Kms wet land marsh habitat supports a large variety of

Water birds which include Garganys, ducks, and Open Bill Storks, Herons, Flamingos & grey pelicans. Tourist facilities. The sanctuary has the following watch towers for sighting the birds. Atapaka 1.5 km from Kaikaluru to see varieties of waterfowl. Murthyraju tank 8 km from Nidamarru East Chodavaram 53 km from Eluru where Open Bill Storks nest in colonies from July - December. The sanctuary is approachable from all four sides of the lake by road, directly to the following places. Herons, Flamingos & grey pelicans. Tourist facilities. The sanctuary has the following watch towers for sighting the birds.It is known the world over for the famous Kolleru Bird Sanctuary. This 673 sq.Kms wet land marsh habitat supports a large variety of water birds which include Garganys, ducks, and Open bill storks, Herons, Flamingos & grey pelicans. The largest fresh water lake in Asia, Kolleru Lake is a virtual paradise for nature lovers. Rich in flora and fauna, it attracts migratory birds from Australia, Siberia, Egypt and Philippines between the months of October and March. During this season, the lake is visited by an estimated 2,00,000 birds, which include Open Bill Storks, Painted Storks, Glossy Ibises, White Ibises, Teals, Pintails, Shovellers, Red-Crested Pochards, Blackwinged Stilts, Avocets, Common Redshanks, Wigeons, Gadwalls and Cormorants, among Others.

Recognizing the rich bio-diversity of Kolleru Lake it was listed as a wetland of International importance in the Ramsar Convention in 1971. The Government of Andhra Pradesh also declared the Lake up to +5 contours as Wild Life Sanctuary vide GO MS 120 dated 4.10.99. Kolleru wetland is an integral part of the local irrigation network and is exploited extensively for paddy cultivation. Kolleru Lake has an area of about 901 sq.km. at +10' MSL contour. The lake serves as a natural flood balancing reservoir for the two rivers. The lake is fed directly by the seasonal Budameru and Tammileru rivers, and is connected to the Krishna and Godavari systems by over 68 inflowing drains and channels.

The lake is an important habitat for a variety of resident and migratory birds, including the Grey or Spot-billed Pelican. A notified wildlife sanctuary, the lake was designated a wetland of international importance in November 2002 under the International Ramsar Convention. The wildlife sanctuary covers an area of 308 sq. km. The Kolleru region is inhabited by a population of 2, 00,147 (1981 Census) distributed in 60 bed habitations and 98 belt habitations spread over in 73 revenue villages including hamlets. The bed villages are island villages and the belt villages are border villages of lake. The major occupation of the people of the belt habitation is agriculture which includes the cultivation of the lakeside areas in the upper reaches. On the other hand, the major occupation of the residents in the bed habitation is fishing with 100% of people involved in it, though agriculture in a small scale is practiced by a few of them. As a result, the fish farming industry in this region has received a big momentum boosting the rural economy of the region. Taking into account huge potential of aquaculture development in Kolleru Lake, known as Carp Pocket of India, a survey was conducted to collect data on various socio and techno economic aspects of the fish culture in Kolleru Lake area to understand the present state of socio-economic conditions of the fish farmers, level of aquaculture technology adoption by them and the resulting output.

Methodology

The authors made personal visits to the sites to take stock of the prevailing situation through close observations of the operations as they were being performed, and through interaction with the farmers, members of fishermen communities, farm managers, entrepreneurs, industrialists, suppliers and providers of various services engaged in different stages of value chain during 2007. Facts have also been gathered from various secondary published sources for an in-depth study.

Pond-owner household was taken as the unit for collecting relevant information. Simple statistical techniques like frequency distribution, classification, categorization etc. were used for analysis and presentation of data.

Salient Observations

The socio-techno-economic investigation on Kolleru Lake involving 221 fish farmers' households has unearthed several important dimensions of aquaculture with social, economic as well as technical values. It is believed that each and every aspect of information is important from social scientist's point of view for rational development of aquaculture in the Kolleru Lake. Salient findings of the study covering different variables are presented below.

Holding size distribution

Holding size is a very important parameter for aquaculture. In order to understand the holding size distribution of sample fish farmers of Kolleru Lake, categorization was done as marginal (<2.5 ac), small (2.5-5 ac), semi-medium (5-10 ac), medium (10-25 ac) and large (>25.0 ac) following established definition applicable for agriculture. It is noteworthy that maximum number of farmers (40%) fall in the medium category having holding size between 10.0 - 25.0 ac followed by large (31%), semi-medium (20%), small (8%) and marginal (1%) which is indicative of the fact that fish

culture in Kolleru Lake is dominated by semi medium, medium and large farmers and these three groups together constitute 91% of total holding size. It is also affirmed that a larger fish firm has higher productivity than the smaller firm. It may also be mentioned that minimum size of pond holding was observed 2 ac and maximum 150 ac with the sample farmers.

Type of ownership of ponds

It is seen that 92% of farmers were themselves owners of their respective ponds as against 8% of the farmers who cultivated fish taking lease from the owner. From social point of view this is a very good trend because in rural areas various types of conflicts arises incase the water bodies are owned by more than one person. This may even result in lack of interest in management of farming practices which ultimately affects productivity and return from the ponds. It is observed that due to this type of conflict some of the productive ponds remain unutilized.

Classification of farmers according to religion

it is found that out of the 221 farmers surveyed, 218 (99%) belong to Hindu community, one (0.45%) is Christian and two (1%) from Muslim community.

Classification of farmers according to caste

Classification of sample fish farmers according to caste demonstrates that General Caste represents 61% followed by OBC (27%) and Scheduled Caste (12%). No farmer was found to be of Scheduled Tribe community in the sample.

Age distribution of fish farmers

Classification of sample farmers according to age indicates that 33% of the farmers belonged to modal age group i.e. 40 - 45 years followed by 40 - 50 years (28%), 35 - 40 years (25%), over 50 years (8%) and less than 35 years (6%).

Family Size

Distribution of fish farmers according to number of children shows that maximum number of farmers (43.43%) are having two children followed by three children (33.03%), four children (8.60%), one child (6.79%) and five children (1.81%). Fourteen (6.33%) out of 221 sample farmers, however do not have any child.

Education level

For sustainable growth, adoption of new technologies and their evaluation in terms of merits and demerits are to be understood by the practicing farmers. Therefore, literacy of farmers is of paramount importance. Education level of sample farmers of Kolleru Lake gives a very good picture of literacy. It is seen that 58% farmers are below metric followed by 19% matric, 14% intermediate, 7% graduate, 1% post graduate and 1% MBBS. There is no illiterate farmer in the sample. Rather it is noteworthy that graduates and even doctors having professional qualification are venturing in the fish culture. This trend appears to be very encouraging for sustainable growth of aquaculture in Kolleru Lake area.

Experience

Individual's success in any venture depends mainly on experience and expertise in that particular activity. Since the carp culture in Kolleru Lake area is of recent origin, therefore need was felt to study the experience of sample farmers of this area. It is seen that maximum (50%) numbers of the farmers are having experience in the range of 10-15 years followed by more than 15 years of experience (22%), less than five years (18%) and in between 5 - 10 years (10%). This suggests the fact that fish farmers of Kolleru Lake are experienced enough to maintain sustainability in production. The observation that 18% of the farmers with less than 5 years of experience in fish culture indicates that new entrepreneurs are also entering into the fish culture leading to its continuation and creation of employment opportunities in rural areas.

Durable assets and farm equipments

The success and economy of fish farming depends on the timely use of farm equipments. Therefore, it was of interest to study the type of assets owned by the farmers of Kolleru Lake. It is observed that almost 94% of the sample farmers were having the water pump which is considered very much essential for timely intake of water from the adjoining canals. Likewise 97.75% of the farmers possessed boats which are also very essential for regular feeding, sampling, and periodical application of fertilizer and medicine, or biomass checking, weed control etc. Almost 80.63% of the sample farmers were having godown located by the side of the farms for storage of feed, fertilizers, gadgets and equipments, etc. Similarly, 62% of the farmers were having water pipes of different sizes for regular intake of water from nearby canals. Over one-fourth (27.48%) of the farmers mostly possessing large water bodies even constructed labour huts on the dyke or adjoining areas as shelters for labours who in turn perform various farm activities at different stages of production and serve as watch and ward for the entire farm. Likewise only 38.29% farmers possessed nets and nylon threads which are bit costly. Only farmers with large and medium size water holding could afford to purchase nets for periodical biomass checking and harvesting. Others hired nets along with fishermen for harvesting. As regard to ownership of more than one asset by the farmers, it is observed that only 12.6% of the farmers were having all the six types of assets like water pump, boat, godown, pipe, labour hut and net. It is gratifying to note that 130 (58.56%) farmers are having the essential prerequisites for farm management like water pump, boat, godown and pipe.

Training

It is quite surprising to note that out of 221 sample farmers only one has indicated to have been trained in fish farming. This is an area where government intervention is necessary for human resource development for proper growth and sustainable development of fish culture as a viable economic enterprise.

Sources of advice

Interview of farmers on the sources from where they have taken needed advice regarding fish culture and related activities unearthed the fact that 58% of the farmers took advices from fellow farmers, followed by 25% managing fish culture based on their own knowledge and experiences, 13% seeking advice from private fish doctors and a minimum of only 4% taking advices from government sources. Most interesting revelation in this context is the emergence of fish doctors to tackle the disease and fish health related problems in aquaculture in the Kolleru Lake area which is probably new dimension in aquaculture sector.

Sources and duration of loan

Survey reveals that out of 221 farmers only 66 (30%) farmers availed short term (10-15 years) loan, to the tune ranging from Rs. 15,000 to 1,40,000, from government banks, and loan from private sources are reported to be nil which is definitely a good trend for development aquaculture sector.

Insurance coverage

Fish culture is a long duration (10-18 months) activity involving a lot of risk. However, contrary to the wide scale practices of insurance coverage for the agricultural crops, the coverage of fish culture by Insurance Company is almost lacking. It is recorded that out of 221 sample farmers surveyed for the purpose only one had insurance coverage by Oriental Insurance. Due attention and enabling policy decision by the government in this area is the need of the hour for sustainable development of this potential sector generating employment and uplifting rural livelihood.

Sources of water intake

Water is very precious and important for aquaculture as the entire culture operation depends on water. Further, periodic intake of water is necessary to replenish the losses due to evapo-transpiration. Enquiry reveals that sources of water for aquaculture in Kolleru Lake are the adjoining canals as most of the farms are constructed and developed by the side of the canals named as Kolleru canal, Chandraya canal, Krisna canal and also the canals developed by irrigation departments. A study on sources of water taken by the sample farmers reveals that majority of the farmers (72%) receives water from Kolleru canal followed by irrigation canal (14%), Chandraya canal (8%), Natural water (4%) and Krishna canal (2%).

Frequency of application of lime

Use of lime is a common practice. Data on use of lime was tabulated to gather information about the frequency of application lime by the farmers. It is seen that farmers used lime 0 - 13 times during the entire culture period. While maximum number of farmers applied lime eight times, 16 farmers did not use lime even for once compared to the maximum of 13 times used by only 3 farmers.

Types of inorganic fertilizer application

It is observed that all the 221 sample farmers used Urea followed by 212 farmers (95.9%) using SSP, 189 farmers (85.5%) applying DAP and 3 farmers (1.4%) using Grow more to fertilise their ponds to enhance natural food productivity.

Organic fertilizer application

Enquiries during survey reveals that all the 221 sample farmers used poultry droppings. Cowdung, in addition to the poultry manure, was used by 107 farmers. These organic manures are used to boost natural fish food productivity of the pond, which ultimately results in higher production of fish biomass. Availability of poultry manure at a very cheaper rate from near-by giant poultry farms might be the reason for its wide use.

Sources of seed

Out of 221 farmers 146 farmers reared fish seed from fry to fingerling in their ponds and the rest 75 farmers purchased seed from private hatcheries. This fact indicates the trend of approaching towards self sufficiency in quality seed production, which is a prerequisite for successful culture operations and a major shift from their traditional dependence on fish seed from West Bengal.

Disease Occurrences

It is suspected that due to use of a variety of inputs such as feed, fertilizer, poultry manure, cow dung, etc. the water quality is gradually deteriorating thereby inviting various types of diseases in the Kolleru Lake area. From an interaction with the farmers on this issue it appears that 40.72% of farmers encountered bacterial disease followed by 40.27% parasitic, 26.69% fungal and 10.41% viral. All the farmers reported some or other kinds of diseases. This matter requires immediate attention of the appropriate corners for sustainable growth and development of the aquaculture industry.

Method of application of medicines

Maximum number of farmers (58%) used medicines in the affected ponds after diluting in water followed by spraying in water (41%) and only 1% out of sample 221 farmers used medicines mixed with feed.

Type of integration

Enquiries reveal that 99% of sample farmers were doing only fish farming. Only one farmer integrated fish culture with agriculture and two farmers integrated dairy with fish farming for better return. Many poultry farms integrating with fish culture were seen around adjoining villages. Similarly series of coconut plants were seen standing on the fallow pond dykes.

Types of feed used

Feed constitutes a considerable proportion of input cost in aquaculture. Therefore, it was felt important to understand the types of feed used by the farmers. Survey on the sample farmers of Kolleru Lake reveals that deoiled rice bran (DORB), cotton seed cake, groundnut oil cake and soybean are the most widely used feed ingredients. Maximum number of farmers (64%) used the combinations of DORB, & oil cake followed by DORB, oil cake, soybean, vitamins and probiotics and gur/jaggery (30%); DORB and soybean (3%); and only DORB (3%). In some cases oil cakes of mustard and sunflower were also used by the farmers.

Method of application of feed

A study on method of application of feed ingredients by the farmers of Kolleru Lake revealed that an overwhelming majority of the farmers (97%) applied feed in perforated bags hanged from bamboo poles into water. This innovative method of feeding assumes paramount importance because it saves wastage of costly feed by preventing it from going into the pond bottom, ensures its optimum use and controls pollution of pond water. This contributes to both economic and environmental sustainability of fish culture operation in this area. A meager 2% of the farmers fed the fish through conventional method of broadcasting the feed in water, and only 1% farmers diluted feed with water and then spread it over water.

Time and frequency of application of feed

For efficient and economic aquaculture operation time of application of feed is also important like the type of feed ingredients used and the feed dispensing method. Survey about the time of application of feed brings out the fact that 98.64% of farmers fed the fishes in the morning hours, followed by only 1.36% used in day time but none used feeding in the evening. From enquiry it could also be known the farmers are using feed daily on regular basis.

Purpose of periodical netting

Netting is generally done by the farmers at the time of harvesting. Besides periodical nettings are also being done which means substantial cost involvement towards hiring charges for nets and manpower. When asked about the purposes of netting, maximum farmers (86%) reported to have done so for biomass checking whereas 0.10% did so for disease checking and 13% reported to have done it for both biomass checking and disease checking.

Types of labour employment

Categorization of labour employed by the farmers reveal that out of 221 sample farmers, 169 (76%) managed the aquaculture activities employing permanent workforce, whereas 10 farmers (5%) engaged labours purely on casual basis depending on the need and the rest 42 farmers (19%) engaged themselves in various farming activities. This is indicative of potential employment opportunities in rural aquaculture.

Sources of household income

Sources of household income from various activities along with fish farming were studied. It is evident that household income is dominated by fish farming only (78.74%), followed by fish farming and business (18.56%). Sources of income from fish farming along with either agriculture or dairy farming or goatry or poultry or service are insignificant.

Production

Aquaculture in Kolleru Lake proves to be promising in terms of both productivity and profitability. Reported productivity ranges from 8 - 12 ton/ha during a culture period of 12 - 14 months, which is good enough to help the farmers earn a handsome income (Anon.2002).

Profitabilitiy

All the sample farmers reported higher productivity and profitability from fish culture compared to agriculture. This of course shows the strength of the technology and points to the scopes of its wide-scale adoption among the farmers and entrepreneurs. It is also estimated that the productivity of large farm is higher than the smaller. Since most of the farmers have semi-medium, medium and large holdings, they earn higher income compared to marginal farmers as they get the advantage of operating with economy of scale. Forward and backward linkage industries Aquaculture in Kolleru Lake area has had significant positive effect on rural and urban fish supply and on income and employment generation. There is a well-established linkage of aquaculture with various forward and backward industries. It is seen that the whole chain of activities beginning from the backward linkage activities of production and marketing of fish seed and fish feed and other inputs like fertilizers, drugs and medicines, and continuing through forward linkage activities of icing and packing, transportation, storage, processing and retailing of the product, have not only made significant contribution to employment and income generation in these linkage activities but also revitalized aquaculture by opening up far off markets. Unlike traditional fishing activities geared mainly to self consumption or for local marketing, fish farming has been mainly conducted as a business venture where production is undertaken primarily to meet the market demand. Increasing demand for fish in domestic markets and the complex network that affects the supply and price of fish is influencing aquaculture production both at national and regional levels. The study reveals that aquaculture practices in Kolleru Lake, although initiated in the eighties have added a new dimension to the economy of the region within a very short period, with the lake recognized by the FAO as one of the fastest growing aquaculture sites in the world.

As a growing economic activity, aquaculture has made it possible to introduce and try new and improved technologies, which have opened up new opportunities such as hatchery production of fish seed, feed production units, input suppliers, aqua-shops, net making units, ice plants, packaging industry and transport networks. In the process it has promoted the growth of a whole chain of activities, from hatcheries to marketing to retailing and export of fish and prawn that have not only added value to the product, but also increased the demand and profitability of aquaculture at the farm level. Kolleru aquaculture, as a business venture is undertaken primarily for the markets located at Kolkata, eastern, north and northeastern states of India and even Bangladesh. The latest survey indicates that progressive farmers are trying to establish post harvest facilities eyeing the markets of Thailand, Singapore, Vietnam and other Asian countries.

Growth of forward and backward linkage industries in aquaculture

Aquaculture in the Keller Lake area has had a significant positive effect on rural and urban fish supply and on income and employment generation. We have observed that there is a well-established linkage of aquaculture with various forward and backward industries. It is noteworthy that plastic tanks and oxygen cylinders are available on the road side in many places on the highway connecting Eluru and Bhimavaram

These are given out on rent for transportation of fish seed in the Kolleru lake area. This business has creating a significant scope for employment. We have also observed that the whole chain of activities beginning from the backward linkage activities of production and marketing of fish seed, feed and other inputs like fertilizers and medicines, and forward linkage activities of icing and packing, transportation, storage, processing and retailing of the product have also made significant contribution to employment and income generation and revitalized aquaculture by opening up far-off markets. Another important cottage industry that has arisen in the area is boat making. Boats are made of discarded telephone poles which are flattened in the industry before being used as a

basic raw material. The majority of ponds in Kolleru Lake area are large in size. Hence boats are frequently used for applying fertiliser, feed, lime and other inputs. The presence of poultry farms around Kolleru offers a supply of poultry manure, a basic organic input for pond fertilization, at a low price. Many fish need centers and aqua-shops have come up in Eluru, Bhimavaram and neighboring areas catering to range of farmers' requirements by providing inputs and information services to them. The large pond size has made it feasible to go for a high degree of intensification. Producers have a high production potential and do have access to the capital or credit, which is required to take up aquaculture on a larger scale. They have also high risk bearing capacity, which the small farmers do not have. Kolleru aquaculturists being resourceful have better purchasing capacity of all the inputs required for production. They are better educated and experienced. Wherever suitable they can even think of integrating the aquaculture with other farming practices for higher profit and optimum utilization of resources. The entire process of aquaculture seems to have been stimulated through interventions at three levels (i) commercialization of production through adoption of scientific technologies and market-oriented approach (ii) ,elopment of forward linkages through improved post-harvest services like packaging, processing, rage, transport, efficient marketing systems and opening up of new markets as well, and (iii) :kward linkages through the provision of inputs like seeds, fertilizers, manures, feed, medicines 1 aquaculture machinery. Wider availability of land resources in both size and suitability and ured supply of water has led to the commercialization of aquaculture production at Kolleru

Lake area. Growth of poultry farms. Growth of boat making industry. Productions from the aquaculture ponds have efficient forward linkages with marketing chains.

The marketing infrastructures are organized and efficient. The harvests are segregated and graded. There are several ice factories and rice mills supplying ice and husks for packaging of fishes. Rice mills around Kolleru lake area have paved the way for successful aquaculture. They supply cheap rice bran, an important ingredient for fish feed. Husk produced at rice mills is used in packaging of fish as an insulator to maintain the cool conditions for a longer time, and can considerably slow the melting of ice. The labourers involved in packaging are efficient in quick packing of the produce whether at the pond site or at the ice factory premises. The organized transport agencies in the locality play a very important role in transporting the fishes to various places of the country quickly and efficiently. Producers sell the produce to the contractors, who in turn send the produce to different wholesalers located at various location Finally the consumers get their fish from the retailers/ venders who have got it from the wholesaler Growth of aquaculture has led to simultaneous growth of ancillary industries such as ice factorie feed mill plants, packaging centers and lorry supply agencies. Due to mass fish production ma ice plants have been established in the local village areas namely Jalipudi, Sreparru, Pedapad Kalakarru, Singapuram, Akiveedu and other places. The ice is being used for packaging and keepi the fish fresh for several days.

The ice plants are observed to employ many people for its operation who earn a good incom The credit for their well being goes to aquaculture development. Ice factories are the forwa linkages of aquaculture. The husks from the rice mills are also used by the packers to prevent t ice melting for hours due to non-conductive nature of the material. Along with huge export potenti the lorry supply agencies have come up in a good number employing many people. On an avera

)0 lorry loads of fish are flagged off daily from Kolleru lake area (Roy *et al.* 2008). Several ermocol and plastic factories manufacturing plastic crates and thermocol boxes for carrying the shes from the production centers to various places have come up in the locality. Thermocol boxes eing light and cheap have relative advantages over plastic crates. Being less costly they can be scarded after a single use and need not be brought back to the packing center for reuse. On the her hand the plastic containers are heavy and costly. The empty crates are brought back in lorries the packing centers for repeated use, which involves substantial transport charges.

reation of rural employment opportunities in sorting, packaging, transporting and arketing

is estimated and reported that for a one-hectare pond vo permanent labourers and 300 man days of casual bour are engaged for aquaculture. The traditional system lowed only casual labour for a period of 4-6 months in year when the peak activities of fish culture go on. In ldition to the direct labour employment in the culture actices, indirect employment in allied activities like ice ctories, packing centers, lorry supply agencies and arketing is noteworthy. Marketing practice is observed be absolutely labour intensive which leaves nployment for labours in various stages like harvesting, sorting, packaging and weighing, icing factory, loading and unloading at wholesale market and retailing in the consumer markets (Roy *al.* 2008).

he shifting land use pattern and the management interventions

It is reported that over the passing decade, the Kolleru crisis gained attention at various management levels also involving frequent political interventions. Before the existing conflict on land-use the entire lake area was cultivated for paddy both as Rabi [is the spring harvest also known as the winter crop] and Kharif [is the autumn harvest also known as the summer or monsoon crop].

Also it is appealing to note that people shifted from agro based economy to carp lture as the record (from forest department statistics and management plan) shows that the farmer rived INR 12,500 per acre from fish tanks as compared to INR-7500 from paddy cultivation.

However, the increasing incidence of aquaculture ponds in the region in past decade was the cause of the unusual hydrological events resulting in flooding of the agricultural land in the sub-catchments. The removal of encroachments was thought as serious step by the Revenue Department for the protection of sanctuary in order to restore its function as a balancing reservoir in the event of flood. In addition the treatments of catchment area the other recommendation of the management was to increase in the carrying capacity of the existing outlet (Upputeru) into the sea to facilitate draini of flood water was also among the listed set of activities; it took awhile for them to be execut when finally in June 2006 the aquaculture demolition came about. As a result a mammoth fund v required to re-locate the community and compensate for the land value.

Econometric approach for estimation of technical efficiency of carp farms

It is reported that technical efficiency (TE) scores for fish farmers of Kolleru Lake ranged fr 0.5051 to 0.9899 with a mean of 0.7260 and modal class is 0.6 to 0.7 that compares well with carp culture ponds. Mean technical efficiencies of carp culture ponds of Nepal, India, Banglad and Pakistan are reported to be 0.676, 0.790, 0.738 and 0.740 respectively for semi-intensive a intensive carp farming and the respective values for extensive farming are reported to be 0.5 0.502, 0.475, 0.625 (Sharma and Leung, 2000a) and between 0.11-1.0 (Jayaraman, 1998). In present study it revealed that over 56% of the farmers operate with efficiency score greater than and no farmer operate below the technical efficiency of 0.50.

Potential for technical efficiency improvement

The result reported suggests that on average the farmers of Kolleru Lake area are able to obt about 72.60% of potential output from the given set of production inputs. Therefore, in the sh run there is a scope of increasing the productivity of carps by around 27.40% by adopting technology and techniques used by best practice farms. Further, it is observed that if the aver farmer in the sample was to achieve the TE level of its most efficient farmer counterpart then average farmer could realize 26.7 per cent cost saving [1 - (72.60/98.99)]. Similarly the m technically inefficient farmer may achieve cost saving of 48.9 per cent [1 - (50.51/98.99)].

Identification of determinants of technical efficiency

Results of maximum likelihood estimates of the technical inefficiency model and determinants technical efficiency estimated based on twenty-five nonphysical socio-economic, demograp

and farm specific factors were included in the model to test the possible impact and association with the technical efficiency of aquaculture farms operating in the Kolleru lake area reveals that of the twenty-five factors considered for technical efficiency model, socio-economic and farm specific factors like religion, age, education, primary occupation, pond size, ownership type, source of water, depth of water, sources of advice taken, method of application of drugs, calamities faced, types of labour, types of feed and disease encountered have positive impact on technical efficiency. On the other hand caste, number of children, experience, renovation, asset cost, source of seed, loan, harvesting technique, method of application of feed, periodical netting for biomass checking and types of organic manure showed negative association with technical efficiency. Of all the farm specific factors showing positive impact mentioned above only pond size and source of water were found significant at ($p<0.05$) and types of labour, religion, method of application of drug, primary occupation and calamities are found significant at ($p<0.10$). Likewise of the variables showing negative impact only periodical netting for biomass checking is found highly significant ($p<0.01$), source of seed and harvesting technique and caste were found significant at lower level of significance ($p<0.10$) demonstrating the influence of these factors in efficiency of carp culture. The stochastic frontier estimate demonstrates the average technical efficiency of 72.60% indicating that there is scope for further increasing the output by 27.40% without increasing the levels of inputs. The empirical results suggest that there are immense possibilities to increase efficiency levels by extending culture operation in large size ponds, provision for supply of fresh water, application of drugs after diluting in water, employing labour on regular basis and leasing out the water bodies to farmers whose primary occupation is fish farming and educating the farmers to procure seed from reliable sources, proper harvesting technique and avoiding periodical netting. From the policy point of view, above mentioned farm specific variables are found to be most promising areas for action, suggesting that policy maker may look into the matter seriously for possible reforms to enhance the productivity without any additional input use.

The post demolition scenario

Kolleru Restoration Cell of Govt. of Andhra Pradesh started demolishing the ponds that ended in December 2006. It is reported that about 25000 acres of pond area has been demolished with a total production loss of one lakh ton annually. A total of 88 Co-operative Societies operated upto 2005 with aquaculture. After demolition the Society members were operating traditional fisheries with prior permission from forest officials who is in charge of Kolleru Lake areas. Through the process of demolition a good number of fishermen were affected creating problem of their livelihood. Therefore, Govt. of Andhra Pradesh has taken rehabilitation of measures for the fisher folk living in and around Kolleru Sanctuary area. Available document Lr.No.844/A5/2006 dated 19.1.2007 suggests that government has sanctioned and released 2.60 crore for implementation of livelihood enhancement scheme for fishermen who are affected due to destruction of illegal fish tanks in the Kolleru Lake Area. Main livelihood programmes initiated by Govt. of Andhra Pradesh are listed below.

- Intensive training programme on post harvest technology
- Short-term training programmes are being conducted to the selected Fishermen/Women on certain value added activities as follows.
- Fish Hygiene - Value addition Fish pickle making, Cutlets, Pakoda, Burger, Boneless fish (CIFA) etc.
- Inland Fish Culture Practices and production of quality fish
- Fish bye products & Marketing Strategies
- Inland Fishing Gear Fabrication and Maintenance
- Awareness on Scampi (Fresh water prawn) culture
- Fish diseases & control
- Kolleru meenotsavam

The fish productivity in natural water bodies declining year after year due to over exploitation, pollution and changing environmental parameters. Hence, fish seed ranching, a globally accepted proven mechanism to improve the fish productivity of natural water bodies and for sustainable production over a period of time. The Fisheries department has taken up 'Seed ranching programme' in a big way and stocked 62 lakhs of fish seed (3" fingerling size) in Kolleru Lake at different stocking points of 7 Mandals of Kolleru Lake extension area in the presence of District Collector. West Godavari and other senior officers of the Department. Ranching is technically feasible and economically viable.

Exposure visits

Awareness Programmes are being conducted through exposure visits to Kerala, West Bengal, Goa, Karnataka and Tamil Nadu States. Till date 260 fishermen were sent on exposure visits to the above States to make them aware of the latest techniques in post harvest and value added technology and other allied fisheries activities.

Supply of bamboos

A list of 15,656 Kolleru Fishermen was submitted to the Forest Department to supply 60 numbers of 1st class bamboos to each fisherman who are identified by the Fisheries Department Officer and village elders as Committee members from seven Kolleru Mandals of West Godavari District.

Establishment of one craft & tackle making plant

The skilled Fishermen/Fisherwomen are to be selected who may be employed in future in the above proposed unit to be established in one of the Kolleru villages.

Establishment of one ice-cum-cold storage unit Site has already been selected at Sriparru for the construction of 20-ton capacity ice-cum-cold storage plant.

Return of migratory birds

The aforesaid livelihood programmes are being implemented for the poor fishermen families of Kolleru under special package by the fisheries department.

Migration of birds

In the demolished areas huge qualities of birds are seen taking food from follow marshy areas.

Rearing of ducks

In earlier reports there is mention of integration of fishery with duckery. During the past demolition of lake area, many migratory duck rearers are found camping in the abandoned areas with flock of ducks having enough space to rear. Their ducks in natural lake having enough natural feed for ducks as they thrive on small fish as well as bottom biota including molluscan shells. It is observed and reported that people rear ducks in natural water bodies, apply supplementary feeding on the shore and pick up eggs from enclosures where ducks are kept in night time for laying of eggs. It is reported that farmers can understand the availability of natural food organisms in water bodies from the sound of the ducks white taking feed. Once the natural feed is exhausted, the owner migrates to other places with all the ducks further rearing.

Grazing of sheep

Another important activity found in the abandoned areas after demolition is the grazing of sheep. During survey it is seen that many people with flock of sheep (50-100 nos.) moving in the marshy areas having enough of suitable edible grasses for the sheep. In the process many people are engaged

Sheep rearing in abandoned lake

in sheep rearing for their livelihood. Earlier the scope of grazing ground organized aquaculture farms were limited.

Shifting of carp culture technology to other areas

As a result of demolition of 25000 acres of aquaculture pond in the Kolleru lake area within 5' contour, a number of farmers have been affected in the process. During the recent survey it is learnt that farmers are in search of areas above 5' contour upto 10' contour level having ensured irrigation for crops construct new aquaculture activities as means of income and livelihood. It is also found that entrepreneurs are in lactic search of suitable land to the neighboring state of Orissa. To the knowledge of the authors, a few aquaculture ventures have been taken Puri district of Orissa.

Conclusion

The investigation has brought to light a number of salient finding relating to important socio-techno-economic characteristics of the practicing aquaculture farmers in Kolleru Lake, Andhra Pradesh known as "Carp Pocket of India". Notable are that the lake is dominated by semi-medium, medium and large farmers (91%) having holding size over 5 ac and 92% of the holdings are owned by a single person leaving no scope for any sort of conflict that arises in case of joint and multiple ownership. Farmer with large holding maintains higher productivity and higher income. It is heartening to note that there is no illiterate farmer. One positive thing is that 61.54% of the farmers belong to the age group of 40-50 years having experience ranging from 10-15 years in fish culture. Most of the farmers are having durable essential assets and equipments like water pump, boat, pipes, godown etc. required for efficient running of the enterprise. Regarding loan it is observed that only 30% of the farmers have taken medium term loan from banks and none took loan from private sector. All farmers used water from nearby canals and also natural sources. All sample farmers made good use of lime, which is very important for pond preparation and water quality maintenance. All the farmers applied both organic manures like poultry manure and cow dung and inorganic fertilizers like SSP, Urea and DAP in order to enhance and maintain the natural productivity of the pond. Presence of many big poultry farms and rice mills around Kolleru Lake facilitated large scale use of poultry manure and rice bran which is available in plenty and at a cheaper rate making the fish practices at Kolleru Lake area economically viable. With regard to feeding, majority of farmers (57.92%) used to feed fishes daily morning using a combination of deoiled rice bran, cotton seed cake and groundnut oil cake. Feeds are kept in perforated plastic bags and then hanged from bamboo poles inside water - an innovative practice which not only avoids waste of costly feed but also checks water pollution thus adding to both economic and environmental sustainability of aquaculture in the region. There is a well established linkage effect (forward and backward) of aquaculture with other supplementary industries like rice mills, ice factories, plastic and thermocol factories, boat making industry, transport agencies, etc. which have created large scale employment opportunities and livelihood options for many. These findings are useful for a number of reasons, and integral toward achieving a sustainable aquaculture development and management plan for the area. It is felt that production activity, undertaken principally for self-consumption, cannot attain the scale needed for efficient resource allocation and technological advancement. On the other hand, aquaculture undertaken as an economic venture on commercial scale provides inputs and services needed for on-farm production, markets for farm products and livelihood opportunities like local seed and fingerling supply, production of feeds using local materials, and fabrication of various materials for farming activities. For all these activities several actors are involved in aquaculture chain e.g., fish farmers, hatchery and nursery operators, seed vendors, commercial fish feed manufacturers, domestic resume based feed producers, fishing equipment manufacturers, fish processors, fish traders and exporters etc. The additional employment and income generated in the value chain create demand for other products and support the growth of other economic activities. The key challenge in aquaculture development in medium and small-scale operation is to reduce poverty, vulnerability and marginalization among people. It is quite satisfactory to note that growth of various forward and backward industries in aquaculture in Kolleru Lake area have helped in strength. The reported alternative livelihood initiative taken by Govt. of A.P. in post demolition scenario with lively illustrations that may have far reaching consequences in restoration of both the wetland as well as aquaculture. Findings may be useful for a number of reasons and will serve as an eye opener for researchers, planners, and social scientists for future policy planning for all round integrated development of similar wetlands of India. From the discussions presented above it appears that this array of information will serve as an eye opener for researchers, planners, social scientists for future policy planning for all round development of aquaculture in Kolleru Lake area specifically and the country as a whole.

Acknowledgements

Authors are thankful to Indian Council of Agricultural Research for sponsoring the project "Technological innovations in aquaculture and its effects on sustainability of farming systems in Andhra Pradesh", and financial assistance for carrying out the investigation.

References

- Anon (2002). Final Report of the Project 'Technological innovations in aquaculture and its effects on sustainability of farming systems in Andhra Pradesh', Central Institute of Freshwater Aquaculture, Kausalyaganga, Bhubaneswar.
- Roy, A.K., K.N. Sahoo, K.P. Saradhi and G.S. Saha (2002). Farm size and aquaculture productivity relationship. Asian Fisheries Science 15(2002) 129-134, Asian Fisheries Society, Manila, Philipines.
- Ramakrishna, R. (2007). Kolleru carp culture in India An aquaplosion and an explosion, Aquaculture Asia, 12 (4)12 -18.
- Roy, A.K., G.S. Saha, P. Kumaraiah, N. Sarangi, B.S.Giri and S.Ayyappan (2008). Socio and techno economic aspects of carp culture in Kolleru Lake, Andhra Pradesh, Central Institute of Freshwater Aquaculture, Bhubaneswar.
- Sarangi, N., P. Kumaraiah, P. V. Rangacharyulu and Bandla S. Giri (2004). Status of freshwater aquaculture in Krishna - Godavari Delta - a Profile. Central Institute of Freshwater Aquaculture, Kausalyaganga,Bhubaneswar-751 002, pp 29.
- Anjaneyulu, M., V. Nagulu, M. Johanson and J.V. Ramana Rao, 1990. Some aspects of management and preservation of water birds and their wetland habitat at Kolleru. Proc. Nat. Symp. Env. Impacts of Environment on Animal and Aquaculture (Eds. G.K. Manna and B.B. Jana) 171-176.
- Anjaneyulu, Y., 2003. Assessment of Environmental Quality of Kolleru Lake and Strategic Management Plans. In Lake Kolleru Environmental Status (Past and Present), Eds. M.K. Durga Prasad and Y. Anjaneyulu, BS Publications, Hyderabad, pp 3-15.
- Anon (1987). Master Plan for Kolleru Lake Development, Government of Andhra Pradesh, Hyderabad.
- Anon, 1987a. Report on fish culture prospects in Kolleru Lake. CIFRI (ICAR), Survey Report No.6 1-6.
- Anon, 1988. Annual Report of CICFRI (ICAR), 1987-88 Barrackpore 32-34.
- Balaparameswar Rao, M. and K. Padmaja (2003). Lake Kolleru - a perspective with special reference to insecticide pollution. In Lake Kolleru Environmental Status (Past & Present). Prasad, M. K. and Y. Anjaneyulu (Eds.), B. S. Publication, Hyderabad-95. Pp. 48-51.
- Chacko, P.I, J.G. Abraham and R. Andal, 1952. A survey of the flora, fauna and fisheries of the Collair Lake. Ind.Com. Journal, Madras. 1274-280
- Chaston, Ian (1987). Marketing Concept. In Marketing in Fisheries and Aquaculture. Fishing News Books Ltd., Farnham, England pp 11-16.
- Das, V. Luther; M. Mohana Rao and S. V. Sharma (2003). Catfish resources of Lake Kolleru. In Lake Kolleru Environmental Status (Past & Present). Prasad, M. K. and Y. Anjaneyulu, (Eds.) B. S. Publication, Hyderabad-95. Pp. 138-140.
- Durga Prasad, M.K. and P. Padmavathi, 2003. Lake Kolleru - the biggest freshwater wetland ecosystem in South India biodiversity and status. In Lake Kolleru Environmental status (Past and Present), M.K. Durga Prasad and Y. Anjaneyulu, (Eds.) BS Publications, Hyderabad, pp 143-157.
- Dutt, S., 1983. On the capture fishery of Kolleru lake. Eco development of Kolleru (a wetland ecosystem) status, position and approach document. INCOR.
- Ghosh, A. K., R. N. Paul, M. Rama Krishnayya, K. Subba Rao and M. B. R. Murthy, 1990, Fish health problems in Kolleru Lake, Andhra Pradesh. J. Inland Fish Soc. India, 22(1) & (2) 980-91.

- Gopal Krishnayya, Ch. and A. K. Lal. 1990. Eutrophication, its causes and consequences with particular reference to Kolleru Lake. Recent Trends in Limnology 145-149.
- INCOR. 1983. Ecodevelopment of Kolleru lake (A wetland ecosystem) status, position and approach document, submitted by INCOR, Department of Geo-engineering, Andhra University, Waltair.
- Jacob. T. J. 1973. Master plan for development of fisheries and uplift of socioeconomic conditions of fisherman in Kolleru Lake, Andhra Pradesh.
- Janardhana Rao, K., K. Venkateswara Rao, T. S. Rama Raju and K. Subba Rao, 1987. Studies on the position and prospects of fisheries of Kolleru Lake with special reference to the management of the resource. J. Indian Soc. Costal Agric. Res. 5(1) 215-221.
- Jayarao, S. S. 1974. Development of Kolleru area. Project report on development of fisheries in Kolleru Lake. Department of Fisheries, Government of Andhra Pradesh.
- Khasim Imam, D and C. C. Panduranga Rao (2003). Status of heavy metal concentrations in fishes from Lake Kolleru. In Lake Kolleru Environmental Status (Past & Present). Prasad, M. K. and Y. Anjaneyulu (Eds.), B. S. Publication, Hyderabad-95. Pp. 41-47.
- KLDA, 1985 & 1987. Annual Report on Kolleru Lake Development. Authority, Kaikaluru, Andhra Pradesh.
- Mangapathi Rao, K. 1987. The nutrient path way in Kolleru lake ecosystem. Proc. National seminar on aquatic biology, Andhra University, Waltair, A.P.
- Munirathnam Reddy, P. and M. Suryanarayan. 1983. The fisherman of Kolleru lake A study of their ecological adaptations. Proceedings, International conference on man and environment, organized by Indian Anthropological Society. Calcutta.
- Murthy, V.S.R. and S. Dutt, 1976. On the fish and fisheries of lake Kolleru, Andhra Pradesh. Mem. Soc. Zool. Guntur 17-27.
- Nageswar Rao, J. (2003). Lake Kolleru and its environs in relation to quaternary geology, geomorphology, and pollution. In Lake Kolleru Environmental Status (Past & Present). Prasad, M. K. and Y. Anjaneyulu (Eds.), B. S. Publication, Hyderabad-95. Pp. 39-40.
- Nagulu, V. and J.V. Ramana Rao, 1980. Survey of south Indian pelicanries. J. Bombay Nat. Hist. Soc., 80(1).
- Neelakantam, K.K., 1949. South Indian Pelicanry. J. Bombay Nast. Hist. Soc. 48(4) 655-660.
- Padmavathi, P. and M. K. Durga Prasad, 2003. Lake Kolleru vis-à-vis the fish ponds - a study on the water quality and zooplankton. In Lake Kolleru Environmental Status (Past & Present). Prasad, M. K. and Y. Anjaneyulu (Eds.), B. S. Publication, Hyderabad-95. Pp. 16-24.
- Padmavathi, P., M.K. Durga Prasad and P.S.R. Anjaneyulu, 2003. Ecology and fish production in a fishpond of Lake Kolleru. In Lake Kolleru Environmental status (Past and Present), M.K. Durga Prasad and Y. Anjaneyulu (Eds.), BS Publications, Hyderabad, pp 103-117.
- Prasad, Durga, M. K and Y. Anjaneyulu, 2003. Lake Kolleru. Environmental Status (Past & Present). B. B. Publications, Hyderabad, 227 pp.
- Ramani, K. V. and Y. Anjaneyulu (2003). Pollution studies of Kolleru Lake by industrial, agricultural, domestic wastes and corrective management plans. In Lake Kolleru Environmental Status (Past & Present). Eds. Prasad, M. K. and Y. Anjaneyulu). B. S. Publication, Hyderabad-95. Pp. 52-60
- Ramaseshaiah, M. and S. Dutt, 1984. Comparative electrophoretic studies of Anabas testudineus and A. oligilep [is (Osteichthys anabantidae) (climbing perch) from lake Kolleru, Andhra Pradesh. Hydrobilogia, 119 57-64.

- Ranga Reddy, Y. and M.K. Durga Prasad, 1982. Lake Kolleru- A limnological perspective. In Souvenir on First All India Conference on Limnology, 3-5 March 1982. Nagarjuna University, India.
- Rao D.V.S & Raju V.T (1997). Economics of Freshwater Fish Culture in Andhra Pradesh, Agricultural Economic Research &Review. Vol. 10(1) PP. 37-44.
- Rao, B.V. Seshagiri, 2003. The impact of fisheries on Lake Kolleru, a wetland ecosystem. In Lake Kolleru Environmental Status (Past & Present). Prasad, M. K. and Y. Anjaneyulu (Eds.), B. S. Publication, Hyderabad-95. Pp 118-123.
- Rao, K. J. and T. S. Rama Raju. 1987. Observations on the carp culture in large freshwater Ponds of Kolleru Lake. J. Aqua. Trop. 4 157-164. Oxford IBH Publishing Co. Ltd., New Delhi.
- Rao, K. Venkateswara; M. Ramakrishnaiah and T. S. Rama Raju (2003). The fish and fisheries of Lake Kolleru. In Lake Kolleru Environmental Status (Past & Present). Prasad, M. K. and Y. Anjaneyulu (Eds.), B. S. Publication, Hyderabad-95. Pp. 124-133.
- Rao, K.J., 1987. Studies on the position and prospects of fisheries of Lake Kolleru. J. Indian Soc. Coastal, Agrl.Res.5 (1) 215-221.
- Rao, K.M. 1987. Master plan for Kolleru Lake development-An integrated ecological approach. Kolleru Lake Development Authority, Kaikaluru, Govt. of Andhra Pradesh.
- Rao, K.S. and K. Lakshmi, 1990. Deteriorating fisheries in the highly polluted Kolleru Lake. Recent trends in Limnology. 165-174.
- Rao, N. Subba, 1986. Economics of Fisheries. Daya publishing House. Pp5-22
- Rao, M.V., K.H. Rao, I.V. Ramana, S.K. Sasmal, S.B. Choudhary and S.K. Bari, 2003. Change detection studies in and around Kolleru Lake using high-resolution data. In Lake Kolleru Environmental status (Past and Present), M.K. Durga Prasad and Y. Anjaneyulu (Eds.), BS Publications, Hyderabad, pp 197-204.
- Roy A. K., K. N. Sahoo, K.P. Saradhi and G.S. Saha. 2001a. Aquaculture in Kolleru Lake Area for Socio-economic Development. Environment and Ecology, 19(4) 927-931.
- Roy, A. K., K. P. Saradhi, G.S. Saha, K.N. Sahoo and A. Antony. 2004 a. CD-ROM Database on Socio Economic Aspects of Carp Culture in Kolleru Lake. In Roy, A.K., A. Antony and M. Rout (Eds.), Bioinformatics and Statistics in Fisheries Research. VOL-I, CIFA, Kausalyaganga, Bhubneswar, India Pp 119-123.
- Roy, A. K., K. P. Saradhi, K. N. Sahoo, G. S. Sahoo and G. S. Saha. 2004 b. Development of Sampling Frame of Cultivable Water Area in Kolleru Lake of Andhra Pradesh. In Roy, A.K., A. Antony and M. Rout (Eds.), Bioinformatics and Statistics in Fisheries Research. CIFA, Kausalyaganga, Bhubneswar India pp 353-356.
- Roy, A.K., K.P. Saradhi, K.N. Sahoo and G.S. Saha. 2000. Statistics of carp culture in Eluru and Kaikalur Mandals of Kolleru Lake in Andhra Pradesh. The Fifth Asian Fisheries Forum Proceedings AFSIB, Mangalore and AoA, Bhubaneswar, India during 17-22 Jan, 2000.
- Roy, A.K., K.N. Sahoo, K.P. Saradhi and G.S. Saha (2001 b). A socio-economic study on aqua culturist of Eluru and Kaikalur mandals of Kolleru Lake area of Andhra Pradesh, In M Sinha, Dhirendra Kumar and P.K. Katiha (eds.), "National conference on eco-friendly management of resources for doubling fish production strategies for 21 century" held at Central Institute of Capture Fisheries Research Institute (CICFRI), Barrackpore, during 22-23 December 1999. Pp 176-179.

- Roy, A.K.; K.N. Sahoo; K.P. Saradhi and G.S. Saha (2002 b). Farm size and aquaculture productivity relationship. Asian Fisheries Science 15(2000) 129-134, Asian Fisheries Society, Manila, Philippines.
- Sahoo, K. N., A. K. Roy, K. P. Saradhi and G. S. Saha. 2001a. A study on marketing channels of fish produces of Kolleru Lake. Journal of Fisheries Economics and Development, 4(1) pp.35-44.
- Sahoo, K.N., G.S. Saha, A.K. Roy and K.P. Saradhi (2001). Impact of technological transformation in carp culture in Kolleru Lake area of Andhra Pradesh. Proceedings of National Workshop on Fisheries Economics Research and Education in India held at Central Institute of Fisheries Education, Mumbai.
- Sahoo, K.N., A.K. Roy, K.P. Saradhi and G.S. Saha (2000). Break Even Analysis of Carp Culture Practice in Eluru Mandal - A Case Study, In Proc. Fifth Indian Fisheries Forum Eds. S. Ayyappan, J. K. Jena and M. Mohan Joseph (2002), held at CIFA, Bhubaneswar during 17 -22 Jan 2000.
- Sankar, U (1997). Econometric applications in Agricultural Economics. In Econometric applications in India (Eds. K.L. Krishna). Oxford publishers pp 1-52.
- Seshavatharam, V. and Venu, P. 1983. Some observations on Ecology of Kolleru Lake - Wetlands Ecology & Management. (Proceedings of the First International Wetlands Conference, New Delhi, India. pp 35 to 44.
- Seshavatharam, V., *et al.*, 1982. An ecological study of the vegetation of Kolleru Lake. Bull.Bot.Survey of India. 2470-75.
- Seshavatharam, V. and Venu, 1982. Some observations on the ecology of Kolleru Lake. In Brij Gopal *et al.*, (Eds.), Wetland Ecology and Management. Part-II 35-44.
- Sharma, S.V. and P.B. Jaya Raju, 2003. Murrel fishery resources of the lake Kolleru. In Lake Kolleru Environmental Status (Past and Present), M.K. Durga Prasad and Y. Anjaneyulu (Eds.), BS Publications, Hyderabad, pp. 134-137.
- Shivaji Rao, T., 2003. Conflict between development and environment of Kolleru lake area. In Lake Kolleru Environmental Status (Past and Present), M.K. Durga Prasad and Y. Anjaneyulu (Eds.), BS Publications, Hyderabad, pp 168-196.
- Sri Ramakrishnaiah, K., 1980. A case study of Kolleru Lake and Upputeru River. Supdt Engineer, Irrigation Circle, Eluru, Govt. of Andhra Pradesh.
- Srivastava, U. K., 1985. Significance of fisheries. In Inland fish marketing in India. Concept Publishing Company, New Delhi. Vol.3, pp 1-15.
- Srivastava, U.K. (1985). Fish ponds Culture practices and Infrastructure In Inland fish marketing in India. Concept Publishing Company, New Delhi. Vol.3, pp.89-201.
- Veeraiah, K and M.K. Durga Prasad, 1996. A study on the organochlorine pesticide residue input into Lake Kolleru through Tammileru river. Ecol.Env & Cons. (2) 83-86.
- Venkateswarlu, V., 2003. Kolleru lake ecosystem A comprehensive study. In Lake Kolleru Environmental status (Past and Present), M.K. Durga Prasad and Y. Anjaneyulu (Eds.), BS Publications, Hyderabad, pp 158-167.
- Vidhyanath, V., 2003. Planning atlas for integrated resource development and ecozoning of Lake Kolleru. In Ibid., 35-38.
- Vidyanth, V., 1980. On the origin and evolution of Kolleru lake (Andhra Pradesh), India. Indian Journal of Landscape Systems and Ecological Studies, 8(1) 60-65 pp.

18

EFFECT OF INCREASED STOCKING DENSITY ON BLOOD OF COMMON CARP (*CYPRINUS CARPIO*)

Ulfat Jan, G. Mustafa Shah, Aijaz Ahmad Bhat

Abstract: Common carp (*Cyprinus carpio*) were reared in the density of 5, 10, 15 and 20 individuals/ aquarium.The aim of the study was to determine the impact of increased stocking density on blood and glucose of *Cyprinus carpio*. Analysis showed that glucose , red blood corpuscles and haematocrit tends to increase,where as white blood corpuscles tends to decrease. Haemoglobin value remained unchanged. Results observed also indicated that although carp has wide environmental adaptations but increased stocking density influences physiology.

Keywords: *Cyprinus carpio*, stocking density, red blood corpuscles,white blood corpuscles glucose, haemoglobin, haematocrit.

Introduction

Living organisms have developed many strategies to survive and cope with environmental changes to which they are exposed. These environmental constraints known as stressor stimuli, induce short-term predictable physiological changes, particularly in fishes that are ubiquirous inhabitants of the aquatic environment of all latitudes. Stressors can include many kinds of environmental changes; PH (Giles, Majewski and Hibden 1984; Tam Birkett, Makarana, Payson, Whitney and Yu-ck 1987) handling and transportation (Robertson, Thomas and Arnold 1988) Osmotic stress (Stairness, Rainuzzo. Sigholt and Jorgenson 1994) and high stocking stress (Vijayan, Ballantyne and Leatherland 1990). The typical primary physiological stress response is increased secretion of catecholamines (epinephrine) and corticosteroids (cortisol) by the hypothalamie pituitary adrenal arteries (Cech, Bartholow, Young and Hopkins 1996). This response is generally independent of the type of stressor, but the quantitative aspect of the response depends on the intensity and duration of the stressor (Mazeaud, Mazeaud and Donaldson 1977). The initial endocrine responses bring

Post Graduate Department of Zoology, The University of Kashmir Srinagar - 190006; Email: aijazafzal@gmail.com several metabolic adjustments including changes in plasma osmolyte, glucose, and lactate concentrations. Cortisol, the most important cortico steroid in telcosts, contributes to the regulation of gluconeogenesis and/or glycogenesis. The high concentration of cortisol gacors glucose mobilization, providing a substrate for rapid threat response (Vijayan and Moon 1992).

Hypersecretion of catecholamines and corticosteroids also induces hematological changes (Mazeaud, Mazeaud and Donaldson 1977). Decreases in haematocrit and hemoglobin have been reported in fishes subjected to an acute cold stress (Chen, Sun, Lee and Chang 1995).

The purpose of this study is to understand fish hematology and physiological response to stress. We used various stocking densities to induce changes in blood glucose, hemoglobin, haematocrit of *Cyprinus carpio*. The species has wide environmental adaptation results indicated that higher stocking density influences physiology.

Material and methods

Common carp (*Cyprinus capio*) were taken from local fisherman Dal Lake Kashmir. These fishes acclimatized in laboratory conditions with minimal stress. Test fishes were divided into four density groups with 5, 10, 15 and 20 individuals / aquarium with 60 L water capacity. All aquaria were uniformly aerated (*Sobo aereter*). Each aquaria drained and filled with fresh water weakly.

Blood parameters

To perform blood sampling, syringe introduced right in the caudal vessel and blood samples (0.5ml) were taken into vials containing EDTA 0.5gm. Blood glucose was determined by O - Toluidine Method Hemoglobin cances in blood (gm/dl) was determined by sahlis method, haemetocrit was determined by microhamatocrit method. Red blood corpuscles and white blood corpuscles counted by neubaur's chamber.

Results

Changes in blood glucose, hemoglobin and haematocrit concentration for stressed fish are presented Table 1 to 3 respectively. Data shows that plasma glucose tends to increase in fish subjected to high stocking density. Hemoglobin and haematocrit tends to decrease with increase in stocking density. Red blood corpuscles increased with the increase in stocking density, where as white blood corpuscles decreased.

Discussion

As temperature decreases the osmotic concentration and blood viscosity increase. Coping with low temperatures requires the ability to maintain homeostasis. Hematological changes are good indicators of this performance (Chen *et al.*, 1995). The increase of blood glucose is the most studied response

Table 1. Effect of stocking density on Red blood corpuscles and white blood corpuscles of carp (*Cyprinus carpio*) in captivity

Month	AQ 1		AQ 2		AQ 3		AQ 4	
	RBCX 10^6/mm	TLCX 10^4/mm	RBCX 10^6/mm	TLCX 10^4/mm	RBCX 10^6/mm	TLCX 10^4/mm	RBCX 10^6/mm	TLCX 10^4/mm
Nov.	2.8	2.9	2.8	2.9	2.8	2.9	2.8	2.9
Dec.	3.0	2.8	3.0	2.8	3.2	2.7	2.8	3.2
Jan.	3.0	2.6	3.1	2.7	3.2	2.6	3.3	2.4
Feb.	3.0	2.7	3.1	2.6	1	3.1	3.2	2.4
March	3.1	2.6	3.2	2.5	3.2	2.3	3.3	2.3
April	3.1	2.5	3.2	2.4	3.2	2.3	3.3	2.3
May	3.1	2.5	3.2	2.4	3.2	2.3	3.3	2.3
June	3.2	2.5	3.3	2.4	3.3	2.3	3.4	2.3
July	3.1	2.4	3.2	2.3	3.4	2.2	3.5	2.2
Aug.	3.2	2.4	3.4	2.2	3.5	2.2	3.6	2.1
Sept.	3.3	2.3	3.4	2.2	3.4	2.1	3.5	2.1
Oct.	3.3	2.3	3.4	2.2	3.3	2.1	3.5	2.1

Table 2. Effect of Stocking Density on Glucose of carp (*Cyprinus carpio*) in captivity.

In captivity (aquaria)				
Month	AQ 1	AQ2	AQ 3	AQ 4
Nov.	50	50	50	50
Dec.	52	63	74	86
Jan.	56	65	76	87
Feb.	58	67	78	89
March	61	73	82	91
April	63	74	84	93
May	66	75	86	92
June	68	76	88	95
July	69	75	89	96
Aug.	70	78	88	98
Sept.	68	78	86	95
Oct.	66	79	87	96

Table 3. Effect of Stocking Density on haemoglobin of *Cyprinus carpio* in captivity.

In captivity (aquaria)				
Month	AQ 1	AQ2	AQ 3	AQ 4
Nov.	7.5	7.5	7.5	7.5
Dec.	7.3	7.2	7.4	7.5
Jan.	7.2	7.3	7.2	7.2
Feb.	7.0	7.1	7.0	6.9
March	6.9	6.9	7.1	7.2
April	7.1	7.2	6.9	6.8
May	7.2	7.3	7.1	6.9
June	7.5	7.0	7.0	6.8
July	7.1	7.0	6.9	6.7
Aug.	6.9	7.3	7.2	7.7
Sept.	7.3	7.2	7.6	7.4
Oct.	7.4	7.4	7.6	7.4

Table 4. Effect of Stocking Density on hematocrit of *Cyprinus carpio* in captivity.

In captivity (aquaria)				
Month	AQ 1	AQ2	AQ 3	AQ 4
Nov.	41	41	41	41
Dec.	42	42	42	43
Jan.	43	43	42	44
Feb.	44	43	43	44
March	44	44	43	45
April	45	44	44	45
May	46	45	43	46
June	46	45	46	47
July	47	47	45	48
Aug.	47	48	47	48
Sept.	46	46	46	49
Oct.	48	49	50	51

Blood glucose (A), Hemoglobin (B) Haematocrit (C) of common carp (*Cyprinus carpio*) subjected to physiological stress of increase stocking density stocking density. Values expressed as mean. D1 = aquarium first, D2 = Aquarium second, D3 = Aquarium third, D4 = Aquarium - fourth.

in this regard. It is the result of the activation of glycogenolysis that is under cortisol control (Vijayan *et al.* 1992). Hyperglycaemia during cold enposure has been reported in many species (Chen *et al.* 1995; Staurnes *et al.* 1994). In cold treated fish, increased plasma glucose is used mainly as an osmolyte. Salmonid fish including *Cyprinus carpio* survive at low temperature by concentrating their plasma and cellular fluid electrolytes (Na and it) and other osmolyte like glucose.

Crowding is considered a common aquaculture related stressor and produces a wide variety of effects on fish such as reduced growth, health and feed conversion ratio's (Wedemeyer 1997). Two factors can affect fish physiology under crowded conditions. The carrying capacity of water to provide oxygen and to dilute metabolic wastes can act as a stressor. Also the requirement for adequate space limits fish health. The rate of oxygen consumption and ammania excretion are the factors that limit carrying capacity of water (Colt and Orwicz 1991). Decreases in haematocrit and hemoglobin could be the result of blood osmo concentration, as shown by an increase in plasma osmolarity. This osmo concentration leads to hemodilution by body cell water. The decrease in hemoglobin concentration in cold treated fish could be also associated with a dimnution of erythrocyte size (Bollard *et al.* 1993). The increase in osmolyte concentration in the plasma leads to water loss from erythrocyte and thus to their shrinleage. Hemolysis has been reported in common carp (*Cyprinus carpio*) during acute water temperature changes from 8c to 4 c. (Chen *et al.*, 1995).

References

- Bollard B.A, Pankhurst NW, and Wells RMG. Effects of artificially elevated plasma cortisol levels on blood parameters in the telcost fish Pagrus auratus (sparidae). Comp Biochem physiol 106A 157-162, 1993.
- Cech JJJR, Bartholow, SD, Young PS and Hopkins TE. Striped bass exercise and handing stress in freshwater physiological responses to recovery environment. trans Am Fish soc 125 308-320, 1996.
- Chen G-R, sun L-T Lee Y-H, and change C-F. characteristics of blood in common carp *Cyprinus carpio* exposed to low temperatures. J. Appl. A quac 5 21-31; 1995.
- Davino R. Caruso C, Carmadella C. Schinna L. Rutiglarno ME, Romano B, Carratore M. Barra V, and Diprisco G. An overview of the molecular structure and functional properties of hemoglobins of the cold adapted Antarctic teleost. In life under extreme conditions edited by Diprisco G. Berlin, Germany springer.Verag. 199. p. 15-53.
- fletcher GI, Kao MII, and Dempson B. Lethal freezing temperatures of arctic char and other salmonids in the presence of icc. Aquaculture 71 369-378. 1998.
- Giles MA, Majewski HS and Hobden B. Osmoregulatory and hematological responses of rainbow trout (*Salmo giardneri*) to extended environmental acidification. Can J. fish Aquat sci 41 1686-1694. 1984.
- Mazeaud MM, Mazeaud F, and Donaldson EM. Primary and secondary effects of stress in fish some new data with a general review. Trans Am fish soc 106 201-212, 1977.
- Robertson L Thomas P, and Arnold Cr. Plasma cortisol and secondary stress response of cultured red drum (*Sciaenops ocellatus*) to several transportation procedure. Aquaculture 68 115-130, 1988.

- Staurnes M, Rainuzzo JR, sigholt T. and Jorgensen I. Acclimation of Atlantic Cod (Gadus morhus) to cold water stress response, osmoregulation, gill lipid composition and gill Na-K- AT Pase activity. Comp Biochem physiol 109A 413-421, 1994.
- Tam WH Birkett I, Markaran R, Payson PD. Whitney DK and YU CK-C. Modification of carbohydrate metabolism and liver vitellogenic function to brook trout (*Salvelinus fontinalis*) by exposure to low PH. Can. J. fish Aquat sci 44 630-635, 1987.
- Tortora GJ and Adams HH. Laboratory exercise in Human physiology. New York, NY. MacMillan, 1989, p, 241-248.
- Vejayan MM and Moon TW. Acute handling stress alters hepatic glycogen metabolism in food deprived rainbow trout (*Oncorhynchus mykiss*) can J fish Aquat sci 49 2260-2266. 1992.
- Vejayan MM, Ballantyne JS and Leatherland JF. High stocking density alters the energy metabolism of brook char, salveliness fortinalis Aquaculture 88 371-381. 1990.
- Colt J, and Orqiez K 1991. Modding production capacity in aquatic culture systems under freshwater conditions. Aqua Engg. 10 1-29.
- Wedemeyer G.A., 1997. Effects of rearing conditions on the health and physiological quality of fish in intensive culture. In fish stress and health in aquaculture, Cambridge University Press, Cambridge p. 35 -71.

19

SEASONAL IMPACT ON WATER QUALITY INDEX OF GROUNDWATER SAMPLES

*Charmaine Jerome and Anitha Pius**

Abstract: Water Quality Index (WQI) is a unit less number of 100 point scale that provides a pointer to the quality of water source. The aim of the investigation is to assess the quality of ground water regarding its use for drinking purpose. A total of 31 samples were collected from selected bore wells within the industrial area and from villages bordering the study area in all the three seasons -pre-monsoon, monsoon and post-monsoon. The samples were analyzed for various water quality parameters by adopting APHA standard methods. From the data obtained, the WQI was calculated by adopting the method developed by Tiwari and Mishra. The WQI values ranged from 49.2 to 502.1, 20.4 to 261.6 and 32.86 to 446.5 in the pre monsoon, monsoon and post monsoon seasons respectively. On careful examination of the data it was found that most of the water quality parameters were beyond the desired limit prescribed by BIS, making the water unsafe for human consumption and hence pre treatment is essential to make it potable water.

Keywords: Groundwater, Physico-chemical parameter, WQI

Introduction

Water is a precious yet non-renewable resource. Water in sufficient quantity and quality is necessary for the well being of all living organisms. Water quality is the physical, chemical and biological characteristics of water in relationship to a set of standards. With rapid growing population, industrialization and improved living standards the pressure on water resources is increasing. Proper management and protection of urban ground water quality has been a major concern in many countries 2-5 because the degradation of urban ground water affects human health and economics.

Department of Chemistry, Mount Carmel College, Bangalore, Karnataka, India,
Department of Chemistry, Gandhigram Rural Institute, Gandhigram-624302 Tamil Nadu, India

Contamination of drinking water may occur by percolation of toxics through the soil to ground water that is used as a source of drinking water, therefore basic monitoring of water quality is essential to observe the demand and pollution level of groundwater. An integral part of any environmental monitoring programme is the reporting of results to both policy makers and the general public. This poses a particular problem in the case of water quality monitoring because of the complexity associated with analyzing a large number of measured variables. One possible solution to this problem is to reduce the multivariate nature of water quality data by employing an index that will mathematically combine all water quality measures and provide a general and readily understood description of water. Water Quality Index has been regarded as one of the most effective ways to assess the quality of water and is communicated on the basis of calculated water quality indices. WQI was calculated by weighted index method to determine the suitability of ground water for drinking purposes.

Geographic Information Systems (GIS) has grown rapidly and finds application in groundwater management and research. The spatial patterns of chemical constituents are useful in deciding the water use strategies for various purposes.The present study is an attempt to generate baseline groundwater quality data and to apply GIS techniques to determine the water quality of the study area during the three seasons.

Materials and Methods

Study area

Bangalore is located at a latitude of 12°58' N and longitude of 77°35'E at an altitude of 921m above sea level. The Peenya industrial area (Fig.1) is located on the north-western suburbs of Bangalore city between 130 01' 42" N and 770 30'45"E about 20km from the heart of Bangalore on National Highway. It is the region's largest enclave of industrial units containing over 2000 industries that sprawl over an area of 40 sq.km. The total population within the zone is 3,50,000.

Spatial Database Construction and selection of sampling wells

Survey of India (SOI) toposheet on 150,000 scale was used to prepare the base map and drainage map to understand the general nature of the study area. The toposheet was scanned and digitized to generate a digital output using MapInfo GIS software. A total of 31 representative locations were selected from the entire study area located within the Peenya Industrial area and he surrounding villages, namely, Laggere, Dassarahalli, Chokkasandra, Shivapura and adjacent areas like, Jallahalli cross, Rajagopalnagar and Mahalakshmi layout. Groundwater from bore wells were collected from these locations during the pre monsoon (April 2008), monsoon (August 2008) and post monsoon month (Jan 2009) spreading over a period of one year. The geographical location of the bore wells were recorded with the help of Global Positioning System (GPS)-GARMIN 60. Sample collection

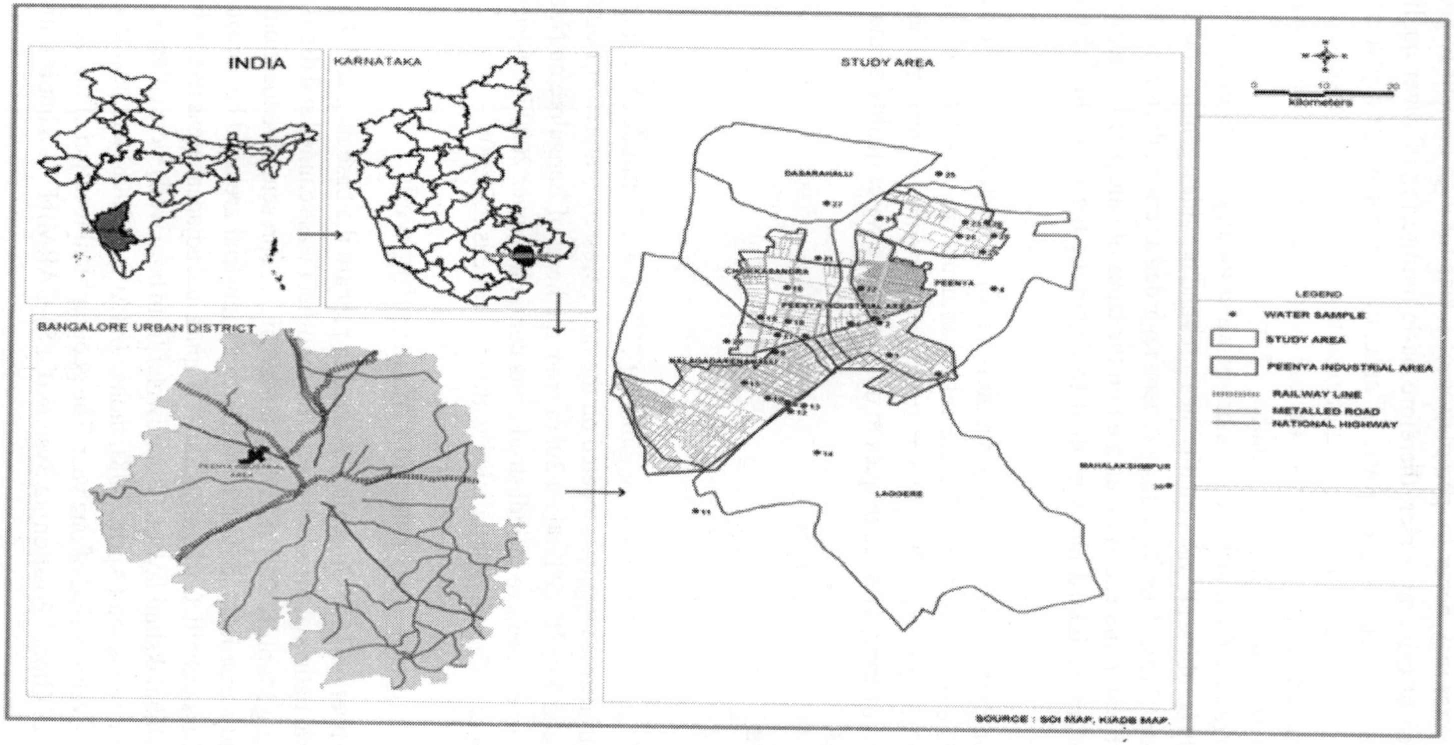

Fig 1. Map Showing Location of Study area with source of water sample

was done in clean, acid pre-washed, high density polythene bottles. The collection, transportation and preservation were done according to standard procedures. The various water quality parameters, like pH, electrical conductivity, turbidity, total dissolved solids (TDS), alkalinity, total hardness (TH), calcium hardness, magnesium hardness, chloride, fluoride, nitrate and sulphate were analyzed in the laboratory according to the standard procedures.

Spatial Distribution Maps

The spatial and the attribute databases generated were integrated for the generation of spatial distribution maps of selected water quality parameters required for the calculation of WQI.

Calculation of WQI

Water quality index was calculated by weighted index method to determine the suitability of groundwater for drinking purposes.

The index developed by Tiwari and Mishra (1985) was used. In the present study nine water quality parameters, namely, pH, alkalinity, TDS, Total hardness, Ca hardness, Mg hardness, nitrate, chloride and sulphate were considered for computing WQI and the unit weight Wi of each parameter is obtained depending upon its weightage, by adopting the following formula

$$WQI = (\Sigma qi\ Wi) / (\Sigma Wi)$$

Where $qi = 100(Vi/Si)$

$$qpH = 100\{(VpH - 7.0) / (8.5 - 7.0)\}$$

$$Wi = K/Si$$

Where qi = Quality rating for the ith water quality parameters (i=1, 2, 3,….N)

Vi = the measured value of the ith parameter at a given sampling location

Si = the standard permissible value for the ith parameter

The standard permissible values for various pollutants for drinking water, recommended by WHO are given in Table 1- for the parameters considered for WQI.

It is well known that the more harmful a given pollutant is, the smaller is its permissible value for the standard recommended for drinking water. So the "weights" for various water quality parameters are assumed to be inversely proportional to the recommended standards for the corresponding parameters i.e.; Wi = K/Si where Wi = unit weight for the ith parameter and K = constant of proportionality. For the sake of simplicity, assuming that K = 1, for pH, assuming the same unit weight as that for chlorides; viz., 0.005. The unit weight Wi, obtained from the above equation with K = 1, are shown in Table 1. According to this Water Quality Index, the maximum permissible value is 100. Values greater than 100 indicate pollution and are unfit for human consumption.

Table1 Permissible levels and unit weight of various parameters

Parameters	Recommended limit (Si)	Unit weight
pH	7.0-8.5	0.005
TDS	500	0.2
Calcium	75	0.0133
Magnesium	50	0.02
T-hardness	300	0.0033
Chloride	200	0.005
Nitrate	45	0.022
Sulphate	200	0.005
Alkalinity	200	0.005

(All units except pH are in mg/l)

Table 2. Water quality parameter data for pre monsoon season

Sample	pH	Alkalinity	TDS	TH	Ca	Mg	Chloride	Sulphate	Nitrate
1	7.5	310	1459	518	482	8.748	494.16	174.49	24.3
2	7.02	350	490	558	228	80.19	288.26	179.9	63.3
3	7.13	200	568	920	760	38.88	426	43.69	78.1
4	6.82	255	1747	396	182	52.002	462.4	97.571	62.5
5	7.96	220	780	248	78	41.31	31.24	16.8	65.35
6	6.8	200	2304	558	512	11.178	494	89.405	30.7
7	7.88	255	742.4	1160	710	109.35	387.1	210.6	32.7
8	7.9	310	899	360	310	12.15	212	38.1	26
9	6.42	565	2020	1824	1457	89.181	910	255.6	84.7
10	6.58	485	2435	1060	984	18.468	795	260.1	18.4
11	7.96	315	1619	296	212	20.412	329.44	58.64	18.5
12	6.34	355	4224	1424	1416	1.944	393.3	216.4	49.6
13	6.01	385	4256	1368	98	308.61	724.4	224.39	136.1
14	7.88	670	2995	784	142	156.01	501.1	67.57	84.9
15	8.18	290	1785	912	290	151.15	213.3	63.83	1.9
16	7.62	325	920	340	296	10.692	348	71.95	43.05
17	6.48	365	1012	418	398	4.86	36.9	17.255	46.4
18	6.84	80	1650	212	74	33.534	74.8	13.476	110.7
19	7.16	520	960	298	160	33.534	375.5	308.12	55.7
20	7.56	540	2135	494	168	79.218	629.06	316.85	94.05
21	7.21	285	1900	1050	542	123.44	411	321.77	99.7

22	7.9	295	1500	1242	94	278.96	379	47.962	95.7
23	7.2	220	675	252	86	40.338	79.1	17.552	169.6
24	7.5	275	1788	202	194	1.944	316.5	13.746	49.6
25	8.01	300	925	518	334	44.712	243.6	47.842	105
26	8.2	255	745	724	176	133.16	278.6	71.97	1.58
27	8.5	225	1658	96	76	4.86	76.68	36.216	6.5
28	8.2	315	2130	528	434	22.842	386.2	42.99	2.3
29	8	400	1082	944	462	117.13	198.8	18.216	159
30	7.6	425	486	486	96	94.77	248.3	48.104	6.95
31	6.8	465	1384	1384	1046	82.134	208	56.148	141

All units except pH are in mg/l

Table 3. Water quality parameter data for monsoon season

Sample	pH	Alkalinity	TDS	TH	Ca	Mg	Chloride	Sulphate	Nitrate
1	7.3	26	345	202	198	0.972	116	53	4
2	7.8	48	74	426	64	87.966	132	84	4.1
3	7.4	86	96	712	316	96.228	344	15	4
4	7.3	91	1036	264	97	40.581	346	64	4
5	7.1	34	96	158	10	35.964	18	9	4
6	7.3	53	1036	422	214	50.544	249	53	ND
7	7.6	32	186	788	198	143.37	224	161	4
8	7.6	41	195	120	100	4.86	98	12	4
9	7.8	12	996	1264	624	155.52	600	102	ND
10	7.9	41	486	698	242	110.81	428	104	2.1
11	7.1	15	486	200	64	33.048	24	26	2.5
12	7.2	90	1612	880	568	75.816	304	90	2.1
13	7.6	42	1500	912	16	217.73	428	94	4.2
14	7.4	54	890	464	69	95.985	346	44	0
15	7.3	102	1063	624	96	128.3	116.9	32	5.8
16	7.5	18	526	198	82	28.188	128	28	ND
17	7.1	32	824	206	197	2.187	16	7	3.8
18	7.4	36	900	196	13	44.469	12	5	2.6
19	7.6	182	486	114	63	12.393	210	182	6
20	7.7	15	1051	268	54	52.002	374	166	0
21	7.1	28	810	692	244	108.86	304	153	2.5
22	7.3	64	720	720	22	169.61	274	26	2.1

23	7.1	16	298	112	17	23.085	18	13	3.4
24	7.3	84	983	96	26	17.01	84.7	4	5.8
25	7.2	94	495	190	124	16.038	69.3	29	4.6
26	7.6	92	325	210	58	36.936	127.4	41	8.5
27	7.28	86	904	58	8	12.15	12.8	22	4.2
28	6.9	112	1051	314	213	24.543	196	34	7.3
29	7.4	244	274	422	204	52.974	114.9	12	8.2
30	6.68	268	208	208	41	40.581	89.6	38	4
31	7.3	295	774	774	548	54.918	111.7	27	14.3

All units except pH are in mg/l

Table 4. Water quality parameter data for post monsoon season

Sample	pH	Alkalinity	TDS	TH	Ca	Mg	Chloride	Sulphate	Nitrate
1	7.38	295	490	484	248	57.35	183.18	108.86	23.01
2	6.8	305	150	536	96	106.9	214.4	108.87	51
3	6	150	390	896	548	84.56	392	31.25	75
4	6	215	1185	340	148	46.66	418	96.064	54
5	8	210	165	196	14	44.23	29.8	13.545	52
6	6	180	1370	522	320	49.09	308	83.246	28
7	7.87	240	390	1100	622	116.2	297	191.37	30.4
8	8	250	470	344	176	40.82	105.1	24.52	21
9	6	500	1565	1636	1294	83.11	793.78	240.96	78
10	6.5	440	1565	884	586	72.41	535.3	234.08	10.1
11	8.5	250	770	252	130	29.65	68.16	48.736	9.92
12	6	320	3765	1240	998	58.81	342	191.38	45.1
13	5.5	330	3920	1208	36	284.8	507	191.37	131
14	8	640	3015	676	96	140.9	423.16	62.068	66.2
15	8.23	260	1215	896	198	169.6	157.62	57.444	0.998
16	7.48	255	592	336	150	45.2	297	62.02	28
17	6	310	645	414	298	28.19	21.2	9.736	31.5
18	6.67	40	1495	180	14	40.34	18.4	9.734	9.4
19	6.5	485	728	216	106	26.73	274	234	35
20	7.18	495	1640	464	104	87.49	568	309.96	67
21	7.16	240	1280	936	298	155	336	309.82	59.2

22	7.61	260	895	1148	40	269.2	344	34.734	69
23	6	215	485	190	30	38.89	38.3	14.517	130
24	7.31	230	1390	184	98	20.9	115	9.73	20
25	8	240	590	316	240	18.47	107.9	31.188	92.5
26	8.08	235	570	600	98	122	198.8	54.71	0.78
27	8.44	205	1390	68	16	12.64	31.2	25.302	2.1
28	8	260	1805	506	260	59.78	247.1	35.278	1.74
29	7.32	385	694	796	320	115.7	157	14.517	74
30	7.09	405	440	440	74	88.94	153.4	40.762	1.12
31	6	420	1116	1116	842	66.58	162	34.738	130

All units except pH are in mg/l

Table 5. WQI for different seasons

Sample No.	Pre monsoon	monsoon	post monsoon
1	171.710	58.175	125.677
2	163.843	77.732	131.777
3	270.582	145.115	235.845
4	142.378	69.289	117.494
5	96.958	27.363	72.144
6	179.219	95.142	139.159
7	279.629	143.940	255.799
8	120.407	36.570	94.578
9	502.102	261.664	446.576
10	309.057	142.654	225.933
11	105.726	39.501	74.070
12	409.939	199.424	333.816
13	334.172	153.576	290.263
14	227.644	88.763	194.256
15	183.359	110.890	166.038
16	131.405	44.714	100.897
17	139.742	55.707	113.942
18	116.026	36.859	39.370
19	132.434	47.813	93.066
20	199.063	66.491	164.661
21	297.131	137.880	225.492

22	267.497	115.913	226.455
23	159.778	22.101	114.339
24	106.248	30.416	65.717
25	193.724	49.810	141.427
26	142.767	51.207	112.933
27	49.210	20.482	32.868
28	153.794	83.063	124.510
29	297.993	96.287	205.563
30	106.039	46.308	88.025
31	404.097	185.401	333.087

The values for the WQI ranged from 49.2 to 502.1 in pre monsoon season, 20.4 to 261.6 in the monsoon season and 32.86 to 446.5 in the post monsoon season respectively.

Results and Discussion

The water quality parameters used in the calculation of WQI for the pre monsoon, monsoon and post monsoon seasons and the calculated values of WQI for the three seasons are presented in Tables 2-5 respectively.

Twenty nine samples recorded values greater than 100 in the pre monsoon season, ten samples showed values greater than 100 in the monsoon season while twenty three samples had values greater than 100 in the post monsoon season. Sample 9 has the highest value of WQI in all seasons while sample 23 shows the least values. Spatial distribution of water quality index for the three seasons are presented in Fig. 2. High Water Quality Index values are seen in the western part of the study area. In monsoon season the WQI is within the permissible limit for a greater part of the area, while in pre and post monsoon seasons the WQI shows higher values for a greater area. This variation can be seen in the stack overlay of WQI for these seasons.

Spatial distribution of different parameters can be compared with the drainage map of the area (Fig. 2). It could be seen that stream network is running towards the west which indicates a westerly slope of the area. The concentration of different parameters is seen to be high in this area. Thus the distribution pattern of the parameters is controlled by the drainage network in this area. Sample 9 which lies towards the western side of the study area shows the highest value forWQI while sample 23 is in the northern part of the study area shows the least value of WQI for all seasons. This can be attributed to a larger number of industries concentrated in the vicinity of sample no.9 when compared to sample No.23 (Fig. 3).

The seasonal variation of WQI at all the sites showed the increasing trend from monsoon to post monsoon to pre monsoon seasons.

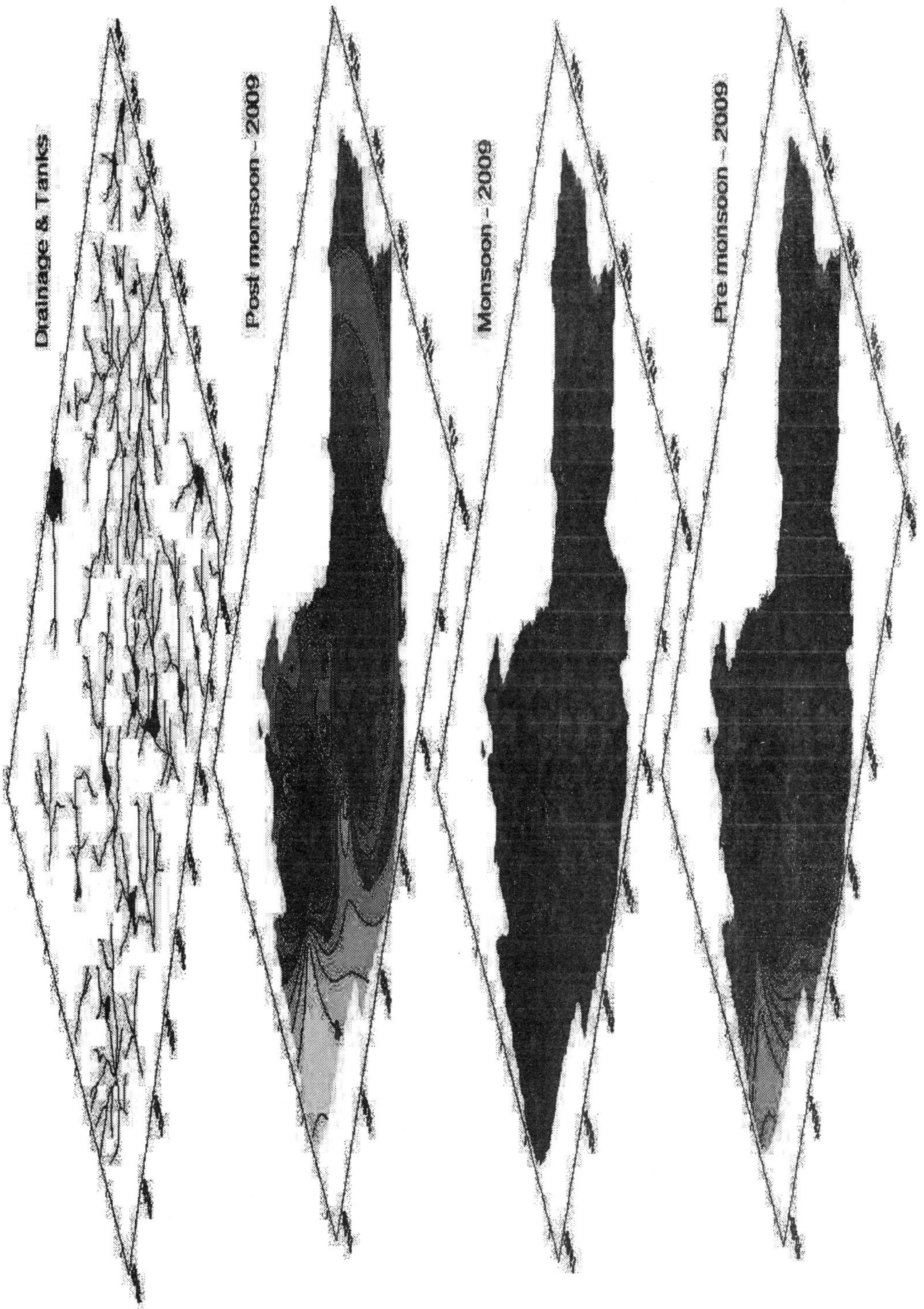

Fig. 2 Water Quality index stack overlay of Pre Monsoon and Post Monsoon.

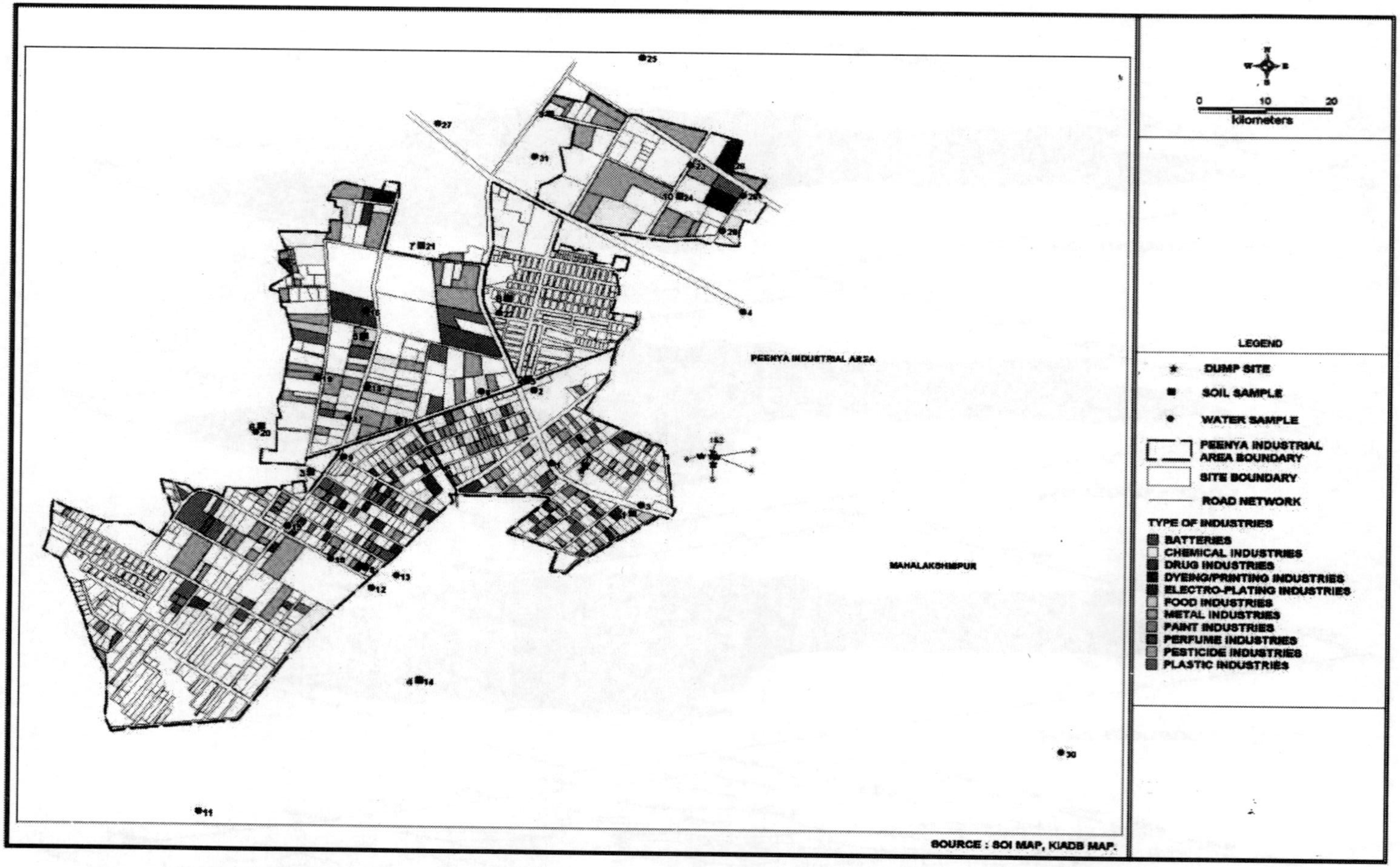

Fig. 3 Map Showing Major Types of Industries

Conclusion

Low values of water quality index during monsoon season may be attributed to dilution effect of rains. Drainage maps indicate a westerly slope which can be correlated to high WQI values of the samples that belong to the western part of the study area.

References

- Kumar, Rajesh, R.D.Singh and K.D.Sharma (1999). Water resources of India.Curr.Sci., 89 (5) 794-811.
- Niemczynowicz, J. (1999). Urban hydrology and water management-present and future challenges. Urban water., 1 1-14.
- Collin, M.L., Melloul, A.J. (2001). Combined land use and environmental factors for sustainable groundwater management. Urban water., 3 226-237.
- Foster, S.S.D. (2001). The interdependence of groundwater and urbanization in rapidly developing cities. Urban Water., 3 185-192.
- Powell, K.L., Taylor R.G., Cronin, A.A., Barrett, M.H., Pedley S., Sellwood, J., Trowsdale, S.A., Lerner, D.N. (2003). Microbial contamination of two urban sandstone aquifers in the U.K. Water Res., 37 339-352.
- Pradhan, S.K., Dipika Patnaik., Rout, S.P. (2001). Water quality index for the groundwater in and around a phosphatic fertilizer plant. Indian Journal of Environmental Protection., 21 355-358.
- Tiwari, T.N. and Mishra M. A (1985). preliminary assignment of water quality index of major Indian rivers. Indian J.Env.Prot. 5 (4) 276-279.
- Bonham-Carter, G.F. (1996). Geographic information systems for geoscientists; Modeling with GIS computer methods in the geosciences. Elsevier Science Ltd.Pergamon, 13 1-50.
- Shivashankara, G.P., Ranga, K., Ramalingaiah and Manamohan Rao (1999). Characteristics of bulk precipitation in industrial areas of Bangalore city. Indian J.Environ.Health 41 229-239.
- APHA (1998). Standard methods for the examination of water and waste water, 20th edn.,American Public Health Association, Washington, D.C.
- Environmental Indicators State of the Environment Report, (2007). Water Resources, IV, 27-60.

20

STUDIES ON WATER POLLUTION WITH SPECIAL REFERENCE TO INDUSTRIAL EFFLUENTS AT DEWAS, MADHYA PRADESH, INDIA

*Sapana Malviya and Manoj Kumar Sisodiya**

Abstract: One of the most serious problems faced by mankind today is the problem of water pollution, billions of galions of water from cities and housing settlements, industries are throw every day more and more of our water resource is slowly becoming unfit for most of the uses. Lack of waste water treatment facilities or their inefficient is the single great test cause of gross pollution of fresh water. After India became independent the scientific studies took new roots and flourished. Studies on water pollution were carried cut by many workers in Indian rivers. The river ecology is very much important in the Indian soil, science it forms a major sowice of water for drinking purpose, other than many other uses recently two resource work on Indian river waters becomes more progressive and many reports are available. The availability of drinking water and its sources are rare, particularly in Malwa region of (M.P.). Many river lack of the state are getting highly polluted. The water quality study of Kshipra River BOD, COD, Solids (T.S.) Nitrogen's , T.K.N. NH4 organic carbon soil sampling water holding capacity, biomass and other quality and quantative results.

Keywords: Unfit water resource waste water, Pollutant water, Degradation

Water pollution is a major problem of the word. It has been suggested that it is the leading world wise calise of deaths and diseases and that it account for the death of more than 14,000 people in addition to the acute problem of water pollution in our countries to struggle with pollution problem are as well. In the most recent national reports on water quality in our day by day decrying water is typically referred to as polluted when it is impaired by Androgenic contaminants and either does hot support a human use like serving as drinking water and undergoes a marked shif in its ability to support its constituent biotic communities such as fish. Natural phenomena such as volcanoes, algae bloom, Storms and earth quakes also cause major changes in water quality and ecological

Asst. Prof. (Botany), Department of Botany, Govt. Bharulal Patidar P.G. College, Mhow, Indore (M.P.)
Department of Botany, Govt. Shri Krishanaji Panwar, P.G. College, Dewas (M.P.)

status of water pollution has many cause and characteristic. Surface water and ground water are generally two categories based on their origin.

Ground water pollution is much move difficult to than surface pollution because ground water can move great distances through unseen. Aquifers non porus aquifers such as clays. Partially purify water of bacteria by simple filtration dilution and in same cause chemical reaction and Biological activity in same cases the pollutant merely transform to soil contaminants ground water that move through cracks and caverns in not filtered and can be transferred as easily.

Analysis of ground water contamination may focus on soil characteristics and hydrology as well as the nature of the contaminant and itself. The specific contaminants leading to pollution in water inculde wide spectrum of chemicals pathogens and physical or sensory changes such as elevate temperature and discoloration. While many of the chemical and substances that are regulated may be naturally occurring example, calcium sodium, iron, manganese etc. The concentration is often the key is determining what is a natural component of water, oxygen - depleting substance may be natural materials such as plant matter as well as man made chemicals. Other natural and anthropogenic substanceless may cause turbidity which block light and disrup plant growth and clogs the gills of same fish species. Many of the chemical substance are toxic pathogens can produce water borne diseases in either human or animal hostes. Alteration of water physical chemical include acidity change in PH. Electrical conductivity, temperature and Eutrophication, Eutrophiation it the fertilization of surface water by nutrients that were previously scare. Other microorganisms sometime forenda in surface water which have caused human health problems include.

The Kshipra river water is also used in fisheries, industries irrigation and other general purposes Kshipra River is a main source of drinking water for human being cattles and other animals. The ghats of river are used dhobies, kombar, brick industries and public for washing and bathing there by adding large amount of soap and detergents in river water.

The major source of pollution in the river is the effluent discharged by factories of Dewas industrial area. The well established industrial complex includes factories like small scale, large scale. Which are related chemical, textile, leather, food processing engineering, Auto parts.

Ground water pollution same time referred to as ground water contaminates is not as easily. Classification as surface water pollution by its very nature, ground water aquifers are sucefible to contaminations from source that may hot directly affect. Surface water bodies may be irrelevant a spice of a chemical contaminant on soil located away from a surface water body may not necessarily create point.

Analyzed Dewas water through several broad categories of methods physical, chemical and biological most method involve collection of samples some methods may conducted *in situ* without sampling such as temperature. Sampling of water for physical or chemical testing can be done by several methods depending on the accuracy needed and the characteristics of the contaminant.

Table 1

S.No	Name	Product	Type
1.	Tata Exports Ltd.	Leather	Tanning
2.	Naveen Chemical	Methgi	Chemical
3.	S.Kumars Enterpries	Cloth	Textile
4.	Prestige food Ltd	Soybean	Food
5.	Singhvi food Ltd	Wead	Food
6.	Ranbaxy Laboratories Ltd	Pharmaceutical	Chemical
7.	Dewas Synthetics	Cloth	Textile
8.	Deepak wools Pvt. Ltd	Blankets	Textile
9.	M.P. woolen mill	Blankets	Textile
10.	Nayan synthetics	Blankets	Textile
11.	Padma synthetics Pvt. Ltd	Blankets	Textile
12.	Gajra Gears Pvt. Ltd	Gears	Engineering
13.	Auto Forgo Pvt. Ltd.	Steel	Engineering
14.	Johnson pedders	Sanitary wares	Ceramics
15.	H & R Johnson	Ceramic	Ceramics
16.	Hind filters	Filter	Engineering
17.	Subham Synthetics	Sanitary wares	Ceramics

Many contamination events are sharply restricted in time most commonly in association with rain water. Scientists gathering data often employ auto sampler devices that pump in cremounts of water at either time of discharge interval. Sampling for Biological testing involves collection of place and animal from the surface water body depending on the type of assessment the organism may be identified for Biosurveys and Returned to the water body.

Water samples may be examined using the principles of analytical chemistry. Many test methods are available for both organic and inorganic compounds. Mostly used method include PH, Biochemical oxygen demand (BOD) Chemical oxygen demand (COD) nutrient (Nitrate and phosphorus compound metal including copper tine, cadmium lead and mercury oil and grease total petroleum Hydrocarbon and pesticides.

Some Dewas industrial facilities generate ordinary domestic sewage that can be treated by municipal facilities industries that generate waste water with High concentrations of conventional pollutant it toxic pollutants heavy metals volatile organic compound on other non conventional pollutant such as ammonia head specialized treatment systems industries generating large volumes of waste water typically operate their own complete on site treatment systems some industries have been successful at redsiging their manufacturing processed to reduce or eliminate pollutants through a process called pollution prevention.

Concluding, we can say that the problem of water pollution is increasing day by day by this Dewas region is also affected. The harmful gases and Poisonous effluent releasing from Industries not only pollutes water but also make the water poisoned which is unusable for humanbing. The graph of negative impact on human health related to it, is increasing day by day.

21

BIODIVERSITY AWARENESS TO ENRICH PEOPLE'S LIVELIHOOD

Sheela Shrivastava

Absrtact: Rich biodiversity has tremendous potential to enrich people's livelihood and biotechnological advancement for food and livelihood security. Now there is a need for us to work for value addition development for favourable conditions to protect this potential bio-resource. Biodiversity is threatened in various ways. The major cause for the loss of biodiversity is global warming which has become global crisis today. Global warming is the result of emission of green house gases as methane, carbon-di-oxide, nitrous oxide, water vapour, carbon mono oxide, nitrogen oxides, aerosol etc. These trace gases influence global warming directly or indirectly. Global warming refers to an average increase in the Earth's temperature which in turn causes change in climate. Alteration in climate is a major factor leading to the depletion of biodiversity as environment has direct impact upon all the living beings and controls their growth, reproduction, survival and dispersal.Though we cannot avoid all the consequences of global warming but this is the time for us to think seriously that what can we do to combat climate change caused by global warming.

Keywords: Biodiversity, Green house gases, Global warming.

Rich biodiversity has tremendous potential to enrich people's livelihood and biotechnological advancement for food and livelihood security. To protect this tremendous potential bio-resource we shall work for value addition development to strengthen the favorable conditions. Human activities are responsible for emission of green house gases which in turn resulting in global warming. One of the major cause of loss of biodiversity is global warming.

What is global warming

Global warming refers to an average rise in the earth's temperature which in turn cause change in climate like alteration in rainfall pattern, rise in sea level, impacts on plants, wild life, ecology and

Professor of Zoology, M. L. C. Govt. Girl's P.G. College, Khandwa (M.P.)

human beings. Most of the warming over the last 50 years is the resultant of human activities which has altered chemical composition of atmosphere through green house gases which have heat trapping properties resulting in to temperature elevation of earth surface. This rise in temperature threatens dangerous consequences like drought, floods, lost ecosystem, rise in sea level, altered weather that can cause intense precipitation or dry spells and more severe storms. Environment has an impact upon all the organisms and controls their growth, reproduction, survival, behaviour and dispersal. At risk are our coastal properties and resources, the productivity of our farms and fisheries and the livability of our cities in summer.

What causes global warming

Earth's atmosphere mainly consists of oxygen and nitrogen. Both of these are transparent to terrestrial radiations and do not contribute to global warming. The other trace gases in the atmosphere basically cause rise in earth's temperature. These gases include carbon dioxide (CO_2), nitrous oxide (N_2O), methane (CH_4), water vapour (H_2O), ozone (O_3), halocarbons, nitrogen oxides (NO_x), sulfur hexa fluorides (SF_6), carbon mono oxide (CO), per fluorocarbons(PFCs), aerosols and non methane volatile organic compounds (NMVOCs). These trace gases influence global warming directly or indirectly.

Sources of Green House Gases

Carbon dioxide - Carbon dioxide is resultant of burning of solid wastes, wood and wood products, fossil fuel etc. On the basis of carbon foot prints the quantity of CO_2 emission per capita of different countries is-

SR NO.	NATIONS	QUANTITY OF CO_2 EMISSION PER CAPITA
1.	America	29
2.	Australia	21
3.	Canada	20
4.	Finland	18
5.	Switzerland	18
6.	Netherland	17
7.	Ireland	16
8.	Cyprus	16
9.	China	3.1
10.	India	1.8
11.	Bangladesh	1.1
12.	Mozambia	0.7

Above table shows that advanced countries are responsible for higher quantity of CO_2 emission.

Methane - Methane is primarily produced through anaerobic decomposition of organic matter in biogenic systems. Agricultural processes like rice cultivation, enteric fermentation in animals and decomposition of animal wastes also emit methane. It is also produced during production and treatment of natural oil and gases. Global warming potential of methane is 24 times higher than that of CO_2. The atmospheric level of methane is reported to have gone up from 0.8 to 1.75 ppmv. Man-made methane contributes 18. 337 % and the rest is of natural origin.

Nitrous oxide - Nitrous oxide is emitted during agricultural and industrial activities and during combustion of solid wastes and fossil fuels. Global warming potential of N_2O is about 170 to 190 times higher than that of CO_2. The present concentration of N_2O is 334 ppmv as compared to 270 ppmv of preindustrial time. Only 4.933% of the total is man-made and the rest is of natural origin. Per-fluorocarbons - result as a by-product of aluminium smelting and also alluring uranium enrichment process. Global warming potential of per-fluorocarbon is about 6000 to 10000 times higher than that of CO_2.

Sulfur hexafluoride - Sulfur hexafluoride is a man-made gas and has a warming potential of 25000 times higher than CO_2. It is mainly used for insulating material for high voltage equipments like circuit breakers and utilities. It is also used for cable cooling systems.

Water vapor - Water vapor is the most significant green house gas which comes from natural source and is responsible for nearly 95% green house effect. About 99.99% water vapor is of natural origin.

Effects of Global Warming

Beside loss of biodiversity global warming affecting world environment in the following way:

- Alarming rise in sea water level to speed up of melting of glaciers and ice caps.
- Intense heat waves generation causing severe health effects.
- Adverse effects on ecosystems and loss of biodiversity due to rise in temperature and due to non adaptation.
- Increase in incidences of drought, cyclones, floods, huricanes etc.
- Breaking out and spread of new diseases.
- Loss of water due to excessive evaporation.
- Increased risk of wild fire.

Global warming may cause biodiversity extinction. Extinction is a hotly debated but poorly understood and employed topic in science. Many times scientists try to forecast the impact of global change on future biodiversity. Some argue that species have managed to survive worse climate change in the past and that current threats to biodiversity are overstated but the climate change is so fast for survival of species. Many biologists say the impacts of climate and resulting shifts in rainfall, temperature, sea level, ecosystem composition and food availability will have important effects on global species richness.

The adverse effects have already begun and this may lead to catastrophic solution unless we put a strict check on the possible causes. The solutions are well in sight but there is a need of strong will to fight the problems. Some of the efforts and alertness can do something to save our environment.

Coffee - One cup of black tea is responsible for emission of 175 kg carbon. By avoiding one cup of coffee one can check emission of above amount of carbon.

Tissue paper - Recycling of 1 kg tissue paper needs 30 litres of water and 3-4 KW electricity. Thus 1 kg tissue paper is responsible for 1.5 to 2.0 tons of carbon emission.

Decaying of food - At global level decaying of potato produces 159000 tons methane gas and cooked rice and pasta are producing 78000 tons and meat producing 160000 tons methane. Thus we should have a strict check on decaying of food.

Cloth -There should be limited use of synthetic clothes. During recycling of cotton cloth we can check 32.5 kg carbon emission, Adoption of car pools, Regular vehicle emission monitoring, Sorting of organic garbage. Plantation, Rainwater harvesting, switch off stand by computers and T.Vs., Use public transport, Save paper, Avoid use of plastic bags. Waste less, Adopt a green way of life.

Above are the some small steps to check global warming. We must adopt them.

22

STUDY OF PAISUNI WATERSHED AREA OF CHITRAKOOT REGION, MADHYA PRADESH & UTTAR PRADESH, INDIA THROUGH REMOTE SENSING AND GIS

Shashikant Tripathi, Sandeep Kumar Soni, Abhishek Kumar Maurya

Abstract: In accordance with their nature, potential and problems with best amount of interference, watershed management approach is aimed at integrated development. For surface water resources and ground water resources, the proper assessment of water potentials helps in additional exploitability at optional level. This objective of watershed management involves geological, hydro-geological consideration and economic, political, financial aspect. For the formulation of proper management programs, a reliable and upto-date information about various factors, viz. size & shape of watershed, topography, soil and their characteristics, land use/land cover, drainage parameters etc., are required. To obtain synoptic view and repetitive coverage, Remote Sensing technique provide a powerful input media, unbiased reproduction of the natural features in the form of photography/imagery and thereby economizing the process of multi-disciplinary approach for planning & decision of land and water resources in a watershed for integrated development. A resource information system is created in a suitable GIS environment for the selected watersheds. This system evolves a sustainable plan for optimum development of the water resources and for finding solutions for different management problems related to natural resources i.e. water resources.

Keywords: Water shed management, Financial aspects, GIS environment

Introduction

Watershed is defined as the entire geographical area drained by a river and its tributaries. It can be small or large. Watershed is associated with each & every stream, tributary or river. Many small watersheds join to form larger watersheds. Watersheds may be delineated very easily through Remote

Pradeep Kr Soni, Remote Sensing & GIS Lab MGCGV Chitrakoot Satna M.P.; Email: sandeepsoni80@gmail.com, avi.maurya@gmail.com

Sensing & GIS Technique. Whereas watershed management is defined as to plan and work toward an environmentally and economically healthy watershed that benefits to all. Watershed management may be defined as process to develop projects, programmes and implementing in to act, to sustain and increase watershed functions that affect the human, plant & animal communities around watershed boundary. Characteristics of a watershed that agencies seek the overall management planning and utilization of watersheds including water supply, water quality, drainage, water run-off & water rights.

At small catchment level, watershed management gain momentum to usher in overall development. There are three general objectives for watershed planning and management i.e. Rehabilitation, Protection and Enhancement. To recognize these objectives, it is necessary to measure soil and water conservation as a part of watershed management strategies. If any area, due to poor distribution of water or lack of sufficient amount of water with high evaporation, it is characterized as arid and semi-arid regions. This calls for managerial solutions involving water conservation and the augmentation of groundwater recharge. There is a basic principle of land and water management that lies in land use practices that ensure maximum rainfall infiltration and control soil and nutrient losses. In many ways, the hydrologic effects of land use change are manifested at different spatial and time scales in all geographical areas.

Due to increasing pressure on land and water, comprehensive water resources and watershed management practices are becoming urgent. Now a day it is great problem of ground water management. Different agencies and water resource managers are working for the adaptation and implementation of technologically advanced tools like Remote Sensing and GIS. The technology is used for reduction of water waste, better and more efficient resource allocation and environmental protection. GIS is used to evolve a sustainable planning for optimum development of the water resources and for finding solutions for water resource management.

Water in the basin arrives in the form of precipitation as part of the water cycle (hydrological cycle). Some precipitation returns to the atmosphere, having been intercepted by vegetation and evaporated from the surfaces of leaves and branches. More is lost to evaporation from the ground surface and transpiration by plants. In arid and semi-arid climates all of the precipitation may be consumed in this way for most of the time; basin run-off occurs only occasionally, following intense storms. Where precipitation exceeds losses to evapo-transpiration, the excess water makes its way through the drainage system. Its rate of progress is not uniform, however; water may be stored in lakes, soils, and as groundwater for considerable periods before it eventually arrives at the outlet, or basin channel, as basin run-off.

Remote Sensing is defined as the science, art and technology to obtain information of any object or phenomenon present on the earth without any physical contact with the target. I t is a modern technology which provide reliable information with respect to identical area. Colwell in 1983 defined Remote Sensing as "the gathering and processing of information about the earth's environment, particularly its natural and cultural resources, through the use of photographic and related data acquired from aircraft or satellite". According to Curran (1985) Remote Sensing can be defined as the use of electromagnetic radiation, sensors to record image of the environment, which can be interpreted to field information.

In the time of Globalization, this technology is accepted universally as a very reliable and powerful technology for monitoring and management of the natural resources. This technology provides repetitive and reliable information in a definite time interval. In this technique electromagnetic radiation (EMR) either derived from natural resources i.e. sun on an artificial sources, acts as the principal media and selective EMR coming from atmospheric windows are used for Remote Sensing, when electromagnetic energy incident on any earth surface features gets shattered, reflected, absorbed or emitted depending upon their physical, chemical and biological characteristics. The differences in atmospheric window allow distinguishing different features on the image. The reflected and emitted radiations interact with the atmosphere while reaching to the sensors. The interacted radiation result in different energy signals, from which we extract valuable information. After it, the signals are sensed and recorded by various sensors mounted on satellite and then sent to the earth's receiving station.

Geographical Information System (GIS) is a sustainable tool for collection, compilation, storation, retrieval, analyzing and manipulation of geographic information, displaying an integrating environmental, economic and social data in signals system. It is a powerful tool and allows entry of many types of data in a single spatial framework. It also facilitates these operations like overlays of data for purpose of comparison, Updation of information, modeling of physical, economic and social processes for the purpose of simulation and prediction.

Remote Sensing and GIS technique is a conventional method for accessing depletion of forest as well as agricultural field. The loss of forest cover as well as agricultural field has been a matter of great concern throughout the world and this is mainly responsible for global environmental problem (Oza, 1992 and Kushalapa, 1992). Information on existing land use/land cover, its spatial distribution and change are essential pre-requisite for planning (Anonymous, 1992).

Significance of the present study

The main aim and objective of the study is to provide watershed area of Chitrakoot region for better planning and decision making towards water resources development and its management.

Location

The study area is located at Chitrakoot Dham Karwi district in U.P. state and Chitrakoot region of Satna district in M.P. state. The Paisuni River flows within two states viz, M.P. & U.P. It is originated from M.P. and flows for a few distances and come in to U.P. The study area is bounded by 24°45'00"N to 25°15'00"N and 80°45'00"E to 81°00'00"E and falls under Survey of India Toposheet No. 63C/16 and 63D/13.

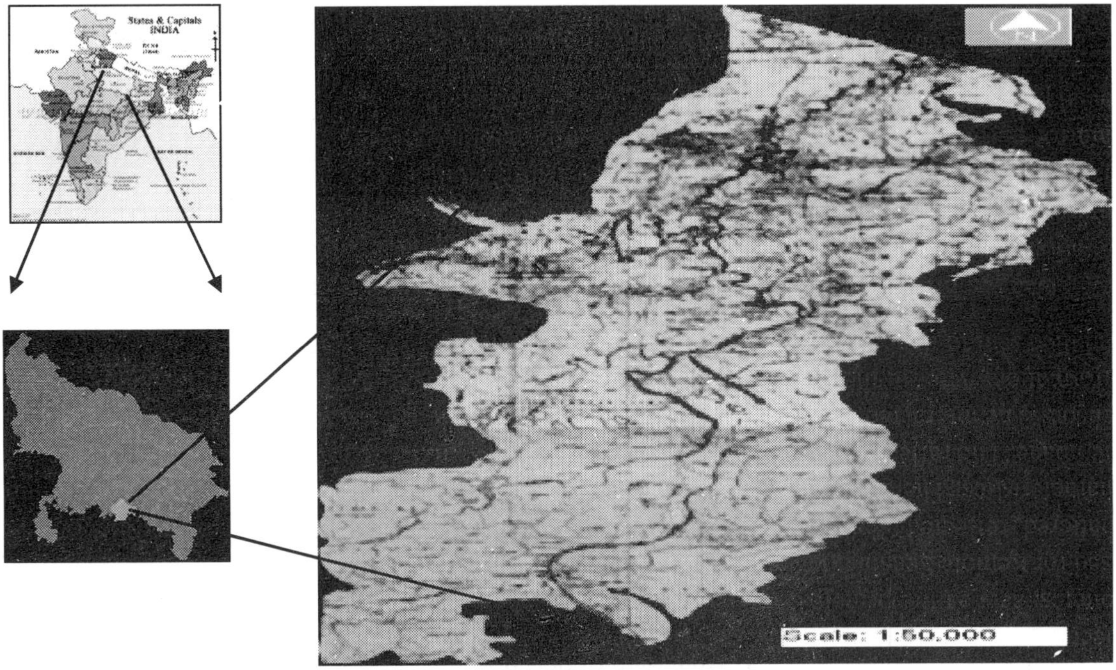

Fig. 1. Sowing Location Map of Study Area

As of 2001 India census, Chitrakoot Dham (Karwi) had a population of 48,853. Males constitute 54% of the population and females 46%. Chitrakoot Dham (Karwi) has an average literacy rate of 67%, higher than the national average of 59.5%; with male literacy of 75% and female literacy of 58%. 16% of the population is under 6 years of age.

Geography

Chitrakoot is a religious place. It is known as "Hill of many wonders". It spreads over Northern Vindhyan Range of Mountains. Chitrakoot Parvat Mala includes Kamadgiri, Hanumaan Dhara, Jankikund, Lakshman pahari, and Devangana famous Religious Mountains.

Drainage

Paisuni River is the main river with complete drainage system, draining their water to the whole area and responsible for agro-economic growth, biodiversity, habitation and livelihood. This drainage system is significant for watershed management. The drainage system is formed with different streams and streamlets.

Work Process

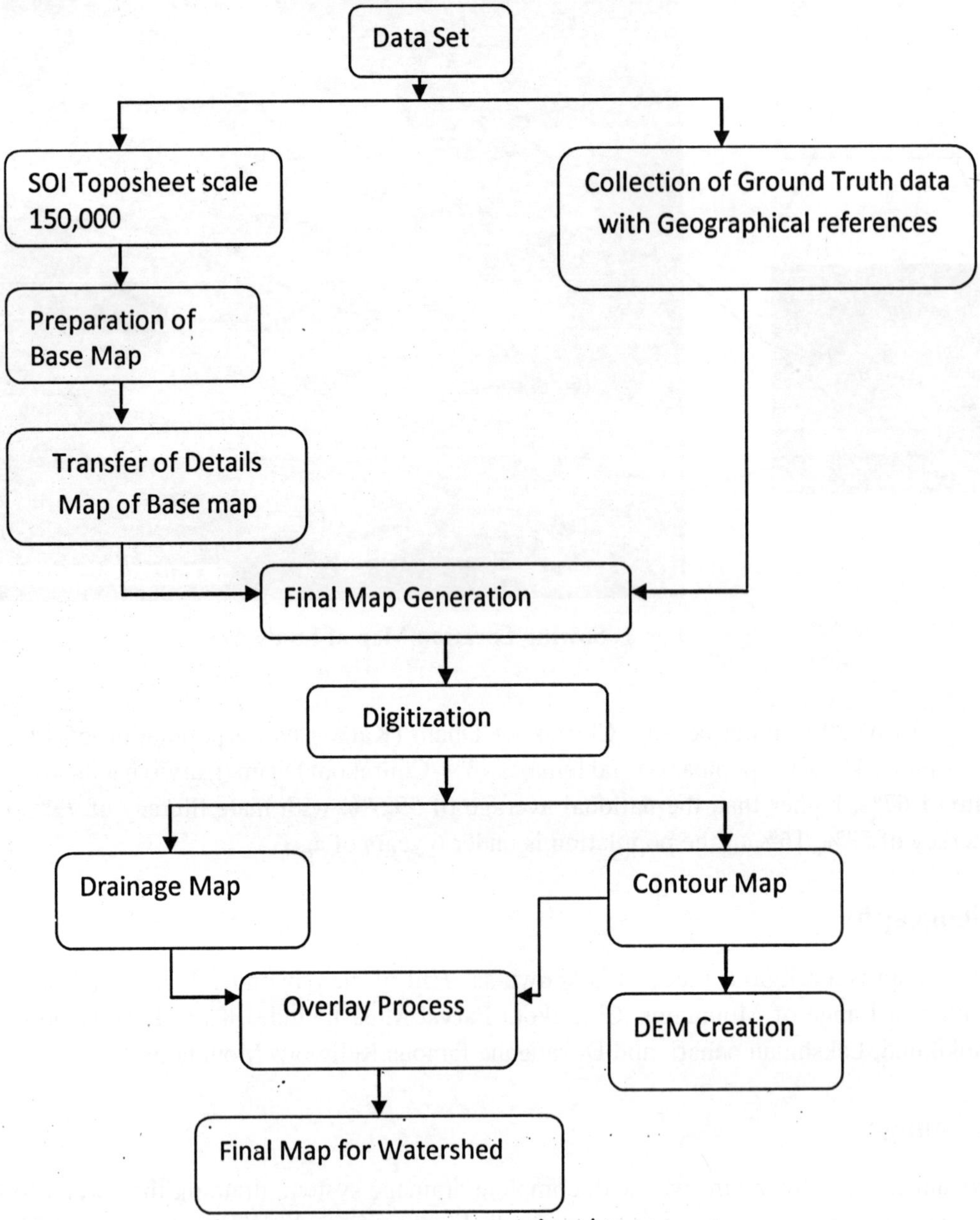

Fig. 2. Flowchart of Work done.

In the working process SOI Toposheet no. 63C/16 & 63D/13 are used. Image processing software ERDAS Imagine 8.5 was helpful for the process.

Result & Conclusion

For the watershed planning and management, drainage map is used as base map. It provides idea about the nature of rocks, present in the area, porosity and permeability of the material present in the area. Contours are also responsible for the formulation of streams and streamlets. These form a complete drainage system which performs a significant role in agricultural and urban & rural area. In the Paisuni watershed region a complete drainage system is found, starting from 1st order to the main river. The drainage pattern of the area is dendritic and the drainage texture of the study area is medium.

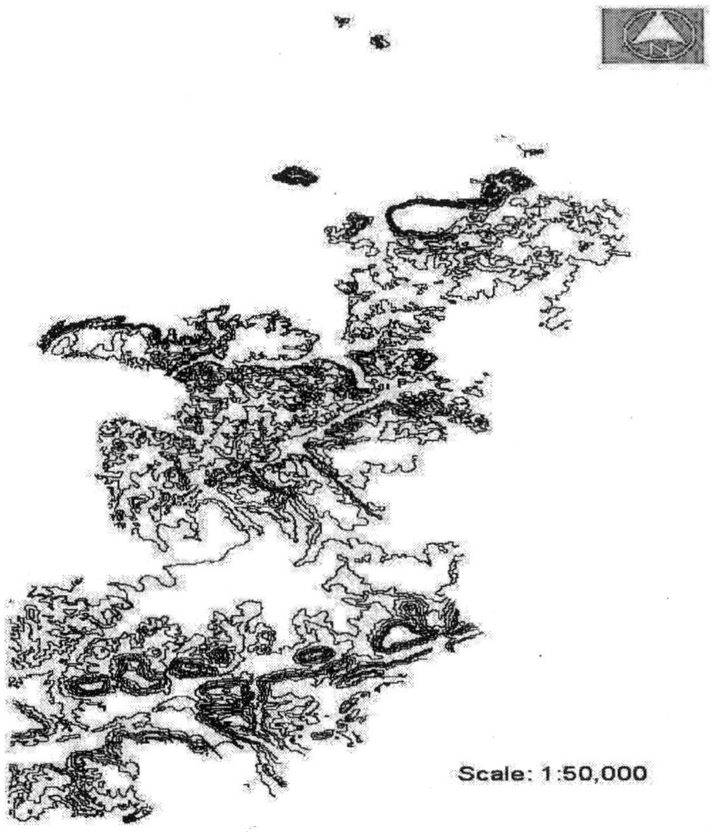

Fig. 3. Showing Contour Map of the Study Area

The interrelated nature of land and water resources calls for a holistic approach towards watershed management. While formulating proper management programmes, some reliable and upto date information are required viz. size and shape of watershed, topography, soil and drainage parameters. These all parameters affect the behavior of a watershed. Remote Sensing & GIS has a capability to obtain synoptic view and repetitive coverage. This technology is used effectively and economically in analysis and inventory, for watershed development and management.

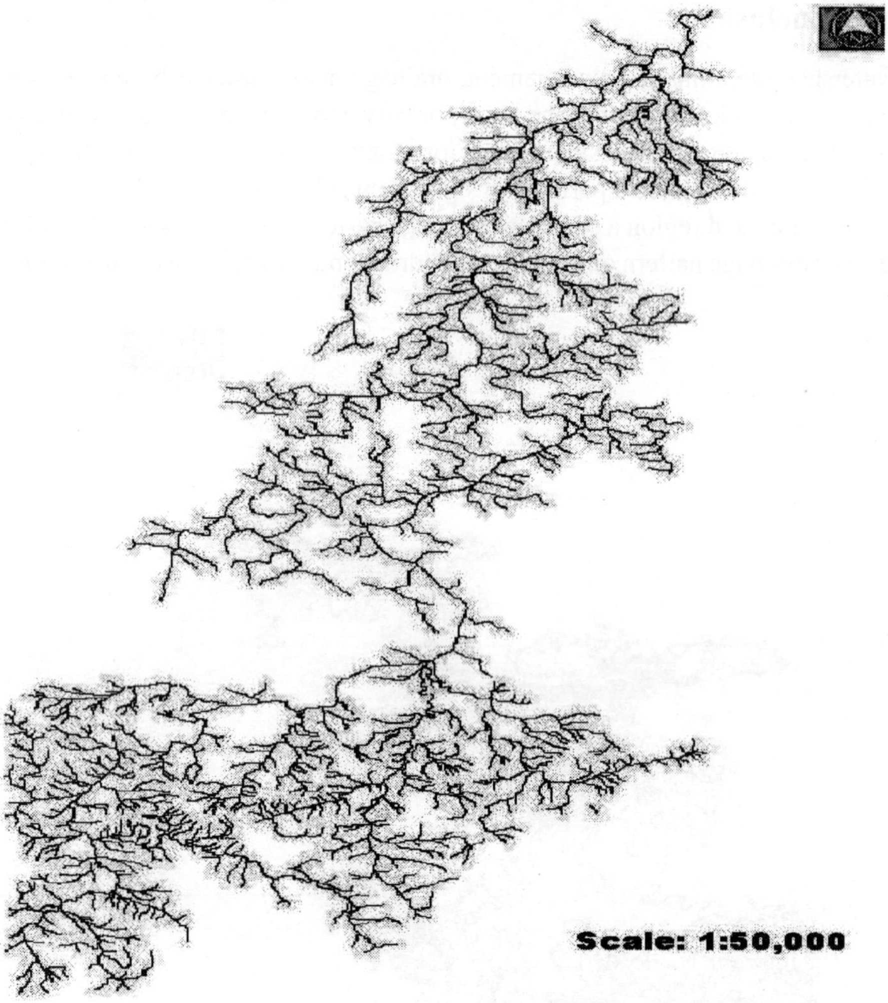

Fig. 3. Showing Drainage Map of the Study Area

Acknowledgement

Authors are thankful to Remote Sensing and GIS Lab, MGCGV Chitrakoot for providing the lab facility for the present study.

References

- Colwell, R.N. (editor), 1983. Manual of Remote Sensing, Second Edition, American Society of Photogrammetry, Falls Church, Virginia, two volumes, 2440 p.
- Curran, P.J., 1985. Principles of Remote Sensing, Longman London, 1989. Remote Sensing of foliar chemistry, Remote Sensing of Environment, 30 271-278.

23

BIODIVERSITY ASSESSMENT OF MACRO-INVERTEBRATES AS BIO-INDICATOR OF WATER QUALITY IN KOTA BARRAGE DAM, RAJASTHAN, INDIA

*N. Sarang, S.P. Singh and L. L. Sharma**

Abstract: Diversity variations in macroinvertebrate and selected limno-chemical parameters of Kota Barrage Dam have been studied on the basis of the samples collected during monthly surveys for a period of one year (November, 2004 to October, 2005) which revealed interesting interrelationship among various factors. This Dam is affected by pollution (industrial specially Thermal Power Plant and domestic effluents) as indicated by low dissolved oxygen. Based on depth of visibility values Kota Barrage Dam could be categorized as "Moderately eutrophic" whereas nitrate nitrogen values of Kota Barrage dam indicated "Mesotrophic" status. On the other hand hardness values of this Dam were "Moderately hard". The species diversity was influenced due to water pollution in Kota Barrage. Despite this the biodiversity of macroinvertebrates was appreciably high with recorded 27 species. Most dominant benthic species encountered were Melanoides tuberculata *Bellamya bengalensis, Gyraulus convexiusculus*, *Indoplanorbis exustus* and *Lymnaea acuminata*. Average density of these macroinvertebrates showed respectively following order of dominance in Kota Barrage.

Average density (No./m^2) = 121 > 117 > 88 >77 > 70

Seasonal average values of benthic fauna at Kota Barrage indicated lower average in monsoon as compared to summer and winter season. In terms of density composition Molluscs formed the dominant group in the Kota Barrage followed by representatives Diptera, Odonata, Coleoptera, Hemiptera and Annelida. The seasonal abundance of benthos in Kota Barrage during investigation period varied between 869 to 1100 No./m^2. *Lymnaea acuminata* and *Chironomus larvae* may be used as pollution indicators in Kota Barrage. Richness of organic matter in this water body might have favoured the species composition and abundance of benthic fauna. Discharge of large quantities of waste causes thermal pollution and it affects the ecology of the ecosystem of Kota Barrage Dam.

Keywords: Biodiversity, Macroinvertebrates, Bioindicator, Aquatic Pollution & Kota Barrage Dam

Indira Gandhi Krishi Vishwvidayalaya, Krishi Vigyan Kendra, Kawardha, District Kabirdham -491 995 (Chhattisgarh)
*Maharana Pratap University of Agriculture and Technology, College of Fisheries, Udaipur-313001(Rajasthan); Email Saranglimno@yahoo.com, sarangfish@gmail.com

Introduction

The benthic biota provides a method of biomonitoring of pollution and acts as a bioindicator. In aquatic system the macrobenthic invertebrate community is more often investigated due to absence of mobility and sensitivity towards physico-chemical stress. Several workers have pointed out that benthic organism provide a valuable indicator of past and present condition of the water quality and prone to be the most useful in assessment of pollution Hynes (1965). Thus the pollution ecology of macrobenthic community becomes a very important biological tool for environmental impact assessment and management. In India the studies on macrobenthos had earlier been carried out by Vasishat and Bandal (1979), Anitha (2005). The studies on the use of this biotic component as bioindicator of aquatic pollution had been carried over Krishnamoorthi and Sarkar (1979). In the lakes of Rajasthan similar studies were conducted by Saxena (2007) and Sharma *et al.* (2007). The bottom fauna play important role in the mineralization and recycling of organic matter. They also serve as the good indicators of water pollution since they form an important food item for fishes.

Discription of study area

The Kota Barrage reservoir located near Kota city is the last structure in the cascade development pattern of Chambal Velly. Kota Barrage Dam is the forth dam in the series of Chambal Velly project, located about 0.8 km upstream of Kota City in Rajasthan. Water released after power generation at Gandhi Sagar, Rana Pratap Sagar and Jawahar Sagar Dam is diverted by Kota Barrage Dam. Ecologically Kota Barrage Dam is different from other reservoirs as there is regular warm water discharge from the adjoining thermal power plant deposition of coal dust all the water levels. The total catchment area of Kota Barrage is 10, 6000 m^2 of which the free catchment area below Jawahar Sagar Dam is just 137 km^2. The live storage is 99 Mm^3. It is an earth filled Dam with a concrete spillway. The dam was completed in 1967 with the primary objective of establishing that Kota thermal power plant. The water of Kota Barrage Dam is used for domestic and public supply, irrigation, industrial supply and thermal power generation. Further details of this dam pertaining to morphometeric feature are indicated in (Table 1).

Material and Methods

The sampling frequency was kept monthly for a total period of over 12 months (from November, 2004 to November, 2005). Water samples were collected from station using a plastic bucket (Plates 1). For the collection and analysis of below stated water quality parameter standard methods such as Adoni (1985) and APHA (1989) were followed. For qualitative analysis of macroinvertebrates standard keys were used Needham and Needham (1962), Pennak (1978) and Subba Rao (1989). However, for quantitative analysis animals were counted individually species wise in the whale sample or sub samples. The number of benthos per unit area was calculated as follows.

$$\text{Benthos No./m}^2 = \frac{N}{A} \times 10^4$$

N = Number of organism per sample A = biting area of samples (15 × 15 cm)

Table 1

S. No.	Items	Kota Barrage
1.	Locations Latitude Longitude	24°18.892N 76°42.069E
2.	Catchment area	10, 6000 sq.m.
3.	Average rainfall (inch)	32
4.	Maximum depth (m)	18.0
5.	Mean depth (m)	10.0
6.	Length of dam (m)	603.0
7.	Height of dam (m)	40.0
8.	Nature of dam	Earth dam and rock fill
9.	Year of impoundment	1960
10.	Purpose	Thermal Power Generation & irrigation
11.	Power units in MW	6 units of Power Generation
12.	Tehsil and district	Kota district

Plate 1 Location of sampling station of Kota Barrage Dam, Rajasthan

Results and Discussion

The results on physicochemical parameters of surface water of Kota Barrage dam are presented in (Table 2). During the study period the temperature of water ranged between 21°C to 39°C minimum in the month of February and maximum in the month of June. This parameter exhibited lowest average in winter followed by monsoon and summer (Fig. 1). In Kota Barrage Dam PH of water indicated alkaline condition with variations between 7.0 (winter) to 9.1 (summer). The mean values of EC representing the total ionic load of water in different seasons varied between 0.222 (winter) to 0.604 mS (summer). The average values for depth of visibility of Kota Barrage Dam were 138.7, 176.0 and 212.5 cm in monsoon, summer and winter respectively (Table 2). Dissolved Oxygen is another vital parameter regulating survival of aquatic life. It was found to be maximum in the month of January (9.2 mg/l) and minimum (4.0 mg/l) in the month of September (Fig. 1). During the study period on 4 occasions dissolved oxygen was found below 5 ppm which is indicative of distressed condition for fish and aquatic biota. Total alkalinity ranged between 84.0 to 190 mg/l (Table 2). The maximum total hardness was observed in the month of February and minimum in the month of June (Table 2). The important nutrients nitrate nitrogen showed the range of 0.001 to 0.020 mg/l. The higher value was observed in July. Orthophosphate recorded in Kota Barrage Dam was fairly high with variations between 0.283 (December) to 1.227 mg/l (May). It is interesting to note that the biodiversity of macrobenthic fauna in Kota Barrage Dam was appreciably high with 27 species (Table 3 and Plates 2, 3, 4, 5, 6 & 7). Most dominant benthic species encountered were *Melanoides tuberculata*, *Bellamya bengalensis*, *Gyraulus convexiusculus*, *Indoplanorbis exustus* and *Lymnaea acuminata* (Fig. 2). The average density of these benthos respectively showed following order of dominance at Kota Barrage Dam (Table 4)

Average density (No./m^2) 121 > 117 > 88 > 77 > 70

Considering various species recorded in Number it is evident that the molluscs were the most dominant with 14 species followed by Diptera (3), Coleoptera (3), Odonata (3), Annelida (2) and Hemiptera (2). The seasonal abundance of macroinvertebrates in Kota Barrage Dam during investigation period varied from 869 to 1100 No./m^2 (Table 4 and Fig. 3).Thus, under the prevailing environmental condition the observed species composition of these benthos could be considered bio-indicators of organic pollution. The values of coefficient of correlation (r) indicate that was a positive relationship between the *Lymnaea acuminata* and PH, Orthophosphate. *G. convexius* culus had positive significant relationship with dissolved oxygen whereas negative relationship with temperature. *Chironomus* larvae had significant positive relationship with carbonate alkalinity (Table 5).

It seen true that the fluctuation brought about by these processes in that water body study, create a very conducive micro as well as macro environment for the healthy growth and multiplication of the molluscan fauna higher abundance of molluscs with increased water temperature and decomposed organic matter has been also reported by Bath *et al.* (1999). PH is another important parameter affecting species diversity and distribution in an ecosystem. The PH value of the water

Kota barrage Dam in the present investigation was alkaline. The alkaline PH was found to be ssociated with more number of species. However, with increasing PH the number of species has een reported to decrease (Venkateswarju, 1969). The richness of molluscs presently observed ay be attributed to the cumulative effect of alkaline nature of water, high calcium contents and acrophytic vegetation which provide both food and shelter, because some of these from are of eriphytic in nature as it has earlier been documented by Tonapi (1980). Pennak (1989) also supported is of view by observing the greater molluscan population in alkaline lakes as compared to acidic kes. Electrical conductivity and dissolved solids are directly proportional to each other (Mittal nd Senger, 1989). As per Rawson (1960) criterion Kota Barrage Dam with electrical conductance ell above 0.222 mS could be considered eutrophic.

The present findings as regard water clarity values could be mainly assigned for biomass, uspended organic matter and other physical disturbances such as boating, bathing and washing in ota Barrage Dam which keep the fine particles in suspension. A close perusal to depth of visibility f Kota Barrage Dam indicates that in general, Occurrence of high algal biomass coincided with w clarity values. Sharma and Durve (1991) proposed a regional classification for assigning trophic tatus to water bodies using average secchi disc values. On the basis of depth of visibility values ota Barrage Dam could be categorized as "Moderately eutrophic". Zhang and Chang (1994) tudied oligotrophic, mesotrophic and eutrophic lakes and reported that in oligotrophic lake hosphorus and nitrogen levels were 0.01 and 0.21 mg/l respectively. Klein (1956) and Sawyer 1960) have classified water of the basis of hardness as follows Soft water (up to 75 mg/l), moderately 75 to 150 mg/l), hard water (150 to 300 mg/l) and very hard water (above 300 mg/l). Based on this lassification also Kota Barrage Dam could be categorized "Moderately hard" water body.

The analysis of adverse affect of sewage on the structure of communities of aquatic organisms vas among the first environmental assessment methods popularly adopted. These organisms are ble to tolerate sewage discharge were categorized as saprobic, these that could only survive in elatively pristine conditional were labeled as oligotrophic and those somewhere in between were now mesotrophic. Ample evidence exists that the macro benthic community acts as a bio-indicator nd has become a very important biological tool for the assessment of water quality (Hynes, 1965; inha and Das, 1993 and Kumar, 1997). Oligochaetes were found at Kota Barrage Dam. Thus, heir ability to withstand considerable oxygen depletion environment is an essential adaptation to heir niche within the community (Brinkhwest, 1974). From the results of present study it is clearly vident that the macrobenthic community is dominated by Chironomus larvae population in Kota Barrage Dam. Chironomus larvae have also been used as pollution indicators by number of workers Gaufin, 1957; Curry, 1962). Thus, the abundance of Chironomus in the benthic population is due o impact of altered nature of substrate due to organic pollution. The presence of *Tubifex* sp., *imnodrilus* sp. and *Limnaea accuminata* in lakes also corroborates with the work of Mason (1981). hese indicator properties of benthos can be used to identify different physical and chemical gradients r eutrophic increases in Kota Barrage Dam and can therefore be employed in environmental nonitoring programmes.

Table 2 Seasonal variations in physico-chemical analysis of water quality parameters of Kota Barrage during Nov., 2004 to Nov., 2005

S. No.	Parameters	Winter		Summer		Mansoon	
		Seasonal range	Average	Seasonal range	Average	Seasonal range	Average
1.	Temperature (°C)	21.0-27.0	23.5	26.0-39.0	33.5	31.0-38.0	34.2
2.	pH	7.0-8.5	7.8	8.0-9.1	8.4	7.3-8.3	7.7
3.	Electric conductivity (mS)	0.222-0.239	0.231	0.22-0.604	0.250	0.228-0.268	0.251
4.	Depth of visibility (cm)	150-300	212.5	100-300	176.0	60.0-260	138.7
5.	Dissolved oxygen (mg/l)	5.2-9.2	7.0	6.0-8.0	7.2	6.0-7.6	6.1
6.	Free carbon dioxide (mg/l)	10.0-32.0	19.5	0.0-14.0	7.0	4.0-16.0	10.0
7.	Carbonate alkalinity (mg/l)	Nil	Nil	6.0-12.0	4.5	0.0-2.0	0.5
8.	Bicarb. alkalinity (mg/l)	90.0-140	116.5	84.0-124	110.0	106-190	132.5
9.	Total alkalinity (mg/l)	90.0-140	116.5	84.0-146	117.0	108-190	133.0
10.	Nitrate nitrogen (mg/l)	0.001-0.006	0.003	0.003-0.014	0.007	0.001-0.020	0.007
11.	Ortho phosphate (mg/l)	0.283-0.957	0.690	0.786-1.227	1.017	0.740-0.867	0.806
12.	Calcium Hardness (mg/l)	520-70.0	59.0	12.0-54.0	41.0	48-56	50.6
13.	Magnesium Hardness (mg/l)	30.0-36.0	32.0	26.0-46.0	36.5	20.0-38.0	30.0
14.	Total Hardness (mg/l)	88-100	91.0	40.0-94.0	78.0	76.0-86.0	80.5

Table 3 Checklist of macroinvertebrates of Kota Barrage during Nov., 2004 to Nov., 2005

S. No.	Name of species	S. No.	Name of species
1.	*Melanoides tuberculata*	15.	*Diplonychus annuletus*
2.	*Bellamya bengalensis*	16.	*Diplonychus rusticus*
3.	*Lymnaea acuminata* (Typica)	17.	*Cybister tocipunetatus asiatieus* Sharp
4.	*Gyraulus cenvexiusculus*	18.	*Sternolophus rcufipes*
5.	*Lymnaea acuminata* (Gracilior)	19.	*Berosus dolerosus*
6.	*Ganiobasis virginica*	20.	*Chironomus* larvae
7.	*Indoplanorbis exustus*	21.	*Tabanus* larvae
8.	*Corbicula striatella*	22.	*Psychoda* larvae
9.	*Parreysia caerulea*	23.	*Helocordulia uhleri*
10.	*Parreysia favidens*	24.	*Sympetrum*
11.	*Scaphula deltae*	25.	*Erythrodliplax* sp.
12.	*Bithynia pulchella*	26.	*Heropobdelloidea lateroculata*
13.	*Lamellidens marginalis*	27.	*Barbronia weberi*
14.	*Physa acuta*		

Table 4 Seasonal quantitative analysis of macroinvertebrates (No./m^2) of Kota Barrage during Nov., 04 to Nov., 05

S. No.	Species of Macroinvertebrates	Winter		Summer		Monsoon		Average density
		Seasonal range	Average	Seasonal range	Average	Seasonal range	Average	
1.	*Melanoides tuberculata*	88-440	209	132-264	99	0.0-220	55	121
2.	*Bellamya bengalensis*	0.0-264	132	88-132	132	0.0-176	88	117
3.	*Lymnaea acuminata* (T.)	88-132	55	176-308	121	0.0-132	33	70
4.	*Gyraulus convexiusculus*	396-440	209	0.0-88	22	0.0-132	33	88
5.	*Lymnaea acuminata* (G.)	0.0-88	22	0.0-220	55	0.0	0.0	26
6.	*Goniobasis virginica*	0.0-176	88	0.0	0.0	0.0-88	22	37
7.	*Indoplanorbis exustus*	0.0-88	22	88-352	154	88-132	55	77
8.	*Parreysia caerulea*	0.0-88	22	0.0-132	33	0.0-132	33	29
9.	*Corbicula striatella*	44-132	44	0.0-176	44	0.0	0.0	29
10.	*Parreysia favidens*	0.0	0.0	0.0-132	66	0.0-264	66	44
11.	*Scaphula deltae*	0.0	0.0	0.0-132	33	0.0-88	22	18
12.	*Bithynia pulchella*	0.0-88	22	0.0	0.0	0.0-176	44	22
13.	*Lamellidens marginalis*	0.0-132	33	0.0-176	44	132-264	99	59
14.	*Physa acuta*	0.0-88	22	0.0	0.0	88-176	66	29
15.	*Diplonychus annuletus*	0.0-44.0	11	0.0	0.0	0.0	0.0	4
16.	*Diplonychus rusticus*	0.0	0.0	0.0-88	22	0.0-176	44	22
17.	*Cybister tocipunetatus*	0.0-88	22	0.0-176	44	0.0	0.0	22
18.	*Sternolophus rcufipes*	0.0	0.0	0.0-88	44	0.0	0.0	15
19.	*Berosus dolerosus*	0.0-132	33	0.0	0.0	0.0	0.0	11
20.	*Chironomus* larvae	0.0-176	44	88-308	99	0.0-88	22	55
21.	*Tabanus* larvae	0.0	0.0	44-88	33	0.0-88	22	18
22.	*Psychoda* larvae	0.0-132	33	0.0	0.0	0.0	0.0	11
23.	*Helocordulia uhleri*	0.0-44	11	0.0-88	22	88-132	55	29
24.	*Sympetrum*	0.0-88	22	0.0	0.0	0.0-88	22	14.6
25.	*Erythrodiplax* sp.	0.0	0.0	0.0-132	33	0.0-132	33	11
26.	*Heropobdelloidea lateroculata*	0.0-44	11	0.0	0.0	0.0-88	22	11
27.	*Barbronia weberi*	0.0-44	11	0.0	0.0	0.0-132	33	15
	Total	660-1496	1078	1012-1364	1100	440-1320	869	1015

Table 5 Correlation and regression coefficients for water quality parameters with Macroinvertebrates of Kota Barrage

Y	X	r	r sqr	a	b
Lymnaea acuminata	pH	0.782**	0.612	-1044.390**	137.9241**
Lymnaea acuminata	Orthophosphate	0.639*	0.408	-85.795	193.8157*
G. convexiusculus	Temperature	-0.591*	0.350	528.000**	-14.6667*
G. convexiusculus	Dissolved oxygen	0.661*	0.437	-453.599**	79.3696*
Lymnaea acuminata	Depth of visibility	0.702**	0.493	-88.560**	00.6785**
Goniobasis urginica	Orthophosphate	-0.638*	0.407	154.166**	-142.2479*
Indoplanorbis exustus	Calcium Hardness	-0.805**	0.648	340.430**	-05.1193**
Indoplanorbis exustus	Total Hardness	-0.704**	0.495	438.373**	-04.4876**
Corbicula striatella	Carbonate alkalinity	0.628*	0.394	11.309	10.2490*
Parreysia favidens	Total alkalinity	0.561*	0.315	-172.437**	1.7224*
Scaphula deltae	Carbonate alkalinity	0.784**	0.615	02.651	9.2771**
Scaphula deltae	Magnesium Hardness	0.591*	0.349	-76.566**	2.9788*
L. marginalis	Bicarbonate alkalinity	0.661*	0.437	-217.735**	2.2399*
L. marginalis	Total alkalinity	0.602*	0.362	-192.938**	1.9976*
Physa acuta	Bicarbonate alkalinity	0.698**	0.488	-149.792**	1.4571**
Physa acuta	Total alkalinity	0.649*	0.421	-137.145**	1.3277*
Cybister tocipunetatus	Bicarbonate alkalinity	-0.555*	0.308	153.976**	-1.1012*
Cybister tocipunetatus	Total alkalinity	-0.580*	0.336	159.688**	-1.1268*
Sternolophus rcufipes	Carbonate alkalinity	0.712**	0.507	08.482	7.6867**
Sternolophus rcufipes	Calcium Hardness	-0.568*	0.322	95.394**	-1.4271*
Sternolophus rcufipes	Total Hardness	-0.565*	0.320	136.960**	-1.4253*
Chironomus larvae	Carbonate alkalinity	0.631*	0.399	31.984	16.6104*
Tabanus larvae	Electric conductivity	0.559*	0.312	-279.884	1214.5091*
Helocordulia uhleri	Temperature	0.654*	0.427	-118.856**	4.7907*
Sympetrum	Nitrate nitrogen	0.611*	0.373	-03.410	4442.6229*
Erythrodiplax sp.	Electric conductivity	0.560*	0.313	-415.922	1785.0136*
H. laterocula ta	Dissolved oxygen	-0.582*	0.339	89.993**	-11.8482*
Barbronia weberi	Calcium Hardness	0.556*	0.309	-58.146**	1.3624*

*, ** Significant at 5 and 1 per cent respectively.

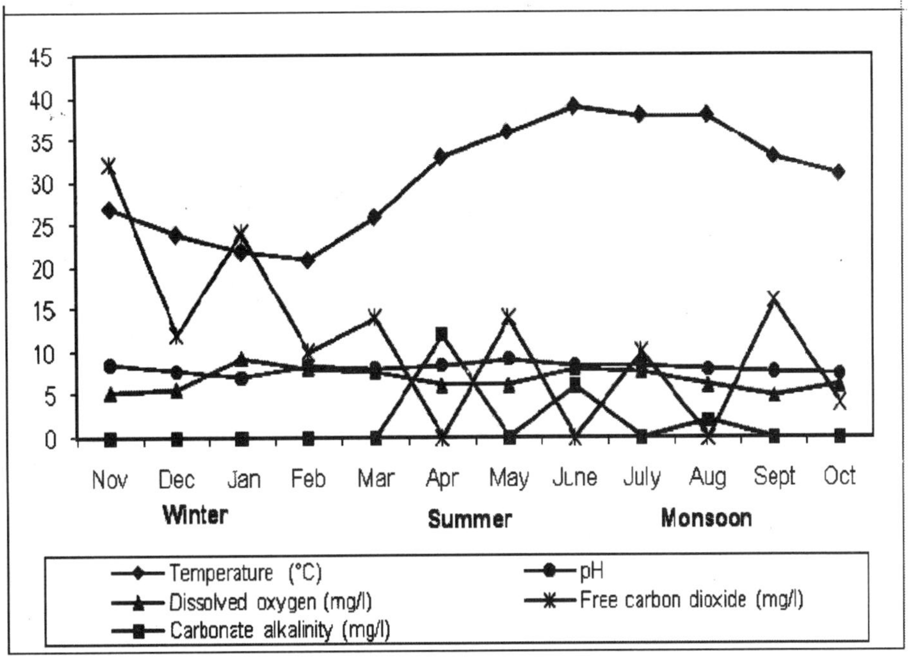

Fig. 1 Variations in temperature, pH, dissolved oxygen, free carbon dioxide and carbonate alkalinity at Kota Barrage during Nov., 04 to Nov., 05

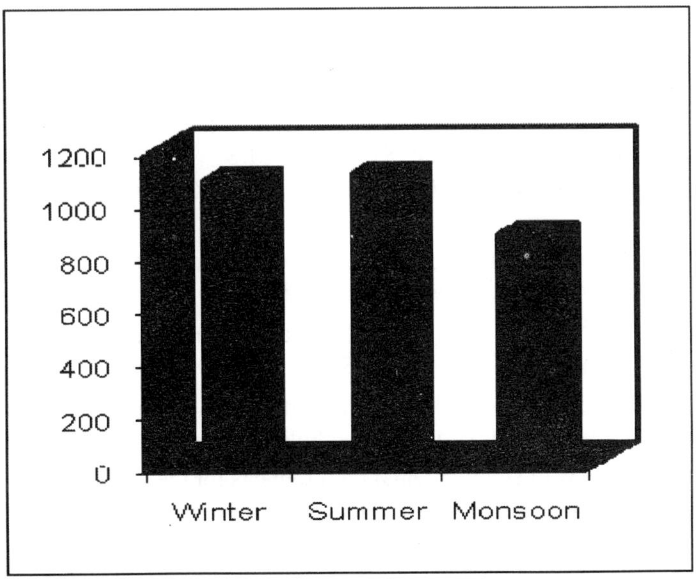

Fig. 2 Average density of five prominent benthos (No./m^2) in Kota Barrage Dam, Rajasthan

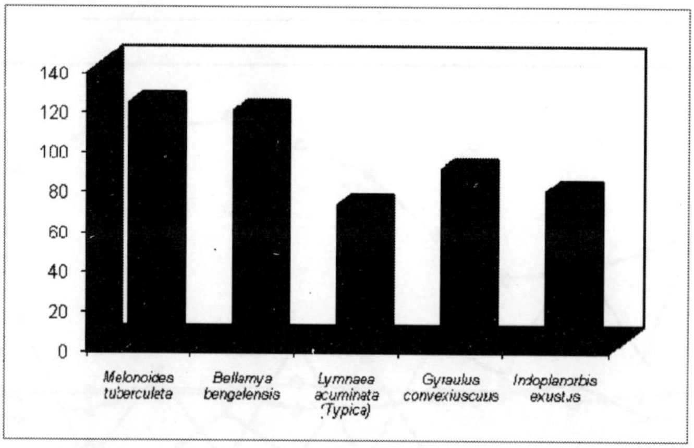

Fig. 3 Seasonal average of benthos (N0./m^2) in Kota Barrage Dam

Acknowledgement

The authors are much grateful to Dr. H.C.L. Gupta, Dean, R.C.A. and College of Fisheries for providing necessary facilities. Thanks are also due to Dr. P.C. Verma, OIC/SO-G, Environmental Survey Laboratory (BARC), Rajasthan Atomic Power Station, Rawatbhata, Kota, India for his kind encouragement and consistent support. We are also indebted to the Director, Zoological Survey of India, Kolkatta and his associates namely Dr. P. Mukhopadhyay, Dr. Animesh Bal, Dr. K. Karmakar, Dr. R. Venkitesan and Dr. A. Misra for identification of aquatic fauna. We are also grateful to Dr. (Mrs) K. Mitra, Principal Scientist, CICFRI, Barrackpue, Kolkatta for the generous cooperation in identification of fauna. Table 1 Morphometeric features of Kota Barrage, Rajasthan

References

- Adoni, A.D. (1985). Workbook in Limnology. Pratibha Publisher, C-10 Gour Nagar, Sagar (M.P.).
- Anitha, G., Chandrsekhar, S.V.A. and Kodarkar, M.S. (2005). Hydrography in relation to macroinvertebrates in Mir-Alam Lake, Hyderabad, A.P. Res. Zool. Surv. India. Occ. Paper No.2351-148.
- APHA-AWW-WPCF (1989). Standard methods for examination of water and waste water (17th ed.). American Public Health Association, Washington, D.C.
- Bath, K.S., Kaur, H. and Dhillon, S.S. (1999). Correlation of mollusks with physicochemical factors at Harike Reservoir (Punjab). India J. Environ. Sci.3159-163.
- Brinkhurst, R.O. and Cook, D.G. (1974). Aquatic earthworms (Annelida, Oligochaeta). In Pollution ecology of freshwater invertebrates (Ed. Harl and Fuller). Academic Press, London.
- Curry, L.L. (1962). A survey of environmental requirements for the midge (Diptera, Tendipedidae). In Biological Problems in Water Pollution. Transactions of 3rd seminar (Ed. Tarzwell), USDHEW, PHS, Robert, A. raft Sanitary Engineering Centre, Cincinnati.
- Gaufin, A.R. (1957). The use and value of aquatic insects as indicators of organic enrichment. Biological problems in water pollution. U.S. Public Health Service, Washington D.C. pp.139-149.

- Hynes, H.B.N. (1965). The significance of macroinvertebrates in the study of mild river pollution. In Biological Problems in Water Pollution. 3d Seminar, U.S. Department of Health Education and Welfare, Cincinnati.
- Klein, L. (1956). The problems of river pollution in industrial areas. Royal Society from Health Journal. 76348-357.
- Kumar, A. (1997). Role of species diversity of aquatic insects in the assessment of pollution in wetland of Santhan Paragnas (Bihar). Journal of Environmental Pollution. 1117-120.
- Krishnamoorthi, K.P. and Sakar, R. (1979). Macroinvertebrates as indicators of water quality. Proc. Symp. Environ. Biology. Academy of Environmental Biology, Muzaffarnagar, India. S.R. Verma, A.K. Tygi and S.K. Bansal (eds.) pp133-138.
- Mason, C.F. (1981). Biology of freshwater pollution. Longman, London pp.209.
- Mittal, S. and Senger, R.M.S. (1989). Studies on the distribution of algal flora in polluted regions on Karwan River at Agra, India. Current trends in Limnological Studies in India in (Press).
- Needham, J.G. and Needham, P.R. (1962). A guide of the study of freshwater biology. Holden Day Inc., San Francisco, p.108.
- Pennak, R.W. (1978). Freshwater invertebrates of the United Sates. 2nd edition, John Wiley and Sons Inc., New York pp.803.
- Pennak, R.W. (1989). Freswater invertebrates of the United States. A Wiley inter science publication, John Wiley & Sonc Inc. P. 628.
- Rawson, D.S. (1960). A limnological comparison of twelve large lakes in northern Saskatchewan. Limnology and Oceanography 5195-211.
- Sawyer, C.H. (1960). Chemistry of sanitary engineers. McGraw Hill Book Co., New York.
- Saxena, M.M. (2007). Trophic sate indices in water quality monitoring. Proc. of NSL 2007 pp. 24-30.
- Sharma, L.L. and Durve, V.S. (1991). Water clarity of 26 waters of Rajasthan in relation to phytoplankton. In Proceedings of the Second Asian Fisheries Forum. Asian Fisheries Society, Manila, Philippines pp.915-918.
- Sharma, L.L., Sarang, N. and Sharma, B.K. (2007). Occurrence of macroinvertebrates in relation to water & sediment characteristics in three seasonal ponds of southern Rajasthan. J. Aqua. Biol., Vol.22 (2)49-54.
- Sinha, R. K. and Das, N.K. (1993). Organic waste and its effect on macrozoobenthos in Ganga at Patna (Bihar), India. J. Freshwater Biol. 533-40.
- Subba. Rao, N.V. (1989). Handbook of freshwater molluscs of India. ZSI, Calcutta, pp.289.
- Tonapi, G.T. (1980). Freshwater animals of India (an ecological approach). Oxford and IBH Publishing Co., New Delhi pp.341.
- Vasisht, H.S. and Bhandal, R.S. (1979) Seasonal variation of benthic fauna in some north Indian lakes and ponds. Indian Journal of Ecology 6 33-37.
- Venkateswarju, V. (1969). An ecological study of the algae of river moosi, Hyderabad (India) with special reference to water pollution-1 physico-chemical complexes. Hydrobiologia. 33 117-143.
- Zhang, H. and Chang, W.Y.B. (1994). Management of inland fisheries in shallow eutrophic, mesotrophic and oligotrophic lakes in China. International Verein Limnology. 24228-229.

24

STATUS OF WATER QUALITY OF NARMADA RIVER AT MAHESHWAR, MADHYA PRADESH, INDIA

*Pramod Kumar Verma and S.K.Mahajan**

Abstract: Water is one of the most unusual natural compounds found on our earth and it is also one of the most and main component for the survival of any living being since life on earth was began in the seas. Rivers are the most important resources in the world and in India in particular. The rivers are being polluted by indiscriminate disposal of sewage and many other wastes of human activities. All water bodies are normally populated by organism and it is almost impossible to find a body of water that is entirely free of life. Any undesirable change in the physico-chemical characteristics of water brings about water pollution. This change is mainly due to human activity such as rapid urbanization and industrialization coupled with injudicious exploitations of natural resources. The discharge of domestic sewage and industrial effluents in to natural water resources results primarily in alteration of their physical and chemical properties which in turn influence adversely the biological population.

The Narmada river originates from the Maikal ranges at Amarkantak, 1057 m. above the sea-level, Now in Shagdol district of M.P. The data revealed that 17 parameters of Narmada river water at Maheshwar. The results indicated that the analysis of physico-chemical parameters of the Narmada river at Maheshwar (MP) during the Summer (April 2006 & 2007), Monsoon (July 2006 & 2007), Post-monsoon (Oct. 2006 & 2007) and Winter (Jan. 2006 & 2007). In the present study of the Narmada river its physico-chemical characteristics viz temperature, transparency, pH, dissolved oxygen, biological oxygen demand, conductivity, total dissolved solids, turbidity, chloride, hardness, free-CO_2, alkalinity, nitrite, nitrate, phosphate, iron and sulphate.

Keywords: Indiscriminate disposal, urbanization, biological population

Govt. P.G. College, Khargone (M.P.)
*Ex-Prof. and Head of Deptt.-Botany, Govt. P.G. College, Khargone (M.P.)

Introduction

In the lotic system, the significance of physical and chemical data for the assessment of water quality has been recognized. The studies consider water quality usually involve physical, chemical and biological variables. The Narmada river originates from the Maikal ranges at Amarkantak, now in Shahdol district of Madhya Pradesh. In its 1312 Km. long journey before joining the Arabian Sea, the Narmada flows through the three states Madhya Pradesh, Maharastra and Gujarat. Direction of the flow of river is from East to West.

Maheshwar is a town in Khargone district of Madhya Pradesh in central India. The Maheshwar town lies on the northern bank of the Narmada river. This town is built on the site of the ancient city of Mahesvari and was the capital of king Kartavirya Arjuna, who is mentioned in the sanskrit epics Mahabharata and Ramayana. Maheshwar was known as Mahissati (Mahishamati in Sanskrit) in ancient times and was the capital of Southern Avanti. There are several bathing ghats and domestic sewage drains from where waste water reaches the river main Hindu festivals are also organized at bathing ghats of the Maheshwar.

Numbers of workers have carried out their work on water quality of different rivers. Khanna *et al.* (1993), Khanna (2003) of river Ganga, Dhindsa and Bhargava (1982) studied the physico-chemical characteristics of river chambal. But no study has been conducted to assess the water quality of river Narmada at Maheshwar.

Material and Methods

In this study, limnological status of river Narmada at Maheshwar was observed for a period of two successive years. During the study period a particular month of the season was selected for sampling. The study period was from April 2006 to January 2008. Physico-chemical characteristics of the water samples were determined following APHA (1998), Adoni (1985) and Golterman (1969).

Result & Discussion

The water samples were collected seasonally from the bathing ghat of Maheshwar. This water receives discharge from several drainages of Maheshwar town.

This investigation was done during the period from April, 2006 to January, 2008. In this connection various parameters of water quality were analysed following the method suggested by APHA (1998). This study indicated that the temperature range of river water is 22 - 27°C, transparency 10.8 - 60.0cm, pH 7.64 - 8.5, dissolved oxygen 6.2 - 14.4 mg/l, biological oxygen demand 3.6 - 7.6 mg/l, conductivity 0.22 - 0.346 mmhos/cm, total dissolved solids 139.8 - 211.06 mg/l, turbidity 78 - 4500 mg/l, chloride 32 - 66 mg/l, total hardness 152 - 224 mg/l, free CO_2 1.2 - 13 mg/l, alkalinity 92 - 138 mg/l, nitrite 0.005 - 0.016 mg/l, nitrate 2.519 - 6.348 mg/l, phosphate 1.174 - 2.842 mg/l, sulphate 38 - 60 mg/l and iron 0.008 - 0.041 mg/l at sampling point.

Table & Graph

SEASONAL VARIATION IN PHYSICO - CHEMICAL PARAMETERS AT MAHESHWAR											
S. No.	Parameters	Summer	Monsoon	Post monsoon	Winter	Summer	Monsoon	Post monsoon	Winter	Max	Min
		Apr-06	Jul-06	Oct-06	Jan-07	Apr-07	Jul-07	Oct-07	Jan-08		
(A) Physical Parameters											
1	Water Temperature (°C)	26	25	26	22	27	26	22	22	27	22
2	Secchi transparancy (cm.)	51.2	10.8	32.5	58.5	52.9	10.8	35.2	60.0	60.0	10.8
(B) Chemical Parameters											
3	pH	8.5	7.64	8.31	8.25	8.18	8.14	8.21	8.43	8.5	7.64
4	DO (mg/l)	12.2	8.4	8.2	9.6	14.4	6.2	8.2	8.4	14.4	6.2
5	BOD (mg/l)	6.8	3.6	3.8	4.5	7.6	3.8	4.9	4.5	7.6	3.6
6	Conductivity (m s/cm.)	0.236	0.294	0.285	0.257	0.228	0.346	0.289	0.229	0.346	0.228
7	TDS (mg/l)	143.96	179.34	173.85	156.77	139.08	211.06	176.29	139.69	211.06	139.08
8	Turbidity (mg/l)	118	4200	1550	78	104	4500	3000	84	4500	78
9	Chloride (mg/l)	46	54	48	41	39	66	43	32	66	32
10	Total hardness (mg/l)	168	218	188	152	166	224	208	178	224	152
11	Free - CO_2 (mg/l)	4.8	2.2	14	1.2	14.4	7.2	13	3.6	14.4	1.2
12	Alkalinity (mg/l)	120	138	122	114	130	110	124	92	138	92
13	Nitrite (mg/l)	0.016	0.014	0.008	0.005	0.007	0.009	0.007	0.005	0.016	0.005
14	Nitrate (mg/l)	2.519	5.742	6.348	2.636	3.541	4.246	3.874	2.754	6.348	2.519
15	Phosphate (mg/l)	1.174	1.248	1.664	1.246	1.947	2.842	1.994	1.678	2.842	1.174
16	Sulphate (mg/l)	46	52	38	60	48	52	40	46	60	38
17	Iron (mg/l)	0.024	0.032	0.041	0.013	0.015	0.008	0.011	0.013	0.041	0.008

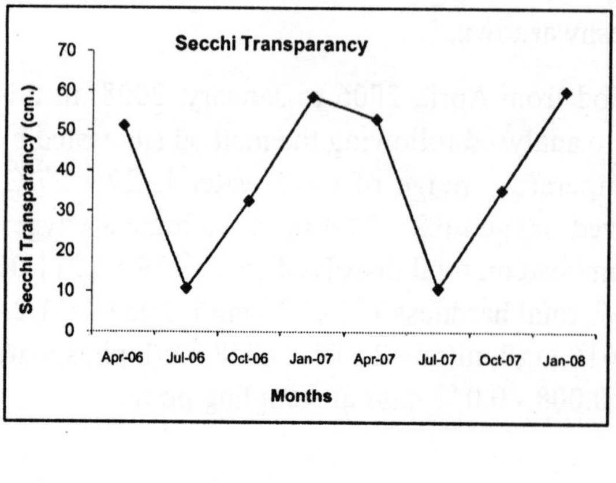

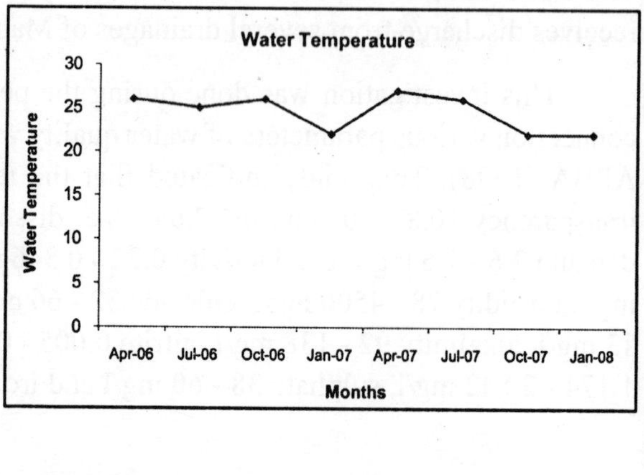

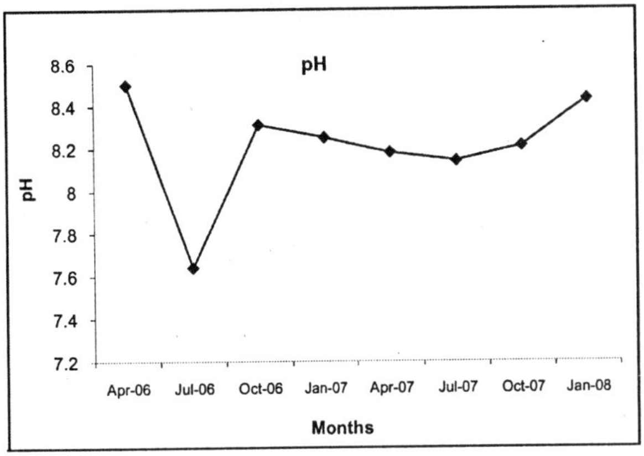
pH
8.6
8.4
8.2
8
7.8
7.6
7.4
7.2
pH
Apr-06 Jul-06 Oct-06 Jan-07 Apr-07 Jul-07 Oct-07 Jan-08
Months

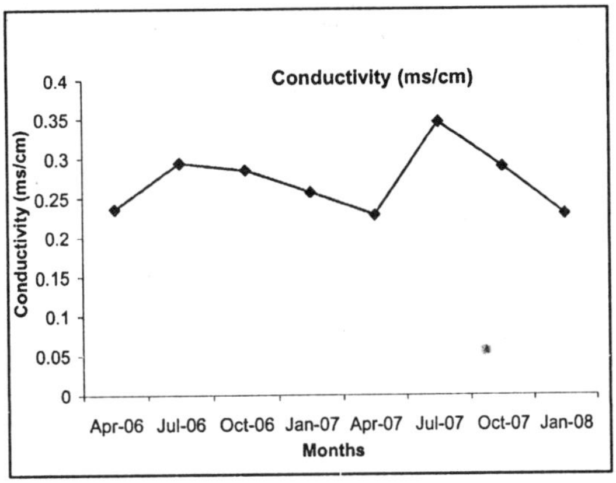
Conductivity (ms/cm)
0.4
0.35
0.3
0.25
0.2
0.15
0.1
0.05
0
Conductivity (ms/cm)
Apr-06 Jul-06 Oct-06 Jan-07 Apr-07 Jul-07 Oct-07 Jan-08
Months

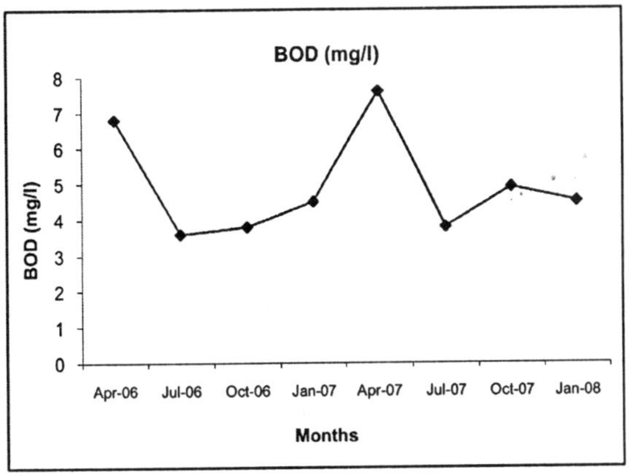
BOD (mg/l)
8
7
6
5
4
3
2
1
0
BOD (mg/l)
Apr-06 Jul-06 Oct-06 Jan-07 Apr-07 Jul-07 Oct-07 Jan-08
Months

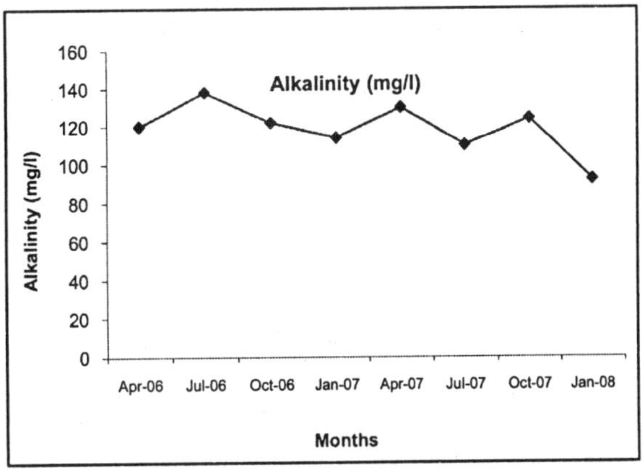
Alkalinity (mg/l)
160
140
120
100
80
60
40
20
0
Alkalinity (mg/l)
Apr-06 Jul-06 Oct-06 Jan-07 Apr-07 Jul-07 Oct-07 Jan-08
Months

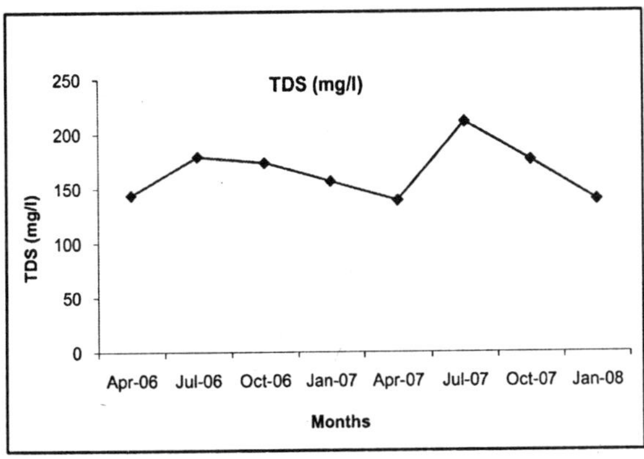
TDS (mg/l)
250
200
150
100
50
0
TDS (mg/l)
Apr-06 Jul-06 Oct-06 Jan-07 Apr-07 Jul-07 Oct-07 Jan-08
Months

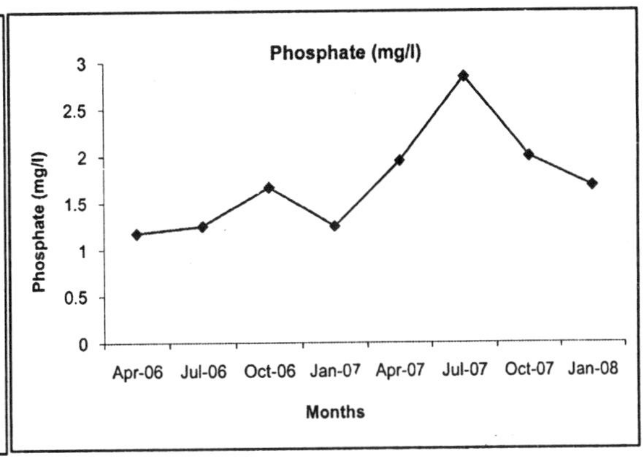
Phosphate (mg/l)
3
2.5
2
1.5
1
0.5
0
Phosphate (mg/l)
Apr-06 Jul-06 Oct-06 Jan-07 Apr-07 Jul-07 Oct-07 Jan-08
Months

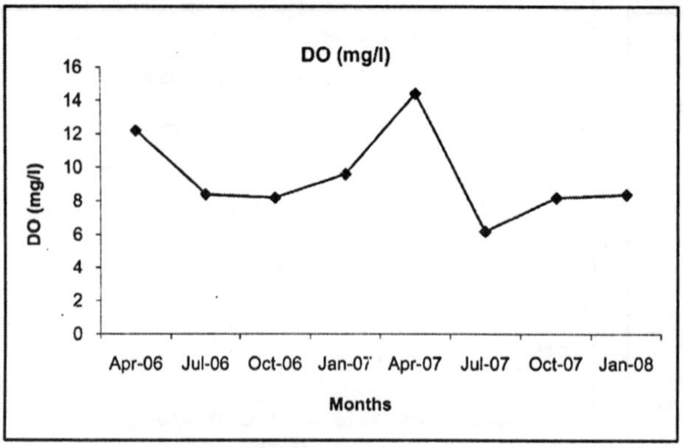
DO (mg/l)
DO (mg/l)
0 2 4 6 8 10 12 14 16
Apr-06 Jul-06 Oct-06 Jan-07 Apr-07 Jul-07 Oct-07 Jan-08
Months

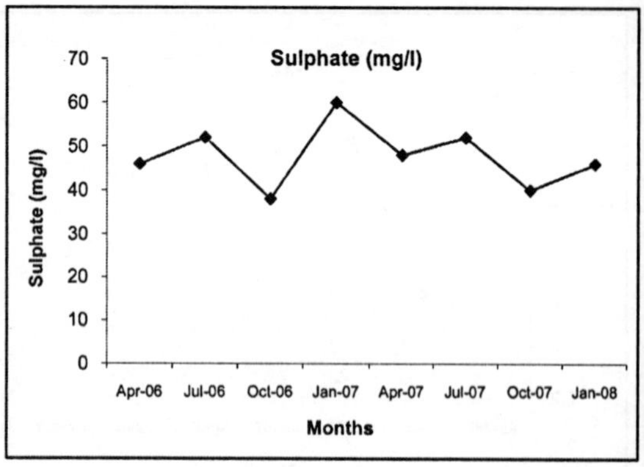
Sulphate (mg/l)
Sulphate (mg/l)
0 10 20 30 40 50 60 70
Apr-06 Jul-06 Oct-06 Jan-07 Apr-07 Jul-07 Oct-07 Jan-08
Months

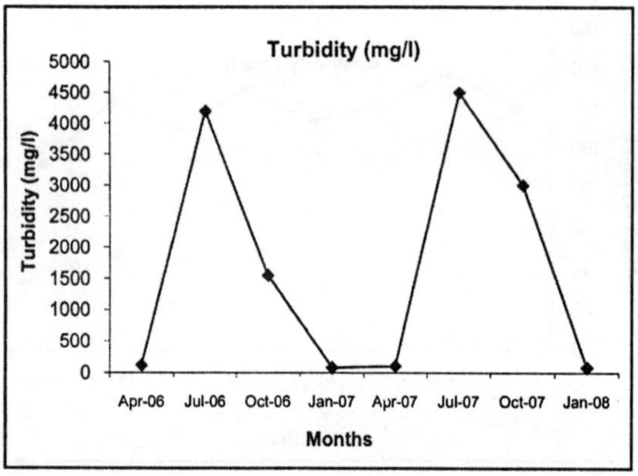
Turbidity (mg/l)
Turbidity (mg/l)
0 500 1000 1500 2000 2500 3000 3500 4000 4500 5000
Apr-06 Jul-06 Oct-06 Jan-07 Apr-07 Jul-07 Oct-07 Jan-08
Months

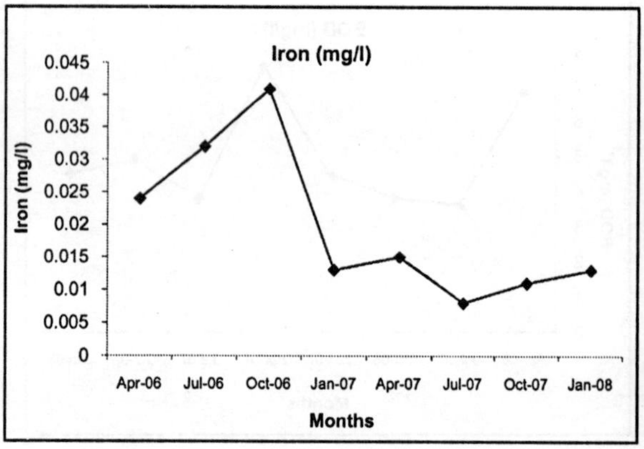
Iron (mg/l)
Iron (mg/l)
0 0.005 0.01 0.015 0.02 0.025 0.03 0.035 0.04 0.045
Apr-06 Jul-06 Oct-06 Jan-07 Apr-07 Jul-07 Oct-07 Jan-08
Months

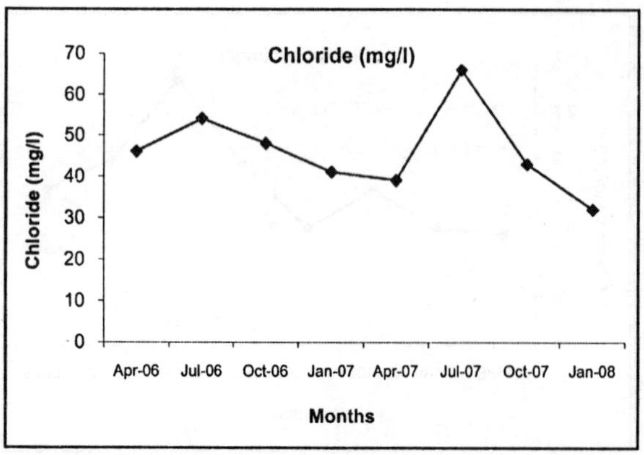
Chloride (mg/l)
Chloride (mg/l)
0 10 20 30 40 50 60 70
Apr-06 Jul-06 Oct-06 Jan-07 Apr-07 Jul-07 Oct-07 Jan-08
Months

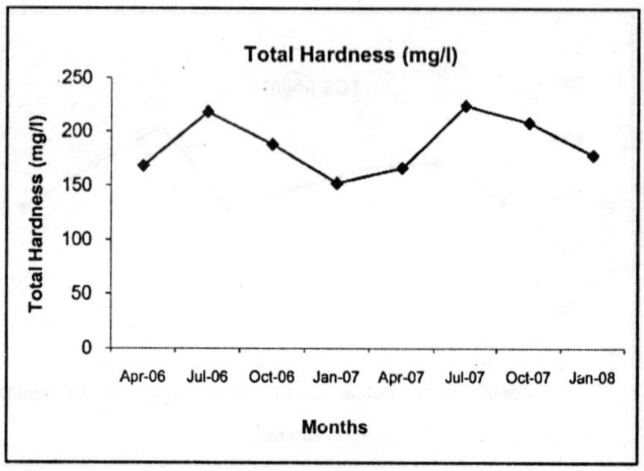
Total Hardness (mg/l)
Total Hardness (mg/l)
0 50 100 150 200 250
Apr-06 Jul-06 Oct-06 Jan-07 Apr-07 Jul-07 Oct-07 Jan-08
Months

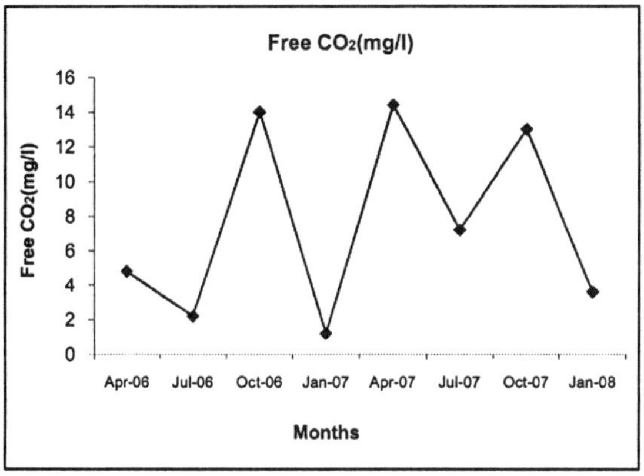

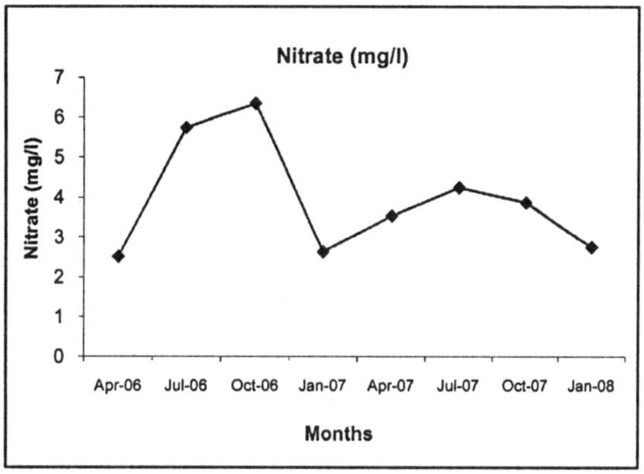

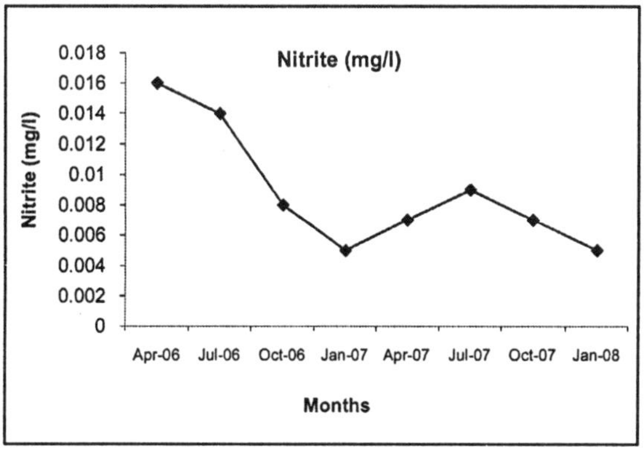

The fluctuation in water temperature has relationship with the air temperature. Similar fluctuation are reported by Khanna *et al.* (1999). Maximum values of the studied parameters were observed during summer season and shock load of first rain in rainy season except DO and minimum values of these parameters obtained during winter, while higher values of dissolved oxygen were observed during rainy period at sampling point which may be due to dilution in rainy period and super saturation of oxygen at lower ambient temperature and less human activities like bathing/ washing etc. in winter. After winter period as temperature of the environment rises bathing, washing and other human activities are also increased which gradually deteriorates the water quality of Narmada river (Maheshwar).

During the study period we found that pH variation lies mostly with in the alkaline range and conductivity values support the fact river is very productive (Rao 1993). Turbidity and transparency are also quite favorable of water ranged between 10.8 and 60.0 cm during the course of study and had a negative correlation with turbidity. Similar results are reported by Dagaonkar and Saxena (1992).

The dissolved oxygen ranged from 6.2 to 14.2 mg/l having moderate fluctuation. The presence of free CO_2 is due to incomplete utilization in photosynthesis and respiratory activity. Free CO_2 and dissolved oxygen showed inverse relationship to one another as similar to Welch 1952, Hutchinson 1857 and Hynes 1970. Chloride were low indicating no organic pollution in river. Chloride concentration was recorded from 32 to 66 mg/l showed a decreasing trend during the study but in the month July it increased up to a comparatively higher level.

Chloride and hardness showed a positive relationship. Similar type of relationship was observed by Khanna *et al.* 1999. Hardness and alkalinity showed negative relationship to one another and their range are quite favorable for fish life as indicated by Trivedi *et al.* (1986). BOD and DO showed a negative relationship as reported by Chopra and Patrick 1994 and other parameters presence of normal range.

Conclusion

It is the duties of concerned authorities as well as people of the area to protect such contaminations and bathing / washing by soaps because Narmada river is a most sacred river and worshiped by the Hindus. Public awareness is most essential specially for those who are responsible to pollute the natural water quality of river Narmada.

The values of various parameters obtained in the present study is found to be higher than the normal permissible limit. Hence the water is gradually becoming polluted and it is not advisable to use for drinking purpose unless it is well treated.

References

- Adoni, A.D., (1985). Work book on Limnology. Pratibha Publishers, Sagar (MP)
- APHA - AWWA WPCF (1998). "Standard methods for the examination of water and waste water." American Public Health Association, Washington DC.
- Chopra, A.K., Patrick Nirmal, J.,(1994) Effect of domestic sewage on self purification of Ganga water at Rishikesh I - Physico-chemical parameters. Advances in Biosciences. 13 (II), 75-82.
- Dagaonkar, A., Saxena, D.N., (1992) Physico-chemical and biological characterization of a temple tank, Kaila Sagar, Gwalior, MP. J.hydrobiol VIII (1), 1-19.
- Dhindsa, S.S. and bhargava, N., (1982). Periodic variation in the Physico-chemical characteristics of the river Chambal at Kota Acta Ecol., 4(2) 21-23.
- Golterman, H.L. and Clymo R.S., (1969). Methods for the chemical Analysis of Fresh water. B.P. Handbook No. 8 Blackwell. Sci.Pub.Oxford. pp 166.
- Hutchinson, G.E., (1975). A Treatise on Limnology vol.I John Wiley and sons, New York London 1050 pp.
- Hynes, H.B.N., (1970). Ecology of Running waters. Liverpool University Press. Liverpool IV impression 1-555.
- Khanna, D.R., (1993). Ecology and Pollution of Ganga river. Ashish Publishing House, New Delhi. pp 1-241.

- Khanna, D.R., Seth, T.R., Malik, D.S. and Singh, R.K., (1999). Fishes and their ecology of the river Suswa at Raiwala Dehradun In D.R. Khanna, Ashutosh, Goutam and Aditya Goutam (eds.) sustainable Ecosystem and Environment. 105-110.
- Khanna, D.R., Singh, Shakun, Gautam, Ashutosh and Singh, J.P., (2003). "Assessment of water quality of river Ganga in district Buland shahar (UP) India" J.Nat.Con. 15(1) 167-175.
- Rao M. Babu, (1993). Studies on the Ecology and Fish fauna of an oligotrophic lake Himayat Sagar, Hydrabad (A.P.) In K.S. Rao (ed) Recent Advances of Fresh Water Biology. vol - II. 73-98.
- Trivedi, R.K., and Goel, P.K., (1986). Chemical and Biological Methods for water pollution studies Environmental Publication Karad.
- Welch, P.S., (1952). Limnology. McGraw Hill Book Co., New York, pp 593.

25

CHIDIKHO - FOREST - A BIRD'S EYE-VIEW

Anuja Gupta, Shamanta Iyengar Surjeet Singh Jat***

Abstract: Forests are one of the largest eco-systems of the world. People living around the forest depend on it for their various needs. Poor management & overexploitation of the traditional forest grow poorer in density & quality. This calls for the study of forest status from time to time to assess their current economic status and future potentials such as generation of employment & revenues. A step, though preliminary, was undertaken to study the forest of Chidikho with this viewpoint. The study deals with the location, the common flora & fauna, the problem and the future prospects. Further scientific study & efforts in this field is expected to generate much needed financial resources for forestry development & bridge the gap between fulfillment of needs of people and awareness of the need for conservation of forests.

Keywords: Overexploitation, Forest status, Reforestation.

Introduction

Forests, have remarkable capacity of self regeneration and proliferation possible to neglect and precious gift. This nature gift has been taken for granted ruthlessly exploited by man for his need and for the animals reared by him. As a result vast tracts once famous for their greenery have taken the deterioration pathway. Forests are of utmost importance as they share a major role in maintenance of balance in nature hence intensive and systematic efforts should be taken for the improvement of the traditional forests. The study area centered around the Chidikho tank which is situated in the center of Narsinghgarh wild life sanctuary. (The name Chidikho is derived from the special shape of the tank). The study area is included in the Narsinghgarh game circle which is a part of the Narsinghgarh will life sanctuary notified in the M.P. state gazette of 15 Jan. 1974 under the will life (Protection) Act 1972.The sanctuary lies between the parallels of latitude 23°35' to 23°40' North

Motilal Vigyan Mahavidhyalaya Bhopal
*Govt. College Barnagar
** Govt. PG College Narsinghgarh

and meridian of longitude 77°2' to 77°10' east. The altitude is between 462.07 & 576.08 meters above msl covering an area of 57.197 sq kms. of projected forests.The terrain slopes moderately except one range of hill-ocks with steep slope running North to South. The hilly range is rocky and bouldory where as the plane varies from good fertile soil to graveily and laterite soil. (Plate-1).

Material & Method

The base line survey included collection of information from the Forest department and local village population. Listing of flora & fauna was done on the basis of morphological characters. Planktons were collected with the help of net & observed under the microscope.

Mammals

Susscrofa (Wild Pig), *Canis aureus*, *Conis lupus* (Wolf), *Felislybicaa* (desert cat), *Maccaca* sp. (Monkey), *Ratufa indica* (Squrril), *Caprolagus hispidus* (Hispid hare), *Felis temminki* (Golden cat), Teropus (Bat).

Reptiles

Crocodilus palustris (Crocodile), *Dermo chyscoriacea* (Lathery turtle), *Varanus griseus* (Menitor lizard), *Python molurus* (Pythons, *Naza-naja* (Cobra), Chamelion, Water snake.

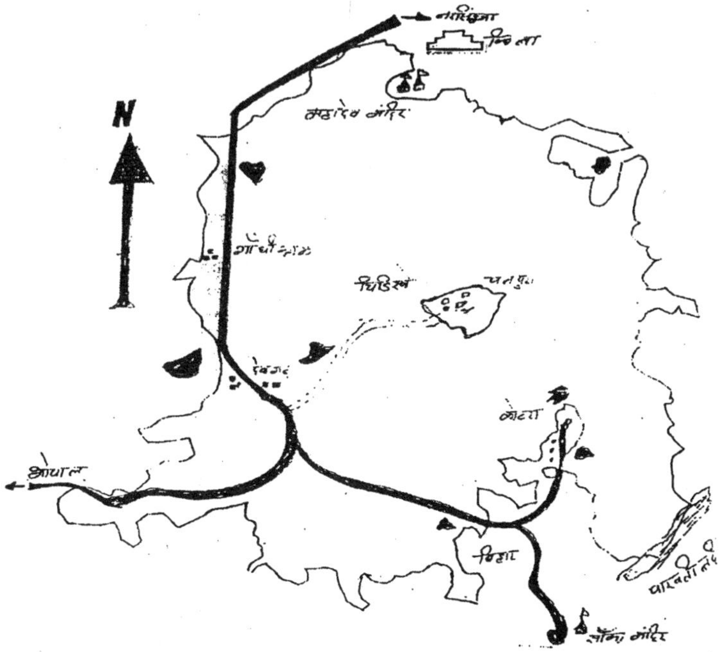

Plate 1. Map of Chidikho pond & Surrounding forest

Results

Major components of flora & fauna

The vegetation is of dry deciduous type. Major plant spp. found are listed below -

Sr.No.	*Botanical Name*	*Local Name*
1.	*Anogeissus latifolia*	Dhawda
2.	*Azadirachta indica*	Neem
3.	*Buchnania lanzan*	Achar
4.	*Feronia limonia*	Kabit
5.	*Ficus bengalensis*	Bar
6.	*Ficus glomerata*	Gular
7.	*Ficus religiosa*	Pipal
8.	*Hardwickia binata*	Anjan
9.	*Madhuca indica*	Mahua
10.	*Mangifera indica*	Aam
11.	*Melia azadirach*	Bakain
12.	*Mitragyna parvifolia*	Kadam
13.	*Phoenix sylvestris*	Khajur
14.	*Eugenia jambolana*	Jamun
15.	*Tamarindus indica*	Imli
16.	*Terminalia belerica*	Baheda
17.	*Terminalia tomentosa*	Sajar
18.	*Acacia leucophloea*	Khejda
19.	*Acacia nilotica*	Babul
20.	*Butea monosperma*	Palas
21.	*Cordia garaf*	Gundi
22.	*Diospyros melanoxylon*	Tendu
23.	*Erythrina sp*	Gadha Palas
24.	*Pongamia pinnata*	Karanj
25.	*Aegle marmelos*	Bel
26.	*Anona squamosa*	Sitaphal
27.	*Balanites aegyptiaca*	Hingot
28.	*Bauhinia racemosa*	Ashta
29.	*Cassia fistula*	Amaltas

30.	*Citrus limon*	Nimbu
31.	*Delonix regia*	Gulmohar
32.	*Euphorbia neriifolia*	Thuar
33.	*Manikara hexandra*	Bhawarsali
34.	*Moringa oleifera*	Surjana
35.	*Opuntia dillenii*	Nagphani
36.	*Pithecolobium dulce*	Vilayati Imli
37.	*Ricinus communis*	Arandi
38.	*Santalum album*	Chandan
39.	*Zizyphus mauratiana*	Ber
40.	*Zizyphus xylopyra*	Ghatoor
41.	*Adhatoda vasica*	Vasak
42.	*Calotropis gigantean*	Aakda
43.	*Carissa opaca*	Karonda
44.	*Helicteros isora*	Marorphali
45.	*Jatropha curcas*	Ratanjot
46.	*Jatropha gossypifolia*	Pilu
47.	*Lantana camara*	Lantana
48.	*Lawsonia inermis*	Mehndi
49.	*Maytenus emarginata*	Baikal
50.	*Thevetia peruviana*	Kaner
51.	*Zizyphus oenoplia*	Makoi
52.	*Calotropis procera*	Aakda
53.	*Datura alba*	Saphad Datura
54.	*Datura metel*	Kala Datura

Fauna - (Based on 2003 data)

Sr.No.	*Animal Name*	*Local Name*	*Namber*
1.	Panthera	Tendua	10
2.	Dancing deer	Sambhar	88
3.	Axis-axis	Chital	308
4.	Baselephus	Nila Gai	44
5.	----------	Lomdi	21
6.	Anteloep	Krishna Mrga	06

7.	----------	Bhadki	17
8.	----------	Chinkara	17
9.	----------	Jarkha	37
10.	----------	Sehi	14
11.	Crocodile	Magar	10

Birds

Terpsiphone pradisee (Asian paradise flycather), *Pavocristatus* (Peafowl), *Amandava formusa* (Blackheaded munia), *Lonchura malacea* (Green Munia), *Passer domesticus* (House sparrow), *Orthotomus sutorius* (Common tailorbird), *Turdoides striatus* (Jangle babbler), *Denodrocitta leucogosta* (White bellied treepie), *Corvus macro rhychos* (Jungle crow), *Acridotheres ginginianus* (Bank myna), *Dicrurus* (Bronzed drongo), *Halcyon pileata* (Black capped king fisher), *Collared scopes, Otus bakkamoena* (Owl), *Francolinus pondiceraianus* (Gray francolin), *Grus leacogeranus* (Siberian crane), *Columba livia* (Blue rockpigeon), *Streptopelia transquebarica* (Red collared dove), *Psittacula krameri* (Rose ringed parakeet), *Centropus sinensis* (Greater coucal), *Coracias bengalansis* (Indian roller), Fishes.

Morphometry of the pond

Area of water body

at maximum water level	312500 Sqm
at full tank level	187500 Sqm
Max. length at full tank level	850 m.
Max. width at full tank level	500 m.
Depth of water body	7 m.

Capacity -

Gross storage capacity	729167 Cum
Live storage capacity	375000 Cum
Catchment Area	2.4 Sq.Km.

Organisms in pond water

Macro phytes: Vallisneria, Potamojaten, Hydrilla, Chara, Nymphea, Wolfia, Nelmbo etc.

Protozoa:Lobomonas, Euglena, Heamatococus, Cryptomona, Pleodorina, Paranema, Anisonema, Amoeba, Diffusia, Paramoecium, Tintinnidium, Vorticella etc.

Other micro animalcules Sponges cyst, Gemules, Bryozoans, Statoblast and Colonial forms etc.

Rotifers Philodina, Brachionus, Trichocera, Keratella etc.

Nemathelminthes: Fasciola hepatica, Diplogasteroides etc.

Round worm platyhelminthes: Planeria etc.

Annelida: Polycheata, Leeches, Earthworms, Sludgeworm, Bristle worm, Manayunkia and Leech etc.

Arthropods: Sigara (Water boathman), Limnochares (Watermite).

Decapods: Shrimp (Mysis), Copepods (Diapotomus, Cyclopes), May flies nymph, Larval forms of midge larva (Chironomous).

Crustacea: Moina, Daphnia, Letodora etc.

Fish: Major carps, Minor carps and cat fishes.

Specially: Cypriniformes, Notopteridae, Cobitidae.

Labeo rohita, *Labeo kalabasu*, *Catla*, *Cirrinus mrigala*, *Channa Stryctus*, *Lepisauteus*, Mahasheer fishes are mainly found in the tank.

Phytoplankton:

Cyanophyceae: Anacystis, Oscillatoria, Anabaena, Lyngbya, Rivularia.

Chlorophyceae Scenedesmus, Selenastrum, Pediastrum, Ankistrodesmus, Zygnema, Spirogyra, Oedogonium, Ulothris.

Bacillariophyceae: Asterionella, Synedra, Gomphonema, Navicula, Nitzschia, Cymbella.

Status of the Area:

No special attention is given for the artificial regeneration of the forest. No special work is being undertaken for preservation of wild life. There is no ban on movements of local villagers at any tie who daily go to the protected areas for collection of fuels and non-wood forest products such as jamun, jack_fruit, ber, awala, tendu fruit, imli, karonda, sitaphal, leaves of ashta, palas, sal, tendu etc. The villagers are farmers, labourers & rear cattle for milk. During scarcity of water and forage supply and otherwise the cattles graze within the protected area.

These practices within the area not only disturb the wild life, they also create the gaps and thin down the crown of the trees. Since self regeneration of traditional forests take time and efforts towards artificial regenaration is nil wide gaps are being created in the area which is leading to soil erosion.

Road crossings, Deogarh to kotra over a length of 5 kms., Deogarh to Chaipura 5 kms. in sanctuary, Kotra Narsinghgarh cart tract 10 kms. along the boundary of sanctuary are also a source of disturbance to wild life.

Future Prospects

Plantation is the need of the hour. Hence a nursery should be established within the forest region. To fill up the gap in shortest time it is important that the fast growing plant spp. are selected. These plants should be protected from the cattles & other vandalizing agents. Enrichments plantations of such forage & fodder spp. of trees, shrubs & grasses should be done that serves as good cover for wild animals and are favoured by the herbivores.The whole forest is at present treated as single zone. Core & buffer zones must be clearly marked out for effective management.

Detailed list of spp. should be prepared for identification of endemic as well as endangered spp. so that special attention can be paid towards them. Artificial dens should also be designed for habilitation of carnivores in the area. In summer season artificial feeding of herbivores should be done until the range reaches self-sufficiency. Efforts should be done towards the building up of reproduction potentials in animals which are growing fewer in number. In the area number of birds are seen. In summer season fire and absence of cover & fruit trees make it difficult for birds to thrive. Till the cover improves by plantation artificial feeding of birds should be done in lean months.

Cattle grazing must be totally prohibited or restricted to definite areas. Crocodiles should be reintroduced in chidikho tank. Study on their habit growth & reproduction should be encouraged. Masasi, Suryavanshi and Futa tank should be taken care of to reduce pressure on the Chidikho tank. Aquaculture can also be done in these tanks. A wide veriety of insect species thrive on threes and generate a number of goods besides honey such as was lac & taser silk. Plantation of trees for production of these can be done in a part of forest.

Social forestry should be encouraged in nearby villages to reduce pressure on existing forest. It is now time to awaken to the fact that without proper management and conservation the forest resources will remain permanently inexhaustible. They are bound to disappear in coming years if they are overexploited. Hence best possible efforts must be putin to redress the damage. In the regard active involvement, co-operation and participation of the people living around the forest is of utmost important as they are strongly linked with these areas. This can be done by silviculture & lac culture. These being labour intensive would create employment opportunities to the villagers and generate revenue.

Acknowledgement

The authors are thankful to the Principal, Govt. P.G. College Narsinghgarh, Forest Department officials, local villagers and collage students of Narsinghgarh. We are also thankful to lab attendants Rajesh Bhilala & Dulichand Nagar.

References

- Ali S., 2002. The Book of Indian Birds, Bombay Natural History Society, Oxford Univ. Press.
- Brandis D., B. Singh & M. Singh, 1971. Indian Trees, Pub. Dehradoon.
- Odum E.P., 1996. Fundamental of Ecology, 1st Indian Edi. Natraj Pub. Dehradoon.
- Prescott G.W., 1969. The Algae A Review, Thomas welson & Son London.
- Raju R.A., 2000. Wild Plants of Indian Subcontinent and Their Economic Use, C.B.S. Pub. & Distri. Dariyaganj, New Delhi.
- Ranga M.M., 2002. Wildlife Management and Conservation, Agrobios (India) Jodhpur.
- Rangothaman G., 2002. Aquatic Ecology, Agrobios (India) Jodhpur.

26

EARLY BREEDING OF CARPS – AN ALTERNATIVE TO INCREASE THE INCOME OF FISH FARMERS

Pankaj Patial and Jagdish Markanday

Abstract: Indian major carps breeds normally once a year during the monsoon season. With the use of improved brood-stock management practices, supplementary feed and intermittent doses of Pituitary gland extract, Indian major carps were bred in early summer (April). A good fertilization, hatching and survival rate was obtained in early breeding. Early breeding ensure the availability of fish seed from April to September which is from July to September in normal breeding. It has been observed that survival of spawn in nursery ponds in month of April, May and June is much higher as compared to July and August. Fish Production will increase by 1000-1500 kg/ha/yr as fish fingerlings attain additional growth of 100-150 gm in 3 months. By adopting this technology, fish farmers and fish breeders can increases their income by Rs. 30,000 to 50,000/ha/yr in the same pond and from same stocking density. The same sets of brooder can be bred again after a gap of 55-60 days in their natural breeding season which further increases their availability of seed and also the income of fish farmers and fish breeders.

Keywords: Indian major Carps, Management, supplementary feed, Pituitary gland and spawn.

Introduction

The appropriate use of various new technologies of carp culture has shown a major success by achieving higher yields necessary to meet increasing demand caused by population growth and an increase in per capita consumption. In India, too, fisheries sector has come long way and has contributed immensely to food basket of the country. Development of the techniques for induced breeding of Indian major carps (IMC) using pituitary gland extract (PGE) by Chaudhury and Alikunhi (1957) has greatly contributed to the rapid development of carp culture and revolutionized carp seed production in the India. Fresh water aquaculture in India has evolved from stages of domestic activity in Eastern States like West Bengal and Orissa to that of industry in recent years, while in

National Dairy Research Institute, Karnal-132 001 (Haryana), India; Email: patialpankaj24@rediffmail.com

States like Andhra, Haryana, Punjab, Uttar Pradesh and Maharashtra the fish culture has been taken as a trade.

It is significant that fish production has been increasing at high rate and about only one third of fresh water in the country has been put to use for aquaculture. The requirement of quality fish has grown manifold after aquaculture has established itself as a highly viable farming enterprise from domestic backyard activity. According to Gopakumar *et al.* (1999), the fish fry requirement for entire country is estimated to be 153626 millions sufficient to stock 2358634 ha of ponds and tanks. In case of Northern states (Haryana, Punjab, Himachal, Rajasthan, Uttar Pradesh and J&K) the fish fry production from all sources is 980.51 million whereas the requirement is 2247.6 million of fish fry which indicates that there is a shortage of 1267.09 million fry. The water available under ponds and tanks in these six states is 377000 ha. It is estimated that these ponds and tanks have a capacity to produce 501300 tons of fish, if stocked properly as per recommendations.

Since resources in term of land and water are limited, so the best option is to breed fish in early summer and continued till September instead of July on the onset of monsoon. With this objective in mind, he present work has been undertaken to develop a technology to induce breed the IMC i.e. Catla (*Catla catla,*), Rohu (*Labeo rohita*) and Mrigal (*Cirrhina mrigala*) in early summer i.e. in April were developed. This will be helpful to meet the fish seed requirement of country and to increase the income of fish farmers.

Material and Method

Pond management

Rectangular pond having an area of 0.2 ha with an average depth of 1 meter was selected at Sultan Fish Seed Farm, village Butana, Karnal. The pond was dewatered, dried and desilted to increase depth up to 2 meters. The pond was ploughed and dried for 10 days and treated with lime (@ 300 kg ha^{-1}. The pond was fertilized weakly with organic and inorganic manure to ensure the proper production of natural food organisms. Details of manure application are given in Table-1. The source of water supply was underground deep tube well. The pond water was changed by 20-25% every 15th days. The seasonal variation in physico-chemical parameters water and meteorological data of air were recorded.

Brood stock management

Immature fishes of Catla (20 pairs), Rohu (20 pairs), and Mrigal (20 pairs) of 2+ age group were stocked at Sultan Fish Seed Farm, Butana in November 2008. The stocking density of the brood fishes was maintained 1300 $Kg.ha^{-1}$. As control, 15 pairs of fishes were also stocked in separate pond of same depth. Supplementary feed containing 35% crude protein, 8% lipids, 2% common salt, 1% mineral mixture (with trace elements and vitamins) and terramycin egg formula (Pfizer) was used. The ingredients of feed are given in Table-2. Supplementary feed was given @ 1% of body weight from November to January, when water temperature remains low and 2-3% during the rest of the period from February to October. Fishes used as control was given groundnut oil-cake and wheat bran in ratio 11 @ 2 % of body weight as supplementary feed. The sampling of the brood fish stock was done fortnightly to assess the growth of the fish.

Table 1 Manuring of Broodstock Pond

Manure	Quantity
Organic Raw dung	15000 $kg\,ha^{-1}yr^{-1}$ (287.0 $kg\,ha^{-1}day^{1}$)
Inorganic manure Urea, Single Super Phosphate Murate of potash	 200 $kg\,ha^{-1}yr^{-1}$ (3.78 $kg\,ha^{-1}week^{-1}$) 400 $kg\,ha^{-1}yr^{-1}$ (7.56 $kg\,ha^{-1}week^{1}$) 50 $kg\,ha^{-1}yr^{-1}$ (0.945 $kg\,ha^{-1}\,week^{1}$)

Table 2 Composition of Supplementary Feed

Ingredient	Amount/100kg
Groundnut-cake	50.0 kg
Rice bran	27.0 kg
Maize gluten	20.0 kg
Common salt	2.0 kg
Mineral Mixture with Trace elements, Vitamin A, D_3 & E	1.0 kg
Terramycin egg formula (Pfizer)	0.001%

Administering intermittent doses of PGE

Catla, Rohu and Mrigal (20 pairs each) were administered doses of PGE from January to April 2008. Details of PGE administered intra-muscular to advanced gonadal maturity are given in Table

Table 3 Hormonal Treatment for enhancing gonadal maturity

Sr. no.	Date	Dose of PG extract mg/fish
1	12.01.2008	1
2	10.02.2008	1
3	05.03.2008	2
4	20.03.2008	2
5	05.04.2008	2

3. Before giving the each dose, the general condition of the fishes and their gonadal maturity were observed. The controlled fishes were given normal saline at rate of 0.5 ml/ kg of body weight instead of PGE.

Induced breeding experiments

Induced breeding experiments were conducted on 19 April 2008 to breed 30 pairs of brood fishes (Catla 8, Rohu 12, Mrigal 10 pair) by injecting PG extract. Details are given in Table 4. Females were given higher doses of fish PG extract @12-16 mg/kg in 3 doses (1st @2 mg/kg, 2nd @2-4 mg/kg, 3rd @ 8-10 mg/kg) depending on maturity condition of fish. The interval of the injections varied from 4-6 hours. Males were given PG extract @ 4 mg/kg in two split doses. First dose to males were given @ 1 mg/kg at the time of first injection to the female. Second dose to males were given @ 3 mg/kg at the time of third injection to the female.

Table-4 Breeding experiment

Fish	Injected	Ovulated	% ovulation	Eggs produced (in lakhs)	% fertilization	Hatchlings produced (in lakhs)	Hatchling %age
Catla	8	7	80	94.83	88	69.26	83
Rohu	12	10					
Mrigal	10	7					
Control	15	0					

Care of spawn, hatchlings, fry and fingerlings

For first three days spawn was kept in circular pool with running water. Three days old hatchlings were nursed in nursery pond for 15 days and fed with egg yolk for first few days. Most of fingerlings were sold after this stage. To check the growth of fingerlings, 10000 fingerlings were transferred to a rearing pond.

Results

By adopting improved brood-stock management practices and administering intermittent doses of PGE, 30 pairs were found to be matured in third week of April. Females were found to be gravid and males were oozing. After achieving the early maturation, induced breeding experiments were carried out. Results of induced breeding are given in Table-4. Out of 30 brood female fishes, 24 brood female fishes bred successfully and produced 94.83 lakhs eggs. Eggs were released 5-7 hours after final injection. The average rate of fertilization was about 88%. and average rate of hatching was 83%. About 6-8 hours after spawning the eggs became hard. Hatching occur within 18-24 hours after spawning. Hatchlings were fed with egg yolk after three days when they start feeding. Fecundity was 1.6, 1.65, 1.45 lakhs eggs.kg^{-1} of body weight for *Catla, Rohu* and *Mrigal* respectively.

The availability of fish seed has been extended now from April to September which was earlier from July to September. It has been observed that survival of hatchlings and fingerlings in nursery and rearing ponds in month of May-June is much higher as compared to July-August. Productivity increased by 1000-1500 kg/ha/yr as fish attains additional growth of 90-120 gm by 3 months. The technology developed will enhance the income of fish farmers and fish breeders Rs. 30,000 to 40,000/ha/yr.

Discussion

IMC (*Catla, Rohu* and *Mrigal*) show gonadal recrudescence as early as March in Assam (la. ca 25°N), June in Orissa (la.ca 20°N) and July in other parts of India depending on their geographical location (Khanna, 1958; and Jhingran, 1975). But by adopting improved brood-stock management practices and administering deferred doses of PGE, the gonadal maturity has been preponed by 3 months in IMC.

In the month of April 2008, average water and air temperature was 24.01°C and 22.7°C respectively The average sunshine was 11.9 hours per day. It is clear from Figure-1, 2 and 3 that temperature, photoperiod and rainfall was increasing from February onward. Since the IMC show gonadal recrudescence from March to June, at the time when both photo period & temperature are increasing, so it is inferred that photoperiod and temperature play an important role in the seasonal maturation of ovary and testes in Indian major carps (Bhattacharyya *et al.*, 2005; Bhattacharyya and Maitra, 2006 and Dey *et al.*, 2005). Verghese (1968) reported early sexual maturation and

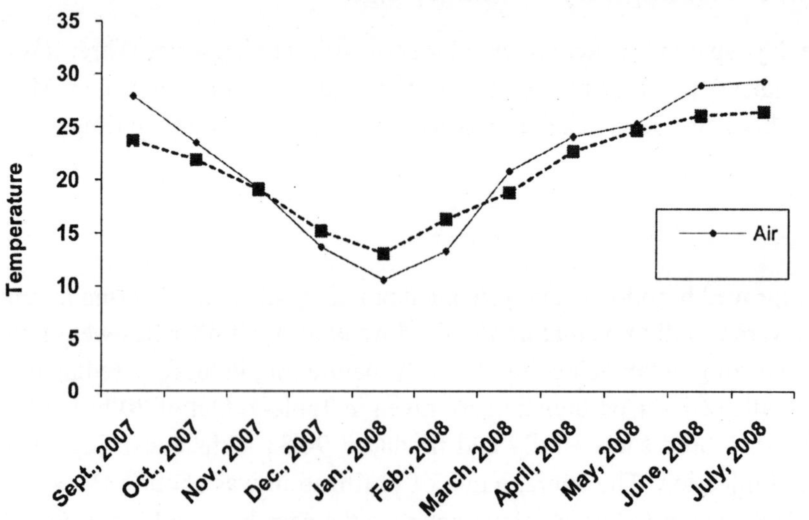

Fig. 1. Average water and air temperature (°C) from Sep 2007 to July 2008

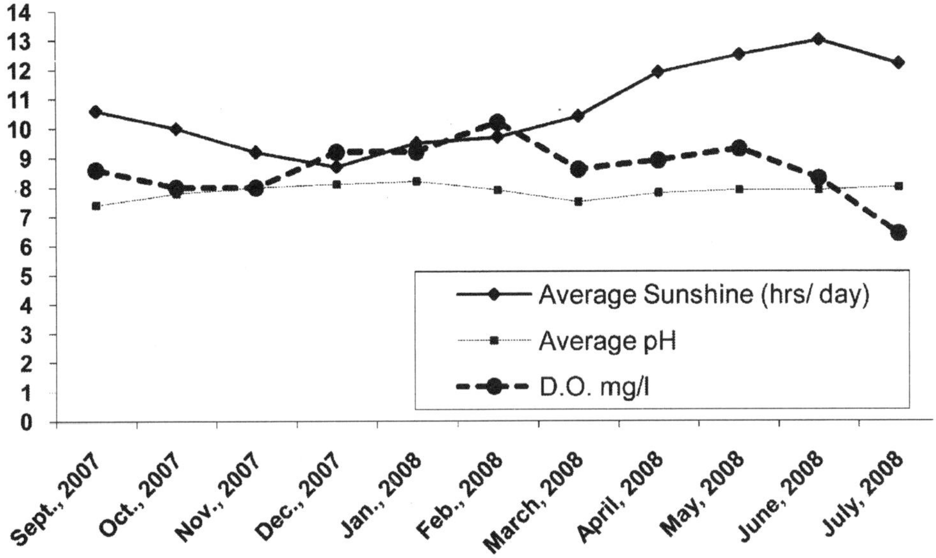

Fig. 2 Details of pH, Dissolved oxygen of pond water and day length during experiment.

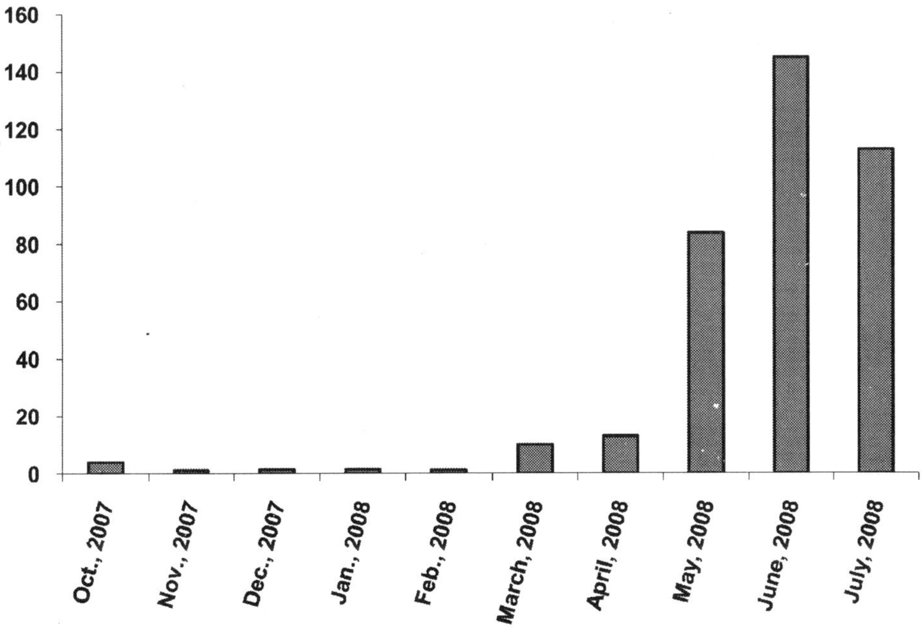

Fig. 3 Rain-fall (mm/month) during experimental year.

spawning in *C. reba* under the influence of artificially controlled light and temperature. The interaction of environmental stimuli (light, rainfall, temperature) affects the endocrine system of carps (Bhowmick, 1977).

Improved brood stock management practices played a key role in early maturation of fishes. The role of management in rearing the brood stock has been highlighted by Alikunhi *et al.* (1960, 1964) and Chaudhary (1960). Both organic and inorganic fertilizers were used to maintain the productivity of pond. Liming of pond was done to maintain the pH of pond between 7.8- 8.2 which is good for fish growth and maturation. Patino *et al.* (2005) also showed that pH ranging from 7.6-8.2 facilitate induced maturation and ovulation. The stocking density of the brood was maintained @ 1300 $kg.ha^{-1}$. Sehgal and Toor (1995) reported that lower stocking densities helps in early ovarian maturation. Supplementary feed containing groundnut oil-cake, rice bran, common salt, mineral mixture with trace elements and vitamins and terramycin egg formula (Pfizer) were given @1-3% body weight. According to Luquet and Watanabe (1986) a restricted diet during early stages of life cycle delays the first maturation and restricted diet during oocyte differentiation reduces the egg number. They also reported that vitamins, fatty acids and minerals are essential for normal reproduction. Terramycin egg formula was given with feed @ 0.001% which stimulates and sustains early and high rate of egg production.

Intermittent doses of PGE were administered @ 1-2 $mg.kg^{-1}$ of body weight from January to April 2007. Fishes were found to be matured in last week of March. Sukumarn and Rahman (1984) bred mrigal by hypophysation in the month of December 1983, bringing the fish into proper maturity by injecting pituitary extracts in the month of November. Pedersen (2003) had also induced the maturation in European eel by weekly injection of PGE for 16-25 weeks. For final maturation and breeding fishes were injected with PGE. Pituitary Gland secretes gonadotrophin hormones (FSH and LH) that play important role in maturation and spawning of fish (Khanna and Singh, 2003). For breeding, circular breeding pool with running water were used. Radheyshyam (2007) reported higher egg production and fertilization rate in breeding pool as compared to breeding hapas. Sinha *et al.* (1974) indicated the role of running /flood water in lowering down the electrolyte level of water causing gonadal hydration for successful spawning. Eggs were released 5-7 hours after final injection. About 6-8 hours after spawning the eggs became hard. Hatching occur within 20-24 hours after spawning. Three days old hatchlings were nursed in nursery pond for 15 days and fed with egg yolk. By this time they grow to 30-35 mm. Then fingerlings were transferred to rearing pond. They attain a size of 90-120 gm till the end of June.

Conclusion

With the application of improved broodstock management and administration of PGE from January to April, the breeding of Indian major carps was advanced by 2-3 months. This will help in mass production of fingerlings before rainy season which is most suitable time for stocking. The technology has direct impact on production, more revenue to farmers and availability of fish seed over longer period. This will play a significant role in meeting the fish seed requirement of the country. The technology will also be useful for multiple breeding as the same fishes can be bred again after a period of 55-60 days.

Acknowledgement

We are thankful to the Director NDRI, Karnal, Deputy Director General (Education), ICAR, New Delhi for providing facilities and financial support. We are also thankful to Sh. Sultan Singh for extending all help.

References

Alikunhi K.H., Vijalakshmanan M.A. and Ibrahim K.H. 1960. Preliminary observations on the spawning of Indian carps, induced by injection of pituitary hormones. Indian Journal of Fisheries 7(1) 1-19.

Alikunhi K.H., Sukumaran K.K., Parameswaran S. and Banerjee S.C. 1964. Preliminary observations on commercial breeding of carps under controlled temperatures in laboratory. Bulletin of Central Inland Fisheries Research Institute, Barrackpore 3 19-20.

Bhattacharyya S., Dey R. and Maitra S.K. 2005. Photoperiodic regulation of annual testicular events in the Indian major carp *Catla catla*. Acta Zoologica (Stockholm) 86 71-79.

Bhattacharyya S. and Maitra S.K. 2006. Environmental correlate of the testicular events in a major carp *Catla catla* in an annual reproductive cycle. Biological Rhythm Research 37(2) 87-110.

Bhowmick R.M., Kowtal G.V., Jana R.K. and Gupta R.K. 1977. Experiments on second spawing of Indian major carps in the same bdreeding season by hypophysation. Aquaculture 12 149-155.

Chaudhuri H. and Alikunhi K.H. 1957. Observations on the breeding of Indian carps by hormone injection. Current Science 26 (12) 381-382.

Chaudhuri H. 1960. Experiments on induced spawning of Indian carps with pituitary injections. Indian Journal of Fisheries 7(1) 20-48.

Dey R., Bhattacharyya S. and Maitra S.K. 2005. Importance of Photoperiods in the Regulation of Ovarian Activities in Indian Major Carp *Catla catla* in an Annual Cycle. Biological Rhythm Research 20 145-158.

Gopakumar K., Ayyappan S., Jena J.K., Sahoo S.K., Sarichar S.K. and Nayal P.K. 1999. National Fresh Water Aquaculture Development Plan. Central Institute of Fresh Water Aquaculture, Bhubaneshwar. pp. 75.

Jhingaran V.G. 1975. Fish and Fisheries of India. Hindustan Publishing Corporation, New Delhi. pp. 954.

Khanna D.V. 1958. Observations on the spawning of the Indian major carps in Punjab. Indian Journal of Fisheries 5(2) 282-290.

Khanna S.S. and Singh H.R. 2003. A text book of fish biology and fisheries. Narendra Publishing House, New Delhi. pp. 403-421.

Luquet P. and Watanabe T. 1966. Intraction "nutrition-reproduction" in fish. Fish physiology and Biochemistry 2(1-4) 121-129.

- Patino R., Bolamba D., Thomas P. and Kumakura N. 2005. Effects of external pH on hormonall regulated ovarian follicle maturation and ovulation in Atlantic croaker. General and Comparativ Endocrinology 141(2)126-34.
- Pedersen B.H. 2003. Induced sexual maturation of the European eel Anguilla anguilla and fertilizatio of the eggs. Aquaculture 224 (1-4) 323-338.
- Radheyshyam. 2007. Spawning efficiency of Indian major carps in hapa and spawning pool in rur area- A comparative study. National Symposium on Ecosystem health and fish for tomorrow, inlan fish society of India and Central Inland Fisheries Research Institute, Barrackpore.55
- Sehgal H.S. and Toor H.S. 1995. Effect of stocking density on ovarian maturation, offspring fitnes and growth of common carp. Aquaculture 129(1-4)113-117.
- Sinha V.R.P., Jingran V.G. and Ganapati S.V. 1974. A review on spawning of Indian major carp Archives of Hydrobiologia 73 (4)518-36.
- Sukumaran N. and Rahman Md. K. 1984. Progress Report of the Tuticorin Sub-Centre for the perio April 1982 to February 1984. Seventh Workshop on AICRP on Composite Fish Culture & Fish See Production, p 144-169.
- Verghese P.U. 1968. Preliminary experiments on modification of reproductive system of inian majc carp, Cirrhinus reba by control of light and temperature. Proc.Indo Pacif. Fish 13 (2) 171-184.

27

STUDIES ON HABITAT DEPLETION AND CONSERVATION STRATEGIES OF DALMA WILDLIFE SANCTUARY, JHARKHAND, INDIA

S.R. Pal

Abstract: The present chapter deals with causes of habitat depletion and conservation strategies of the Dalma wildlife sanctuary. The Dalma wildlife sanctuary is located in East Singhbhum district and partly Saraikella-Kharsawan district of Jharkhand state containing diverse habitat due to its specialized location in south Chhotanagpur plateau. The paper also describes some important faunal and floral species which are under threat of different categories due to habitat loss by various factors. Author has also suggested various conservation measures to check habitat and biodiversity loss of the sanctuary.

Keywords: Habitat, Depletion, Wildlife, Conservation

ntroduction

'he depletion of habitat including biodiversity is a global phenomenon. Its causes are varied, nany and different from realm to ream, continent to continent, country to country, state to state and listrict to district. Important legislations have been framed and steps are being taken at global and ountry level to conserve and to protect as well as to restore and regenerate the habitat and the iodiversity for prosperity of human race and for maintaining the ecological balance of varied cosystem of the earth. But the inferences as should be are yet to be obtained practically. In other vords it can be said the result is not up to the mark. The habitat and bio-diversity scenario is fading lay after day from macro to micro level because legislations, rules and regulations which are made o conserve and protect the biodiversity and steps proposed to be taken for the same are mostly neant for global level and country level and are insuitable for a micro area of an other geographical rea. As such the causes of vanishing of habitat are different in Dalma Sanctuary to that of other

)eptt. of Zoology, S.P. College, Khasmahal, Jamshedpur - 831002; E-mail: Satyaranjan_pal@yahoo.com sanctuary of the state. Here I have tried to find out the causes of depletion of habitat in this sanctuary which I have been observing since last 25 years.

Materials

Dalma wildlife sanctuary, inaugurated by the late Sanjoy Gandhi in 1975 is located 10 kms from Jamshedpur town (22°48′N to 86°12′E) Jharkhand. The sanctuary extends over the thickly forested slopes of the 3000 ft high Dalma Mountain Range. The total area of the Sanctuary is about 19: sq.km. The forest offers excellent cover for its inhabitants which include elephants, barking deer sloth bears, leopards, porcupines and tigers. Several hideouts have been identified and marked out in the sanctuary to help wildlife lovers sight the wild animals in their natural habitat. The steep slopes of the sanctuary provide excellent trekking opportunities and attract many trekkers throughout the year. Another attraction within the sanctuary is a temple dedicated to Shiva, which is visited by a large number of devotees during Shivaratri. The sanctuary provides tourist accommodation in the Tata Steel and Forest Department Guest houses. October to June is the best time to visit.

Observations

Observations on depletion of habitat and biodiversity in the district have been made since 1984 to uptil now. The factors of depletion of the biodiversity in the plateau are:

Low rate of literacy among the ecosystem people (specially among the aboriginals, allied to aboriginals & semi-hinduized aboriginals)

Lack of awareness of the importance of Environment among the Ecosystem people.

Non-existence of ecosystem people forum or village committee (s) for conservation of the forest.

Illegal felling of trees of the forest coverage by antisocial elements.

Lack of employment opportunities for the ecosystem people. No serious attention has been paid by State Govt. & Tata Steel on this aspects uptil now.

Paucity of alterative arrangement of fire wood. More than 75% people use wood for cooking purposes & 90% Ecosystem people depend on forest bush, shrub, and trees for cooking and heating purposes.

Appointment of corrupt and insufficient forest officials.

Illegal collection of forest product by non-licensed holder persons.

Improper implementations of forest legislation (s) by the authorities.

Misutilization of fund meant for the management of forest conservation and habitat protection

Observation of "Sendra Festival" in the Dalma region in mid April every year by aboriginals and allied to aboriginals.

Construction of Road (s) and Building (s) in the forest coverage.

Illegal quarrying (s) in the forest coverage by every money earners.

Afforestation of exotic plant spp. after destroying the bushes in the forest coverage.

Lack of abundant irrigation facilities and water resources in the forest coverage.

Lack of awareness campaign (s) regarding the significances or needs of habitat in daily life of mankind.

Lack of literacy campaign (s) in the remote hilly tracks where the very primitive tribes inhabiting and lack of economic development and communication facilities in those areas.

Increase of population and insufficient shelters for the people.

Poverty among the aboriginals, allied to aboriginals and Semi-hinduized aboriginals.

Killing of wild animals by pouchers.

Remedial steps to be taken to conserve the habitat in the sanctuary

More employment opportunities for aboriginals, allied to aboriginals and Semi-hinduized aboriginals.

Vigorous afforestation programmes which should be implemented honesty.

NGO (s) should work sincerely and honesty to conserve the habitat and to control the birth rate in the district.

Illegal felling of trees and pouching of wild animals should be completely stopped.

More education facilities should be given to aboriginals allied to aboriginal & Semi-hinduized aboriginals, who are living in the inaccessible areas.

Awareness campaign of the importances of habitats and biodiversities should be done at least once in a month in Panchayat level through government organization and NGO(s) existing in the district.

Education of peoples on the rights of originals, allied to aboriginals and semi hinduized aboriginals in forests.

Educating people regarding the concealed economic and environmental costs of over consumption of habitat resources.

Implementation of *in situ* and *ex situ* policies of biodiversity conservation.

Banning of useless and harmful politics on the biodiversity and habitat conservation and environmental restoration.

Meeting all the genuine social needs and lagitimate aspirations of people by blending economic development with environmental and biodiversity impertives to remove poverty.

Discussion

The sanctuary covers a vast area and abounds in natural vegetation cover. The sanctuary comprises of various species of plants trees, which are hard to find somewhere else. The sanctuary is a home to various species of animals that mainly comprises of elephants, gaurs, black buck, chital, wild dogs, sambar, gharials, muggers and many more.

During summer various herds of elephants come to this place in search of water. Gaurs remain close to elephants as they can feed on the clumps that are pulled down by elephants.

At the beginning of its establishment it was very rich in the faunal and floral composition. But recently the biodiversity and habitat scenario of the sanctuary has depleted beyond imagination due to unkind and unsympathetic behaviour of the people living in its nearly places.

Thus the habitat and the biodiversity in the sanctuary is not in a good state and it is necessary to regenerate it for the betterment & healthy as well as prosperous life for all living beings.

References

- Govt. of Jharkhand (2003): Hand book of forest Statistics, Forest dept.
- Mahata, M.C. (1982-83): Wild life in Chotanagpur Plateau Arnyak, Tata College Prakashan Sansthan, Chaibasa 89-107.
- Proter S.H. (1955): The book of Indian Animals, Bombay Natural History Society.
- Prabhat Khabar(2010): Dalma Me Sendra Festival (Jangli Janwar ka Shikar) (Hindi Daily Newspaper
- of April).
- Richard, Ed, Giles H. (1982): A manual of wildlife conservation, Natraj Publications, Dehradun.
- Shukla, R.S. (2000): Forestry for Tribal Development, Wheelers publishing, New Delhi 224.
- Sinha , R.K. (1966): Biodiversity global concern, Common Wealth Publishers New Delhi.
- Tara Datt (2001): Tribal Development in India with special reference to Orissa. Gyan Publishing house. New Delhi.-237 Role of NGOs in Promoting Environmental.

28

ROLE OF NGO'S IN PROMOTING EDUCATION AND AWARENESS

*Hilal Khaliq, Dinesh Parmar, N.C. Sharma Najeeb Ahmad*and Umar Farooq***

Abstract: Nature has been enough sweet and generous in creating this universe and in particular our planet "Earth" and in this regard everyone has been assigned a specific function. But the lack of co-ordination and a shift of paradigm towards vested interests have created a fracas and disequilibrium which may act as peril to ones survival. Some non-governmental organizations have shouldered the responsibility of equilibrating this gap to finally maximize our survival. Their basic work remains to address regional, national and international environmental issues and then discuss the governmental policies among administrators, different NGOs, interest groups, academicians and communities. Dissemination through publications, broadcasting or telecasting the issues, conducting meetings, organizing school actions, environmental day celebrations, parades and to carry out many other efficient activities for implementation of strategies to minimize the cause and affect of environmental deterioration in broader perspective remains the popular effective practices in this regard. Such awareness and strategic implementation would be instrumental in bringing a general shift away from activities focused on opposing something to activities supporting something and a shift from a small group of people with a heightened awareness of issues engaged in activities to ordinary and illiterate citizens joining a broad range of activities. Thus, this civil awareness will pave a way for our long and soulful survival. This further discusses the important issues and recommendations for implementation of an effective program at micro levels as a test trial prior to further implementation of awareness strategies in different socioeconomic groups.

Keywords: Environment, Awareness, Non-governmental organization (NGOs), Human peace, Society development.

Department of Biochemistry and Genetics, Barkatullah University, Bhopal-462026

* Department of Limnology, Barkatullah University, Bhopal-462026

** Govt. Degree College Bemina, Srinager J&K.; Email hilalkhaliq@gmail.com

Introduction

The recent advances in every walk and sphere of life have shown our scientific powess but at the same time have been a witness to many grave problems which have led to the environmental degradation in different ways and a blow to our continuous survival. India being a populous country, struggled more at this front as the health care services are already not up to the mark and the majority of people who live more than half a million villages and in the proliferating slums of our cities or riparian owners add to the agony of illiteracy, uneducated and unaware burden. Although different governments tried their best to ward off this burden but succeeded more or less to a small extent. World Development Report 1997 The State in a Changing World called on the state to fulfill at least five fundamental tasks establish a foundation of law, maintain sound economic policies, invest in social services and infrastructure, protect the vulnerable and protect the environment, without which "sustainable, shared, poverty-development is impossible."

Charitable and voluntary organizations since immemorial times have been contributing significantly towards the awakening and educating of different communities about health care, environmental hazards etc. at the same time Non-Governmental Organizations (NGO's) have equipped themselves and came up enthusiastically to provide services like relief, education, awareness and help to governmental agencies in effective implementation of their policies. NGO's are viewed as voluntary and nonbureaucratic, as representing the private initiative of citizens taking affairs into their hands, contesting the state and holding it accountable (Rostrup, 2002). They are also seen as providing alternative to state, using a self-help initiative to fill in gaps left by the deficiencies of state.

As a result, all the sects of the society have realized the potential of these NGO's and their success compared to others because of their motivation, dedication, sympathy and determination towards the upliftment, conservation and solution of the problems of society.

Thus, at present there is need to create a concern for our environment that will trigger pro-environmental action including simple day to day activities to protect it and contribute to it's over all longevity with the objectives as under

To figure out most decisive factor in environment awareness.

Thorough discussion of environmental issues and governmental policies among NGO's, administrators, academicians and the community heads to develop general strategy for environmental education and awareness.

Determine the focal problems at all points.

To accesses the role of media in creating environmental education and awareness.

All gender involvement to support the development of values that respect the objectives through education communication and institutional changes.

Activities Undertaken by NGO's Regarding Environmental Education

Waste Management

Waste is valuable raw material at wrong place. It can be converted into useful products by use of appropriate technology. Recently this has developed as a new concept of Zero waste management. Waste includes both municipal solid waste and Biomedical waste (Chitra, 2003). NGO's are addressing the problem of handling and disposing of this waste and its possible solution. They educate people to help in collection of garbage from individual households to secondary points rather then dispersing on roads, streets leading to blockage of drains, which in turn results in overflowing of liquid wastes and excreta on the streets nearby residential areas facilitating the spread of foul smell and many diseases. Thus, this is a way of relocating waste rather than a managing it. But it helps in minimizing the bad effects of the waste.

Water Conservation

Water is basic and foremost need of all living organisms. But the present scenario is not encouraging as very often we hear about the suicides, fist fights and quarrels about its unavailability or quality. NGO's particularly fought for this injustice and in this connection Waterways Monitoring Programme was formed in 1991, as a joint Programme with several NGO's and dedicated individuals with a sole purpose of developing clean and pollution free water available to people (Chitra, 2003).

To keep it at bay with the promises NGO's conducted a series of meetings with various governmental agencies and made it a success in its preliminary phase. With this awareness and concern a man named V. Salunke initiated movement known as "Pani Panchayat" to conserve water in drought prone area of Pune and then whole area followed (Bharucha, 2005).

Rain Water Harvesting

Many cities in India suffer from very serious perennial water problems every summer like Bhopal. Therefore it is is important that all avenues of water source be tapped. NGO's have propagated the system of rain harvesting in several residential areas and people have responded well by contrasting open roof top tanks, Medhbandi, Saza Kuva and Johads etc (Bandhu *et al.* 1999, Chitra, 2003, Bharucha, 2005, David, 2007). World Bank study of 27 Asian cities with population over 10 lakhs, Chennai and Delhi are ranked as worst in terms of water availability per day, while as Mumbai is ranked 2nd worst performer. Thus it becomes imperative to find a solution and an alternative to this problem.

Sanitation

Inadequate sanitation facility is an important environmental depleting problem. It poses a threat to human health especially in low income societies or slum dwelling areas. NGO's concept of self help is best displayed by the community sanitation improvement projects in same or adjoining catchments. There role has also led to the reduction of pollution rates in some areas as well e.g. Giriappa road in T. Nagar in Chennai (Chitra, 2003).

Plantation

Different NGO's have held seminars, parades and other functions to stress and stem the importance of plantation on world plantation day or other days as well. The results are instrumental as trees are being planted for the purpose of beautification and abating noise along roads, parks, playgrounds, burial grounds etc with the larger perspective of environmental protection and sustainability in view. Many NGO's stressed the afforestation of river banks as people now understand that it prevents soil erosion and is necessary for our well being.

Roof Gardening

With the cultivable or agricultural land virtually shrinking rapidly and deforestation eloping the minds our minds. A new shift in agricultural practices is needed, which came with roof gardening. Different professionals, individuals, labs and NGO's have been propagating and training people in setting up such gardens. Recently we see people hobbying such things and save on the environmental and economic front as well.

Pollution Control

India has large number of industries and the task of pollution control becomes more complexed with rise in the number vehicles coupled with poverty and population explosion. Thus a comprehensive pollution control approach is the need of hour which includes prevention, availability of technical solutions, fines and more NGO involvement about publicizing the pollution control measures.

Chitra (2003) has reported that "Paryavaran Vahinis" were constituted in 184 districts involving the local people to play an active role in preventing poaching, deforestation and environmental pollution. An environmental information system network has been setup to disseminate information on environmental issues. India has a large network of NGO's which are involved in spreading the message of sustainable development to public.

Law/Policy making and Legislations

Although both state and central governments have formulated different laws/acts at different times and then timely amended to suit to the needs and desires of the situation. These laws/acts not only have constitutional implications but also an international background. The govt. formulated many laws (water Prevention and Control of Pollution act in 1972, Forest Conservation Act, 1989, The

Air Prevention and Control of Pollution Act 1981, Environment Protection Act 1986) but they lack teeth and enforcement as laws are only as good as the ones that can enforced. Thus mere legislations, laws or policies or projects won't work but the proper enforcement and a strict punishment for defaulters will make them of any use. NGO's like Green Peace activists or PETA always bark about the same.

Their implementation at proper time will prove a saviors as the reports around the globe have been encouraging. Capleton *et al.*, (2005) reported that proper implementation of the European Environment and Health Process on UK environment and health policy, plans and practice has strengthen relationship and promoted cooperative working between government departments responsible for environment and health issues. It has further raised awareness and a positive impact of the political profile of environment and health issues in the UK in turn led to the stimulation of NGO interest in the relationship between environment and health.

Environment Ethics

As per Bharucha (2005) there are two aspects that are closely connected with ethical issues that are related to our environment. These are based of valuing nature and appreciating the beauty of nature and treasuring the magnificence of the wilderness.

We have certainly forgotten that earth is our mother and are ruining it with every interest. Earlier forests and trees were treated as sacred but now more advanced shopping malls and recreation parks, casinos and pubs are preferred and the forests fell prey to such dreams and desires.

We hardly have time to appreciate nature or gaze at the scenic sunset or spend time to sit in the incredible silence of forests or listen to the songs of birds, roars of rivers or the sound of wind rustling through the leaves.

But now we need to bring back an ethnic concept based on environment education and awareness. Different NGO's along with some hopeful and dedicated individuals have already started it like W. Mathei, Sunder Lal Bahugna etc.

Recent Issues of Climate Change, Global Warming, Acid Rain, Ozone Layer Depletion, Nuclear Accidents, Oil spills and Holocaust

During recent times the above mentioned words have found a wide usage and made us shiver about their consequences. But less clearance has lead to more confusion and chaos regarding their nature and understanding. Thus a proper awareness is required and NGO's suit the need well as they hang around the banners outside Copenhagen, Kyoto etc during climate summits and urge us to support them for a better future for our next generations.

Role of NGO's in Promoting Environmental Awareness

The basic goals in environment al education more defined as creating environmental awareness, impart general knowledge for a basic understanding of environment acquiring environmental friendly attitudes and values and to generate new pattern of behaviors environment. (UNESCO, UNEP, Tbilisi-1977). While as United Nations Conference on environment and development in Rio de Janeiro, Brazil in 1992 popularly known as earth summit adopted an action plan for sustainable development Agenda 21 and its chapter 36 states that "education is critical for promoting sustainable development and improving the capacity of people to address environment and development issues".

Education without communication is simply impossible (Hannah, 2002) communication needs an appropriate medium to have more benefits as for as environment awareness outside educational establishments is more concerned mass media and traditional media like family, neighbors and colleagues have the role of the medium that enables people to communicate successfully (Hannah, 2002). India being a developing country, the roles of print and electronic media seems to out class all other forms of awareness as they have wide coverage and are easily accessible.

Mass media

The mass media used for awareness includes newspapers, radio, and television due to the fact that these are most widely used easily available and accessible in both urban and rural areas.

Newspapers

They have played an important role in independence awareness and their number has increased ever since. But none of these is having an additional or special addition on environment or related issues as they are the back bone of news in India. But certainly from sometime due to the active participation of some NGO's environmental issues find some space in columns and editorials of some heading dailies needed e.g., by highlighting the negative consequences of use of pesticides by farmers or sewage disposal from an industry to the adjoining rivers. Environmental magazines need to be published highlighting the problems and possible solutions and should be made available at lower costs.

Radio

It is the cheapest most easily available with a signal coverage almost in the whole country. So governmentally motivate and publicly accepted programmes like, Khait Khaliyan, Ao Dilli Savarin and Gami Bhaion Handi Baapat special programmes about the recently achievements and applicability of science and technology should be broadcasted more often. NGO's specially working on these issues has made a huge difference with the help of interested individuals. NGO's have developed interest among peoples by implementing wide pre-programme publicity.

Television

In recent times television has replaced all other sources of entertainment with the availability of DTH's and satellite revolutions. Now it is perceived as most authoritative and eco-friendly media by a vast percentage of people. Secondly science has proven that audio/visual education has a lasting impact then anyone alone and we have over children highly inclined toward TV serials and soap operas. Like as, discovery channel animal planet, national geography have added a lot to our understanding and interest generating among all kinds, race and religions. Programmes financed by government and dispersed through different NGO's have played an important role in many countries (Chitra, 2003and Wacker *et al.*, 1999).

Environmental friendly advertisements like no use of cracker on the occasion of religious festivals or immersion of God's, drainage of slaughter houses should be prominently highlighted by NGO's working in the related media. The active campaign of policy of the Ministry of Environment and Forests and the true activism of NGO's has more ably promoted the environmental awareness (Hannah, 2002) then any one alone. A survey by (Hannah, 2002) has shown that more than half of the interviewers declaring that the information provided by media has been most important in making them aware of the environmental problems. While as more awareness about possible solution is being recently given by different NGO's separately and in collaboration all over the globe.

Institutional/Campus Awareness

In 1986, Nation Policy on education mad environmental education an integral component for the first time after independence beginning from childhood to all ages. But its coverage and development of skills for the study of environment and environmental ethics, attitudes and actions need strengthening (Bandhu, 1999).

To bridge this gap we need wide participation of NGO's along with governmental agencies and in this regard a new approach to address environment issues at school level is the eco-club scheme that aims to bring NGO's more closely to the school system (Hannah, 2002). The introduction of environmental studies as a curricula area at primary stage (1st to 5th) has given a further impetus to it. Different NGO's have arranged seminars, parades and essay writing competitions in schools about "use and disuse of crackers" "artificial and natural colors" or "uses and abuses of polythene bags" etc. Further they have organized rallies on environment day with cardboard readings "Say no to tobacco and Cigarette smoking". In a similar study Hannah (2002) reported that overall age groups 67% of the interviewers reported that they been taught environmental issues at school in contrast to 245 who had not been taught at school while as 9% could not remember.

Traditional Media

NGO's have used traditional media like direct interaction and communication of an individual with his family, friends, neighbors or environmental activists in enhancing environmental awareness. To support the work of NGO's dealing with environmental issues the Ministry of Environment and forests started the National Environment Awareness campaign in 1986 with a yearly amount of 300 to 400,000 rupees and every year nation wide 5000 NGO's are selected and granted10,000 to 25,000 rupees to conduct campaigns on the topic of year that is chosen by ministries project team (Hannah, 2002).This cooperation between National Environmental Awareness campaign and NGO's have proved to be an efficient and result delivering instrument to target the rural and less educated part of the population.

Additional Factors in Explaining Environmental Awareness

Although media has bee a dominant partner in promoting environmental awareness but in the recent times cultural and religious teachings have also been started for this awareness. Recently environmental issues are being specifically highlighted in political agendas like speeches or five year plans which is an indication of prevailing level of environmental awareness. While as its total insensitivity in early times at the bureaucratic/administrative level has given the nation a very heavy backlog of pollution and ecological degradation (Bandhu, 1999).

NGO's have stressed at governmental level more Environmental Training Institutes need to be constructed and provided with the necessary infrastructure and have collaborative projects about environment, sustainable development with other countries that have already such experience. In this direction Tata Energy Research Institute (TERI) has launched a major project "Growth with Resource Enhancement of Environment and Nature (GREEN India 2047). In addition many NGO's are fighting for setting up regulatory standards and norms that industries need to adopt clean technologies and bring about implementation in management practices (Chitra, 2003). Such regulation should help in avoiding any other incidence similar to Bhopal Gas Tragedy (1984) and the after effects of the environmental contamination which are a real threat to human peace. Thus this shift from complete ignorance to a bit of care and concern in itself is a victory or success of the respective NGO's.

Limitations of NGO's

Recently NGO's have been used as tools to mask causes of sufferings or to be simply an internal or foreign policy tool that creates rather than counters human or environmental sufferings. The poor regulatory framework in many countries makes it difficult to control or eliminate NGO's whose e methods or activities are questionable or even illegal. Thus they continue to poison the environment.

Sometimes it is difficult to coordinator governmental and NGO activities especially over longer period and becomes more difficult when NGO's are small, engaged in fragmented activities. Thus they struggle to get appropriate support and legitimacy by local civil society (Zurila, 2002).

Some NGO's are facing credibility crisis with a number of cases of embezzlement and scandals. Chitra (2003) reported that Central Bureau of Investigation (CBI) found that out of seven NGO's receiving funds to help women in self employment, four had not done a thing and the rest existed only on paper. About 400 such NGO's have been blacklisted by the Council for the Advancement of Peoples Action and Rural Technology (CAPART) and the Central Social and Welfare Board (CSWB) has blacklisted 3000 NGOs for their poor performance in different fields and areas of our country (Chitra, 2003).

In recent times NGO's have been used as advertising agents and to fulfill the desires of financing boards/bodies, trusts. Many people own more than one NGO at a time and use them to pocket wealth by getting a lot of profit.

Many NGO's have become money making machines or a source of business. NGO's which don't exist in real and have a good rapport with press and media get more coverage while as NGO's doing genuine and positive work don't make good news to press/media.

Recommendations

The role of NGO's needs to be raised and their activities should be widely shown by media or any effective way especially in rural areas. Governments should cooperate with NGO's which are working in environmental field by proper screening of their validation and then making them accountable. Media should be encouraged to play a more active role. Participation of women should be made mandatory and they should be educated about the awareness and the preservation of environment.

Acknowledgements

We would like to thank the whole team of the NGO "AWARDS" (Alfred Wagner Advanced Research and Developmental Society) especially Dr. Ayaz Naik., and our friends Tahir M., Barpete S., Prasad B., Lalit A., Fayaz A., Parvez A., Swaleha, Aliza, Shakeel A., Wagmare A., Gayoor A., Suman Gupta., Farooq H., for their valuable suggestions.

References

- Bandhu, D., Dyal, R. (Eds.) 1999. Environmental Education for a Sustainable Future, Indian Environmental Society, New Delhi.
- Bharucha, E. 2005. Social Issues and the Environment. Environmental Studies, Universities Press (India) Pvt. Lim. 158-200

- Capleton, A.C., Stevens, J., Harrison, P.T.C. 2005. The Impact of the European Environment and Health Process on UK Environment and Health Policy, Plans and Practice What Difference Has It made? European Journal of Public Health. 15; 546-551
- Chitra, A. 2003. "Role of NGO's In Protecting Environment And Health" in Martin J. Bunch, V. Madha Suresh and T. Vasantha Kumaran, eds., Proceedings of the Third International Conference on Environment and Health, Chennai, India, 15-17 December, 105-112
- David, R. 2007. Environmental Education. Frank Bros. & Co. (Publishers) Ltd., New Delhi.
- Hannah, H. A. 2002. Comparative Study on Environmental Awareness and Environmentally Beneficial Behavior in India. CMS ENVIS Centre, Centre for Media Studies, Research House, Community Centre, Saket, New Delhi-110017.
- Rostrup, M. 2001/2002. The Role of Nongovernmental Organization in Providing Health Care. Annual World Bank Conference on Developmental Economics. 175-186
- Wacker, C., Vairo A., and Wolf M. 1999. Partnerships for Urban Environmental Management the Roles of urban Authorities, Researchers and Civil Society. Environment and Urbanization, 11; 113-126.
- Zurita, B. 2001/2002. Comment on "The Role of Nongovernmental Organization in Providing Health Care." By Rostrup, M. Annual World Bank Conference on Development Economics. 187-188.

29

ASPECTS OF GLOBAL WATER RESOURCE AVAILABILITY, USE AND ITS CONSERVATION

Rekha Verma, R.K.Prajapati and Anjali Mandway***

Abstract: Water is a vital substance in all parts of environment. The Indian Ocean contains 97% of water of which 77×106 km^2 of the water is completely unfit for human consumption and industrial and domestic uses because of its high salt contents.The survey of literature reveals that is locked as ice-cap and 0.3% is available for human consumption. The demand of water increasing approximately 2.4% per year due to high rate of population growth. The developed countries like America require of water 600-1200 liters/day/persons and developing countries like India and Pakisthan consumes 92-120 liters/day/persons as per their living standard. The increasing human population impact of critical availability scarcity of water by discharge 60% domestic sewage created water pollution. The study envisaged that clean and pure water is not available 2 people for drinking purposes out of 10 people in the world. The study further reveals that approximately 22 lac Peoples are dying every year due to water born disease like cholera, hepatitis, typhoid, constipation etc. this effects was observed due to mismanagement of water (17-40%) in metro and cosmopolitan cities like Mumbai, Delhi, Kolkata etc. In capital of state of Madhya Pradesh, Bhopal 30% water is rendered waste due to disturbed and leakage pipe line. The study suggests conserving natural resources of water through public awareness in rural and urban area and by the rain water harvesting. The study also establishes some sort of balance between the population growth and urbanization of natural resources.

Keywords: Domestic use, Mismanagement of water, Balance

Introduction

Water resources are sources of water that are useful or potentially useful to humans. Uses of water include agricultural, industrial, household, recreational and environmental activities. Virtually all of these human uses require fresh water. Approximately 97% of water on the Earth is salt water,

*School of Environmental Biology, Awadhesh Pratap Singh University Rewa-486003 (M.P.) **R.V.P.S. Govt. College, Umaria (M.P.); Email; rekha_ve17@rediffmail.com, rajeshprajapati37@yahoo.com leaving only 3% as fresh water of which slightly over two thirds is frozen in glaciers and polar ice caps.[1] The remaining unfrozen freshwater is mainly found as groundwater, with only a small fraction present above ground or in the air.[2]

Though the volume of global water is very large, serious environmental and water-supply problems in many parts of the world have been caused by wasteful exploitation of fresh water, reckless release of deleterious and even dangerous water-borne wastes into rivers, lakes and the ground, and careless management and mismanagement of watersheds, recharge areas and aquifers. Most of these problems, which relate to both the dry and densely populated areas of the world, could have been forestalled with proper foresight, based on understanding of the hydrological cycle (UNESCO, 1971). Water availability is of major importance both for the functioning of aquatic species and also for the availability of water for human use water that is only available in a few wet years must not be considered renewable. Because most global runoff comes from areas of very low climatic variability, the total global runoff is generally of low variability.

Water pollution is one of the main concerns of the world today. The governments of many countries have striven to find solutions to reduce this problem. Many pollutants threaten water supplies, but the most widespread, especially in underdeveloped countries, is the discharge of raw sewage into natural waters; this method of sewage disposal is the most common method in underdeveloped countries, but also is prevalent in quasi-developed countries such as China, India and Iran. Sewage, sludge, garbage, and even toxic pollutants are ali dumped into the water. Even if sewage is treated, problems still arise. Treated sewage forms sludge, which may be placed in landfills, spread out on land, incinerated or dumped at sea.[5] In addition to sewage, nonpoint source pollution such as agricultural runoff is a significant source of pollution in some parts of the world, along with urban storm water runoff and chemical wastes dumped by industries and governments. The study of water balances is the application in hydrology of the principle of conservation of mass, often referred to as the continuity equation. This states that for any arbitrary volume and during any period of time, the difference between total input and output will be balanced by the change of water storage within the volume.

Global water balance and Its Use

Water is widely distributed on Earth as freshwater and salt water in the oceans. The Earth is often referred to as the "blue planet" because when viewed from space it appears blue. This blue color is caused by reflection from the oceans which cover roughly 70% of the area of the Earth. The oceanic crust is young, thin and dense, with none of the rocks within it dating from any older than the breakup of Pangaea. Because water is much denser than any gas, this means that water will flow into the "depressions" formed as a result of the high density of oceanic crust. (On a planet like Venus, with no water, the depressions appear to form a vast plain above which rise plateaux). Since the low density rocks of the continental crust contain large quantities of easily eroded salts of the alkali and alkaline earth metals, salt has, over billions of years, accumulated in the oceans as a result of evaporation returning the fresh water to land as rain and snow.

First to present a general scheme of the global hydrological cycle was Bruckner.[8] Assuming that the total amount of water on the earth is constant, he expressed the annual water balance as follows

$$P_0 = E_0 - D_0 + D_1 = E_1 - R$$

$$P_1 = E_1 + D_0 - D_1 = E_1 + R$$

Where the subscripts $_0$ and $_1$ refer to oceans and land areas respectively; D_0 is the annual amount of water vapour transported by the atmosphere from the oceans to the land; D_1 is the annual transport of vapour from land to sea; and R is the annual discharge of rivers to the oceans.

During the twentieth century large-scale water balances have attracted wide attention among scientists. Original studies based on the physical nature and mechanisms of the processes of evaporation, precipitation and runoff are based. [9, 10, 14, 15, 16, 25, 26,] Nace prepared a summary water balance, based upon an extensive review of scientific literature.[27]

World water balance was initiated by the E.Ya. Brikner of USSR.[8] The method of estimating world water balance suggested by him has probably a greater importance than Lhe actual estimate of balance. The next step was made by R. Fritzsche, [13] who was the first to obtain values of river runoff close to the present ones. Improved upon it by adding runoff from Polar glaciers to river runoff. Both results of G. Wust's estimation [16] of the world water balance are characterized by low values of precipitation and evaporation of the Ocean. Attention was drawn to this by W. Meinardus [15] who probably prepared the first reliable map of atmospheric precipitation for the Ocean table 1. [20]

In 2000, the world population was 6.2 billion. The UN estimates that by 2050 there will be an additional 3.5 billion people with most of the growth in developing countries that already suffer water stress.[3] Thus, water demand will increase unless there are corresponding increases in water conservation and recycling of this vital resource.[4] The United Nation Educational, scientific and culture organization and state hydrological Institute, World Resource Institute and many other instate prepare a graphical representation of the global water withdrawal and consumption by humans from natural water resources or reservoirs for various purposes like agriculture, industry or domestic purposes and also predict future forecast consumption from the 1990s to 2025 in Asia, North America, Europe, Africa, south America and Australia and Oceania. They were discussed the more withdrawal and consumption of water in Asia and very low in Australia and Oceania (figure 1).

One estimate of global water distribution indicates oceans and seas & bays are greater container of water 1,338,000,000 Cubic kilometers approximate 96.5% of the total global water which is salt water and Only approximately 3 % is the fresh water out of which only 0.3 % of water is available

Table1 World Water Balance from the data of different authors (cm)

Author	year	Precipitation			Evaporation			Total river runoff from land
		Land	Ocean	Total	Land	Ocean	Total	land
E. Brikner	(1905)	85	98	94	67.3	105	94	17
R. Frizche	(1906)	75	98	91	54.5	105	91	20.5
G. Wust	(1920)	75.2	74	74.3	50.2	84.3	74.3	24.9
W. Meinardus	(1934)	67	114	100.2	42	124.2	100.2	2.5
G. Wust	(1936)	66.4	82.3	77.7	41.6	92.5	77.7	24.8
M.I. Lvovitch	(1945)	72	114	101.5	47.6	124	101.5	24.4
M.I. Lvovitch	(1964)	73	114	102	48.7	124	102.0	24.9
M.I. Budiko	(1956)	70	102.4	92.8	44.6	112.7	92.8	25.4
F. Albrecht	(1961)		104.6			114.4		22.4
R. Nace	(1968)	67	88.5	82.5	47.0	97	82.5	28.5
M.I. Lvovitch	(1969)	73.5	113.7	102.0	48.5	124.0	102.0	25.0

Sources: World water Balance (General Report) online at iahs.info/redbooks/a092/093019.pdf

Table 2 World Water Balance

Water balances elements	Volume km	Depth mm
Peripheral part of the land		
Precipitation	102 100	873
River-runoff	37 400	320
Evaporation	64 700	553
Closed (inland drainage) part of the land		
Precipitation	7 400	231
Evaporation	7 400	231
World Ocean		
Precipitation	410 500	1 137
River water inflow	37 400	103
Evaporation	447 900	1 240
The Earth		
Precipitation	520 000	1 020
Evaporation	520 000	1 020

1. Including the evaporation of 755 km^3 of river-runoff.

Sources: World water Balance (General Report) online at iahs.info/redbooks/a092/093019.pdf

Table 3 The distribution of renewable river water across the Earth's surface is very uneven. Earth 28% Water 72%

Continent or Region	Renewable river water (KM^3)	Percent of world total
Sub-Saharan Africa	4,000	9.20
Middle East and North Africa	140	0.32
Europe	2,900	6.70
Asia (excluding Middle East)	13,300	30.6
Australia	440	1.01
Oceania	6,500	14.9
North America	7,800	17.9
South America	12,000	27.6

for human consumption (Table 4). The distribution of renewable river water across the Earth's surface is very uneven. Earth 28% Water 72% (Table 3).[24]

Even within these regions, there can be huge variations. For example, as much as a quarter of Australia's limited renewable fresh water supply is found in almost uninhabited Cape York Peninsula[6]. Also, even in well-watered continents, there are areas that are extremely short of water, such as Texas in North America, whose renewable water supply totals only $26km^3$ in an area of $695,622km^2$, or South Africa, with only $44km^3$ in $1,221,037$ km^2

Water conservation

No matter how freshwater is used-whether for agriculture, industry, or municipalities-there is great potential for better conservation and management. Water is wasted nearly everywhere. Until actual scarcity hits, most countries and most people take access to freshwater for granted.

"We have to stop living as if we had unlimited water supplies and start recognizing that we must deal with serious water constraints," Falkenmark and colleagues have warned. The relevant question about freshwater is not "how much water do we need and where do we get it?" Rather, it is "how much water is there and how can we best benefit from it?" That is, we must do better at managing the demand for water rather than continuing to focus on "supply-oriented" water management. [21]

On the demand side, a variety of economic, administrative, and community-based measures can help conserve water immediately. In the long run, slowing population growth will slow the increase in demand for water and help buy more time to develop better water conservation and management strategies. [22] The ways of the water conservation and management on the indoor and outdoor are as fallows-

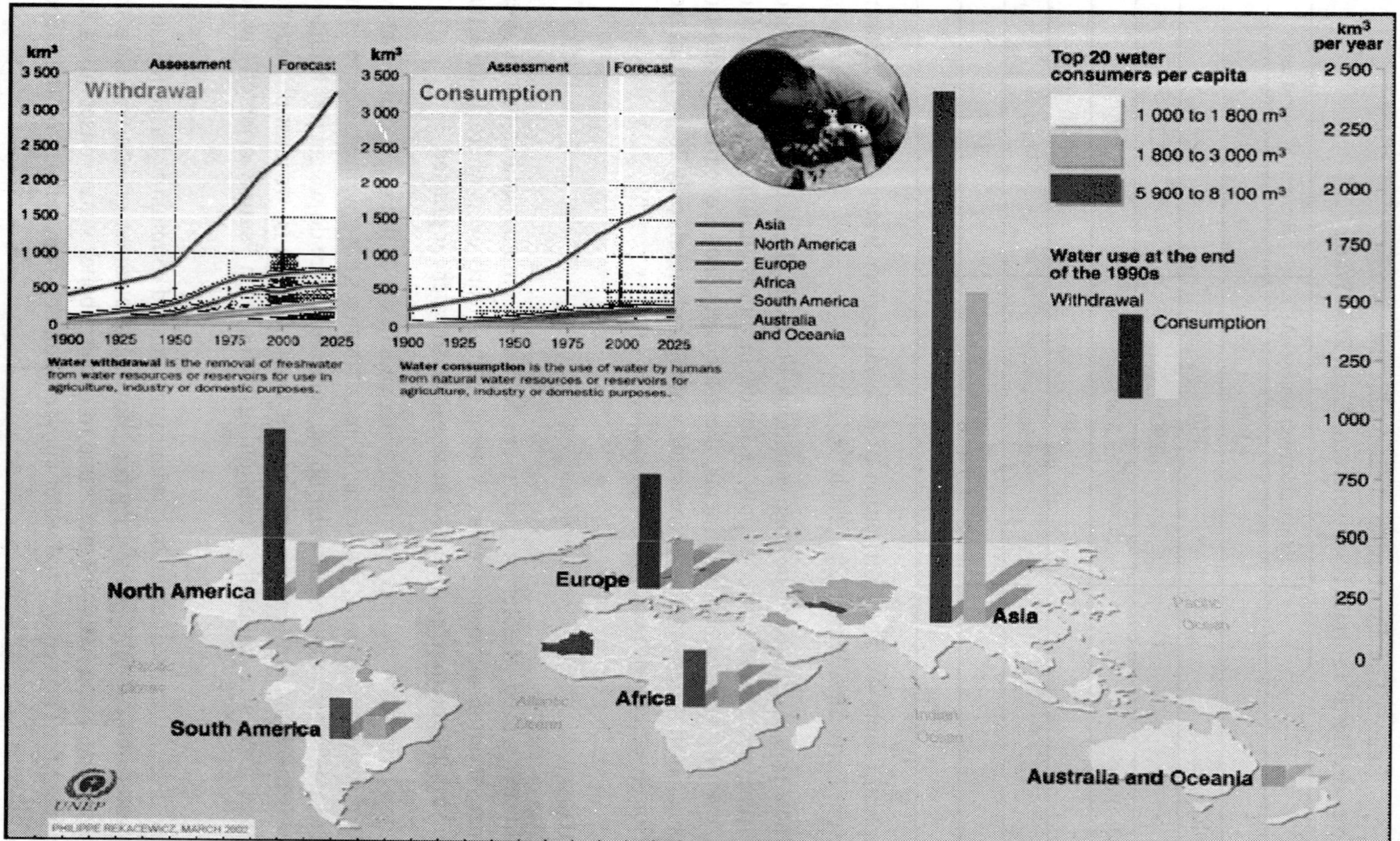

Fig. 1 Global Water Withdrawal and consumption

Sources: Igor A. Shiklomanov, State Hydrological Institute (SHI, St. Petersburg) and United Nations, Educational, Scientific and Cultural Organization (UNESCO, Paris), 1999; World Resources 2000-2001, People and Ecosystems The Fraying Web of Life, World Resources Institute (WRI), Washington DC 2000; Paul Harrison and Fred Pearce, AAAS Atlas of Population 2001, American Association for the Advancement of Science, University of California Press, Berkeley

Never put water down the drain when there may be another use for it such as watering a plant or garden, or cleaning.

Verify that your home is leak-free, because many homes have hidden water leaks. Read your water meter before and after a two-hour period when no water is being used. If the meter does not read exactly the same, there is a leak.

Repair dripping faucets by replacing washers. If your faucet is dripping at the rate of one drop per second, you can expect to waste 2,700 gallons per year which will add to the cost of water and sewer utilities, or strain your septic system.

Check for toilet tank leaks by adding food coloring to the tank. If the toilet is leaking, color will appear within 30 minutes. Check the toilet for worn out, corroded or bent parts. Most replacement parts are inexpensive, readily available and easily installed. (Flush as soon as test is done, since food coloring may stain tank.)

Table 4 One estimate of global water distribution

Water source	Water volume, in cubic miles	Water volume, in cubic kilometers	Percent of fresh water	Percent of total water
Oceans, Seas, & Bays	321,000,000	1,338,000,000	--	96.5
Ice caps, Glaciers, & Permanent Snow	5,773,000	24,064,000	68.7	1.74
Groundwater	5,614,000	23,400,000	--	1.7
Fresh	2,526,000	10,530,000	30.1	0.76
Saline	3,088,000	12,870,000	--	0.94
Soil Moisture	3,959	16,500	0.05	0.001
Ground Ice & Permafrost	71,970	300,000	0.86	0.022
Lakes	42,320	176,400	--	0.013
Fresh	21,830	91,000	0.26	0.007
Saline	20,490	85,400	--	0.006
Atmosphere	3,095	12,900	0.04	0.001
Swamp Water	2,752	11,470	0.03	0.0008
Rivers	509	2,120	0.006	0.0002
Biological Water	269	1,120	0.003	0.0001
Total	332,500,000	1,386,000,000	-	100

Source: Gleick, P. H., 1996 Water resources. In Encyclopedia of Climate and Weather, ed. by S. H. Schneider, Oxford University Press, New York, vol. 2, pp.817-823.

Avoid flushing the toilet unnecessarily. Dispose of tissues, insects and other such waste in the trash rather than the toilet.

Use the minimum amount of water needed for a bath by closing the drain first and filling the tub only 1/3 full. Stopper tub before turning water. The initial burst of cold water can be warmed by adding hot water later.

Don't let water run while shaving or washing your face. Brush your teeth first while waiting for water to get hot, then wash or shave after filling the basin.

Operate automatic dishwashers and clothes washers only when they are fully loaded or properly set the water level for the size of load you are using.

When washing dishes by hand, fill one sink or basin with soapy water. Quickly rinse under a slow-moving stream from the faucet.

Kitchen sink disposals require lots of water to operate properly. Start a compost pile as an alternate method of disposing food waste instead of using a garbage disposal. Garbage disposals also can add 50% to the volume of solids in a septic tank which can lead to malfunctions and maintenance problems.

Consider installing an instant water heater on your kitchen sink so you don't have to let the water run while it heats up. This will reduce heating costs for your household.

Insulate your water pipes. You'll get hot water faster plus avoid wasting water while it heats up.

Never install a water-to-air heat pump or air-conditioning system. Air-to-air models are just as efficient and do not waste water.

Install water softening systems only when necessary. Save water and salt by running the minimum amount of regenerations necessary to maintain water softness. Turn softeners off while on vacation.

Check your pump. If you have a well at your home, listen to see if the pump kicks on and off while the water is not in use. If it does, you have a leak.

When adjusting water temperatures, instead of turning water flow up, try turning it down. If the water is too hot or cold, turn the offender down rather than increasing water flow to balance the temperatures.

If the toilet flush handle frequently sticks in the flush position, letting water run constantly, replace or adjust it.

Don't over water your lawn. As a general rule, lawns only need watering every 5 to 7 days in the summer and every 10 to 14 days in the winter. A hearty rain eliminates the need for watering for as long as two weeks. Plant it smart, Xeriscape. Xeriscape landscaping is a great way to design, install and maintain both your plantings and irrigation system that will save you time, money and water. For your free copy of "Plant it Smart," an easy-to-use guide to Xeriscape landscaping, contact your Water Management District.

Water lawns during the early morning hours when temperatures and wind speed are the lowest. This reduces losses from evaporation.

Don't water your street, driveway or sidewalk. Position your sprinklers so that your water lands on the lawn and shrubs... not the paved areas.

Install sprinklers that are the most water-efficient for each use. Micro and drip irrigation and soaker hoses are examples of water-efficient methods of irrigation.

Regularly check sprinkler systems and timing devices to be sure they are operating properly. It is now the law that "anyone who purchases and installs an automatic lawn sprinkler system MUST install a rain sensor device or switch which will override the irrigation cycle of the sprinkler system when adequate rainfall has occurred." To retrofit your existing system, contact an irrigation professional for more information.

Raise the lawn mower blade to at least three inches. A lawn cut higher encourages grass roots to grow deeper, shades the root system and holds soil moisture better than a closely-clipped lawn.

Avoid over fertilizing your lawn. The application of fertilizers increases the need for water. Apply fertilizers which contain slow-release, water-insoluble forms of nitrogen.

Mulch to retain moisture in the soil. Mulching also helps to control weeds that compete with plants for water.

Plant native and/or drought-tolerant grasses, ground covers, shrubs and trees. Once established, they do not need to be watered as frequently and they usually will survive a dry period without any watering. Group plans together based on similar water needs.

Do not hose down your driveway or sidewalk. Use a broom to clean leaves and other debris from these areas. Using a hose to clean a driveway can waste hundreds of gallons of water.

Outfit your hose with a shut-off nozzle which can be adjusted down to fine spray so that water flows only as needed. When finished, "Turn it Off" at the faucet instead of at the nozzle to avoid leaks.

Do not leave sprinklers or hoses unattended. Your garden hoses can pour out 600 gallons or more in only a few hours, so don't leave the sprinkler running all day. Use a kitchen timer to remind yourself to turn it off.

Consider using a commercial car wash that recycles water. If you wash your own car, park on the grass to do so.

Avoid the installation of ornamental water features (such as fountains) unless the water is recycled. Locate where there are mineral losses due to evaporation and wind drift.

If you have a swimming pool, consider a new water-saving pool filter. A single back flushing with a traditional filter uses from 180 to 250 gallons or more of water.

Create an awareness of the need for water conservation among your children. Avoid the purchase of recreational water toys which require a constant stream of water.

Encourage your employer to promote water conservation at the workplace. Suggest that water conservation be put in the employee orientation manual and training program.

Report all significant water losses (broken pipes, open hydrants, errant sprinklers, abandoned free-flowing wells, etc.) to the property owner, local authorities or your Water Management District.

Encourage your school system and local government to help develop and promote a water conservation ethic among children and adults.

Support efforts and programs to create a concern for water conservation among tourists and visitors to our state. Make sure your visitors understand the need for, and benefits of, water conservation.

Encourage your friends and neighbors to be part of a water conscious community. Promote water conservation in community newsletters, on bulletin boards and by example.

Conserve water because it is the right thing to do. Don't waste water just because someone else is footing the bill such as when you are staying at a hotel.

Try to do one thing each day that will result in a savings of water. Don't worry if the savings is minimal. Every drop counts. And every person can make a difference. So tell your friends, neighbors and co-workers to "Turn it Off" and "Keep it Off".

Water conserve through the rain water harvesting method.

Develop watersheds for the management of water resource and grow more trees they help in recharging the groundwater and attracting the precipitation, where more forest areas more rainfall occur.

The minimization of loss or waste, care and protection of water resources and the efficient and effective use of water. The adaptation and implementation of a strategy by a water institution or consumer to influence the water demand and usage of water in order to meet any of the following objectives economic efficiency, social development, social equity, environmental protection, sustainability of water supply and services, and political acceptability.

Conclusion

Water is the most valuable gift of Mother Nature to earth, without water life is impossible on planet earth; it permeates marine estuarine, fresh water (rivers and lakes) and ground water environment across coastal and inland areas. Contribution made by the different workers tin the field of water resource oceans and seas & bays are greater container of water 1,338,000,000 Cubic kilometers approximate 96.5% of the total global water which is salt water and only approximately 3 % is the fresh water out of which only 0.3 % of water is available for human consumption. The distribution of renewable river water across the Earth's surface is very uneven Earth 28% Water 72%. In India also 12% of people get clean drinking water the rest 88% quench their thirst from polluted water due to more than three million people get affected or die of entire disease in the world every year. The Study indicated the more consumption of water in the Asia and low in the Australia and Oceania. Every one should made contribution in the conservation and management of water, if he wants to save water or other natural resource of our future generation.

References

- Earth's water distribution, United States Geological Survey. http//ga.water.usgs.gov/edu/waterdistribution.html. Retrieved 25.04.2010.
- Scientific Facts on Water State of the Resource. GreenFacts Website. http//www.greenfacts.org/en/water-resources/index.htm#2. Retrieved 2008-01-31.
- "World population tc reach 9.1 billion in 2050, UN projects". Un.org. 20.04.2010. http//www.un.org/apps/news/story.asp?NewsID=13451&Cr=population&Cr1. Retrieved 10.05.2010.
- Groundwater - the processes and global significance of aquifer degradation. Google.com. 28.04.2010. doi10.1098/rstb.2003.1380. http//www.google.com/url?sa=t&ct=res&cd=1&url=http%3A%2F%2Fwww.pubmedcentral.nih.gov%2Farticlerender.fcgi%3Fartid%3D1693287&ei=q9vnRsOEJpmSwQHny8TfBQ&usg=AFQjCNHS5RKSHrKezWO_PBtqp06oMAZs8Q&sig2=becO6aqfyo1izUxdiBcILg. Retrieved 15.05.2010
- Ocean dumping of sewage sludge is prohibited in the United States by the Marine Protection, Research, and Sanctuaries Act (MPRSA).
- Brown, J. A. H.; Australia's surface water resources. ISBN 064402567X.
- Brikner, E,. A. (1905) Water cycle balance on the Earth, Pocliuouedeiiye, vol. 7, N 3.
- Bljdiko, M,. I. (1956) Heat balance of the carth surpace, L.
- Lvovitcnh.,T. I (1945) World water regime elements, Sverdlovsk-Moscow.
- Lvovitcnh., T. I (1964) River runoff. Water regime types of rivers (of the world). Physico-geo-graphical atlas of the World, sheets 58-61, M.
- Lvovitcnh.,T. I. (1969) Water resources of the future, L.
- Fritzschc, R. (1906) Nicdcrsclilag, Abfluss und Verdunstung auf den Landfiicheii der Erde,Zeitschr. für Gewässerkunde, Bd. 7, H. 6.
- Nace, R,. (1968) Water of the World Geological Survey, US Department of the Interior, New release, July.

- Meinardus, W. (1934) Eine neue Niedercchlagskarte der Erde, Petfernicrnris Geogr. Mi/t.8, 0, 1-IV.
- Wust, G. (1920) Die Verdunstung auf dem Meere, Berlin.
- Albrecht, F. (1961) Der jahriiche Gang der Komponenten des Warme tind Wasserhaushaltes der Ozcanc. Berichte des Deutschen Wetlerdinstes, Nr. 79 (6, 11).
- http//www.mass.gov/dep/water/resources/watercon.htm#conserve
- Global Water Withdrawal and Consumption website http//www.un.org/Pubs/chronicle/2003/issue1/0103p40.pdf Retrieved 14.05.2010.
- Lvovitch, I.M. (1970) World water Balance (General Report) Website iahs.info/redbooks/a092/093019.pdf Retrieved 14.05.2010.
- Falkenmark, M., Lundqvist, J., and Widstrand, C. (1989) Macro-scale water scarcity requires micro-scale approaches Aspects of vulnerability in semi-arid development. Natural Resources Forum 13(4) 258-267.
- Gardner-outlaw, T. and Engleman, R. (1997) Sustaining water, easing scarcity A second update. Washington, D.C., Population Action International, p. 2-19.
- Save Water 49 Ways Saving Tips, American water and energy saver, Website http//www.americanwater.com/49ways_saving_tips.php Retrieved 14.05.2010.
- Gleick, P. H., 1996 Water resources. In Encyclopedia of Climate and Weather, ed. by S. H. Schneider, Oxford University Press, New York, vol. 2, pp.817-823. Website http//ga.water.usgs.gov/edu/waterdistribution.html Retrieved 18.05.2010.
- Zoodobenolk, I. (1956) Water balance of continents and oceans, Dolilotli A.N.S SSR,v ol. 108, N 5.
- Albercht, F. (I 961) Der jahriiche Gang der Komponenten des Warme tind Wasserhaushaltes der Ozcanc. Berichte des Deutschen Wetlerdinstes, Nr. 79 (6, 11).
- Keller, R. (1967) Gewisser und Wasserhaushalt des Festlandes, Lcipzig.

30

IMPACT OF IDOL IMMERSION ON THE PHYSICO – CHEMICAL PROPERTIES OF UPPER LAKE OF BHOPAL, MADHYA PRADESH, INDIA

Charanjeet Kaur Arti Patel* Vinita Vipat** and Amit Dubey***

Abstract: Idol immersion in surface water bodies is an age old practice. Increase in the number of idols & change in the materials used for idol making has a negative impact on the water quality of the water bodies in which they are immersed. The issue of water quality deterioration due to idol immersion is more concerning to the Bhopal city because the traditional site for idol immersion viz. the Upper Lake, is also the major sources of drinking water for more than 40% of the residents of Bhopal city, besides being a very important ecological hub. Therefore, any deterioration in the water quality of this water resource directly affects the residents of the city. Upper Lake (Lat. 23°12'-23°16'N Log.77°18' - 77°23'E) situated in Bhopal was constructed by Raja Bhoj in 11th century. The water spread area of the Lake is 31 km^2 and it has a vast catchment of 361 sq.km. The Upper Lake is a major source of potable water & used for the purpose of idol immersion also (Kulshrestha *et al.*, 1988). The immersion of idols in water body has resulted in deterioration of the water quality, accumulation of toxic chemicals & sediments, shrinkage of lake area & above all a loss of aesthetic value. For the purpose of the study samples were collected before, during & after idol immersion from the Prempura immersion ghat. The water samples collected from the epilimnetic & hypolimnetic layers of the water body were analyzed for different physico-chemical parameters. The comparison of water quality before during & after idol immersion reveal that there is significant deterioration in the water quality of Upper Lake due to idol immersion. The chapter discuss in detail the impact of idol immersion on the important water quality indicating parameters.

Keywords : Water quality deterioration, Epilimnion, Hypolimnion.

Introduction

The city of Bhopal is known as the city of Lakes by virtue of large number of water bodies present in and around the city. Upper and Lower Lakes comprising the Bhoj Wetland are the lifeline of the

*Sri Sathya Sai College for Women, Bhopal
**Lake Conservation Authority of M.P., E-5, Arera Colony, Bhopal

city as they cater the needs of more than 40% of the total population of the city. However, the water quality of these water bodies is also under threat due to idol immersion activities. Immersion of idols in water bodies have a negative impact on its water quality and increasing immersion of idols in water bodies has become one of the important non-point sources of its pollution. Idol Immersion adds into water the different constituents that the idols are made up of viz. plaster of paris, bamboo, husk, paints and other materials like poojan samagri, coconut etc which are immersed along with idols. The addition of these materials in the water quality deteriorate its water quality as the biodegradable matter like bamboo, husk etc increase the BOD in the water and depletes the oxygen from the water column. Similarly the inorganic matter like plaster of paris increase the concentration of Calcium and Magnesium in the water body thereby increasing the salinity, TDS and hardness of water. The paints used in the decoration of idols add toxic heavy metals into the water, increasing the toxicity levels in the water.

Thus present study is an attempt to assess the impact of idol immersion on the water quality of the Bhoj Wetland of Bhopal.

Materials & Methods

Idol Immersion in water bodies is an age old practice. Ganeshutsava & Durgautsava are celebrated with great enthusiasm & zeal throughout the country. These festivals mark the worship of idols of Lord Ganesha & Goddess Durga for a few days and the festival ends with the immersion of idols in the water. The size and decoration of idols is also increasing along with the increase in the number of idols every year. The materials used in the making of idols have also changed over past fewars. Earlier, clay and soil were used for making idols but now they have been replaced by newer material like plaster of paris which is easier to mould and easily available.

But this increase in the number of idols and the change in the materials used for idol making has had a negative impact on the water quality of the water bodies in which they are immersed. Environmentalists all over the country are emphasizing the deterioration in the water quality due to idol immersion activity and efforts are on to minimize the effect of idol immersion on the water quality of these precious aquatic resources.

Study Area

The study area is Upper and Lower lakes of Bhopal. Upper lake (Lat 23°12' E - 23°16'N Log77° 18' - 77° 23' E) situated in Bhopal, was created by Raja Bhoj in 11 century. The Lake has large catchment area of 361 sq. km. and at present has water spread area of 31 sq.km. The traditional site of idol immersion in the Upper Lake was the Sheetal Das ki Baghiya but due to its proximity with the water intake points, the idol immersion activity was shifted from this place to Prempura, which is situated in the downstream of Upper Lake at the Bhadbhada Spill Channel. Therefore due to natural gradient the water from Upper Lake flows in the direction of the spill way. Therefore, any pollution caused in the down stream does not directly affect the water quality of the main water body. Keeping this in mind the idol immersion hat at Prempura was constructed as a conservation measure under the Bhoj Wetland Project & ever since has been the main idol immersion site for the city.

Parameters Analysed

For studying the physico-chemical properties of the Upper Lake samples were collected before(pre), during, & after(post) Ganesh idol immersion. Parameters analysed are pH, turbidity, conductivity, TDS, DO, total hardness, calcium hardness, magnesium hardness, total alkalinity, carbonate alkalinity, bicarbonate alkalinity, chloride, total phosphorus, orthophosphate, nitrate, BOD, & COD.

Results & Discussion

The present investigation reveals that the pH was recorded in mild alkaline range at the surface, in tune of 7.96 & 7.52 in the bottom samples. However the immersion of Ganesh idols adds organic debri in the release of CO_2 and its subsequent dissolution in water. The value of pH during idol immersion at the surface was recorded as 6.57 and at the bottom was recorded as 6.51. After a few days the organic debri settles down to the bottom and as a result is decomposed to a greater degree. This decomposition releases high concentration of CO_2 which further lowers down the pH to 6.24 and 6.18 in the surface and bottom samples respectively.The concentration of free CO_2 also recorded an increase from nil - 0.8 mg/l in the samples collected after idol immersion. DO is one of the most important water quality indicating parameters.The concentration of which by and large regulates the life activities in water bodies. The concentration of DO, pre idol immersion was recorded in the moderate range vis 9.6 at the surface and 6.4 at the bottom,which reduce to 2.4 at the surface and 1.6 at the bottom. During idol immersion which further reduced to 1.2 at the surface and 0.8 at the bottom.The concentration of DO is regulated by two activities -

- The photosynthetic evaluation by the aquatic autotrophs.
- Useup by all organic matter in decomposition activity.

During idol immersion a large amount of suspended particulate matter release in the water column which hampers the passage of light there by reducing the photosynthetic activity in the water column. Further more the addition of organic matter due to idol immersion depletes the oxygen from the water column.Thereby further reducing the concentration of DO.These two factors combine & reduces the concentration of DO in the samples collected during and post idol immersion.

Variation in Water Quality at Prempura Idol Immersion Ghat - Ganesh Idol Immersion

S. No	Parameter	Pre				During				Post			
		Surface	Bottom	Ref. Surface	Ref. Bottom	Surface	Bottom	Ref. Surface	Ref. Bottom	Surface	Bottom	Ref. Surface	Ref. Bottom
1	pH	7.96	7.52	7.89	7.52	6.57	6.51	6.81	6.81	6.24	6.18	6.74	6.25
2	Free CO_2 (mg/l)	Abs	0.80	abs	1.20	1.20	2.40	1.00	2.60	1.80	2.40	1.20	1.60
3	Dissolved Oxygen (mg/l)	9.60	6.40	7.60	4.40	2.40	1.60	10.00	1.20	1.20	0.80	9.60	6.80
4	Turbidity (FAU)	8.00	16.00	5.00	8.00	12.00	24.00	13.00	11.00	8.00	18.00	11.00	8.00
5	Conductivity mS/cm)	0.18	0.21	0.14	0.16	0.23	0.24	0.19	0.19	0.32	0.41	0.25	0.29
6	Total Dissolved Solids (mg/l)	109.80	128.10	85.40	97.60	140.30	146.40	115.90	115.90	195.20	250.10	152.50	176.90
7	Total Alkalinity (mg/l)	84.00	88.00	76.00	82.00	120.00	124.00	114.00	102.00	136.00	148.00	132.00	114.00
8	Total Hardness (mg/l)	84.00	96.00	120.00	114.00	146.00	150.00	144.00	150.00	156.00	212.00	168.00	172.00
9	Ca Hardness (mg/l)	56.60	65.40	76.00	82.30	100.80	115.50	84.00	94.50	102.00	124.90	86.90	95.60
10	Mg Hardeness (mg/l)	27.40	30.60	44.00	31.70	45.20	34.50	60.00	55.50	54.00	87.10	81.10	76.40
11	Chloride (mg/l)	19.98	16.96	15.96	15.98	32.97	24.98	24.98	20.98	36.98	44.98	32.96	42.96
12	Total Phosphorus (mg/l)	2.32	3.24	1.20	1.57	8.60	4.05	3.20	3.87	9.96	10.23	4.56	6.54
13	Phosphate (mg/l)	0.15	0.24	0.47	0.57	2.04	2.23	0.81	1.51	3.32	3.96	2.25	2.74
14	Nitrate (mg/l)	0.25	0.32	0.16	0.25	0.62	0.74	0.64	0.69	1.23	1.56	1.42	1.69
15	BOD (mg/l)	6.00	8.00	2.00	4.00	14.00	18.00	6.00	8.00	22.00	28.00	12.00	16.00
16	COD (mg/l)	20.00	22.00	8.00	14.00	44.00	56.00	22.00	32.00	68.00	88.00	42.00	52.00

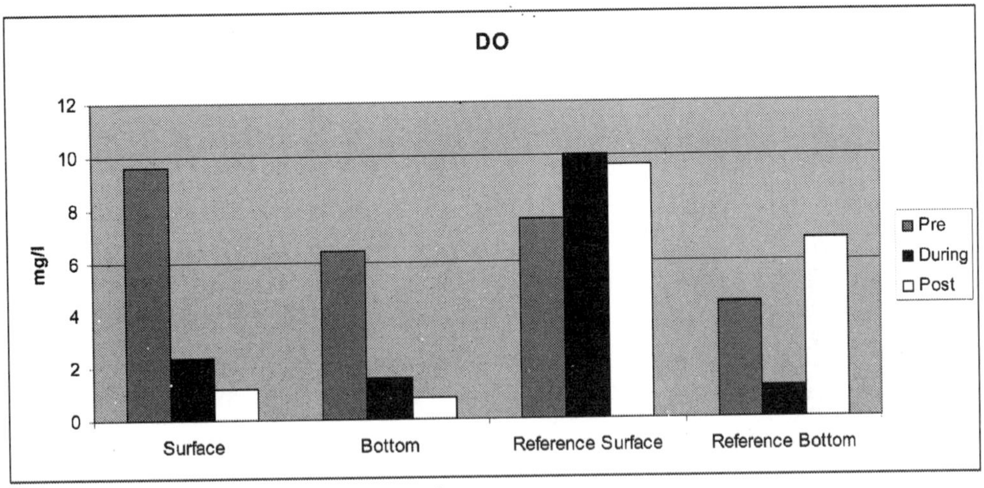
DO
12
10
8
6
4
2
0
mg/l
Surface
Bottom
Reference Surface
Reference Bottom
Pre
During
Post

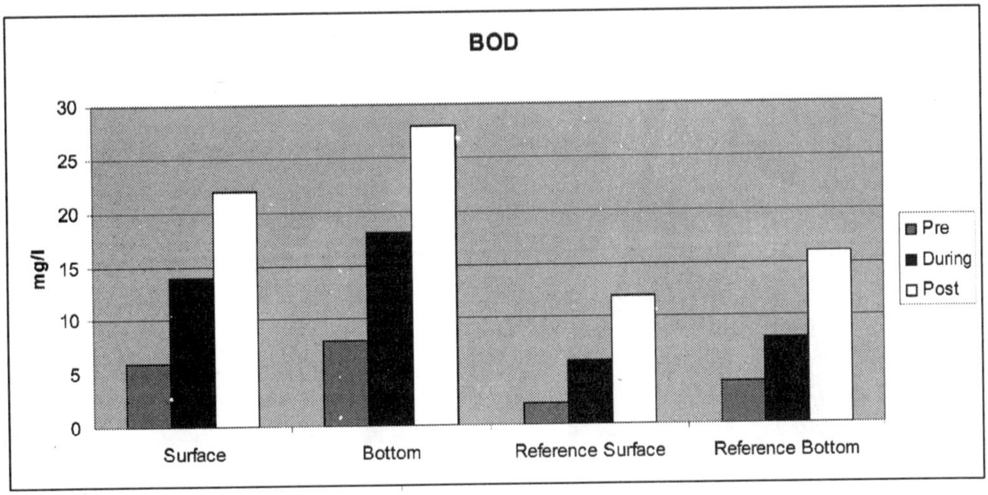
BOD
30
25
20
15
10
5
0
mg/l
Surface
Bottom
Reference Surface
Reference Bottom
Pre
During
Post

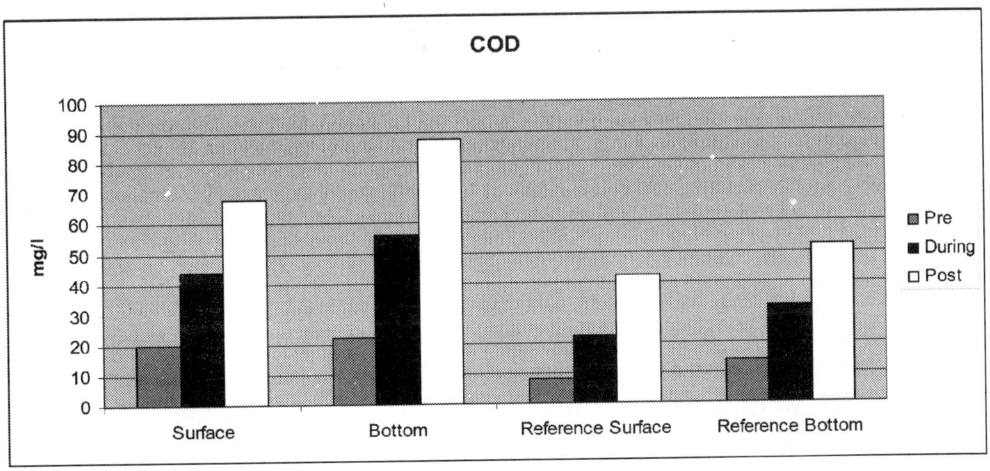
COD
100
90
80
70
60
50
40
30
20
10
0
mg/l
Surface
Bottom
Reference Surface
Reference Bottom
Pre
During
Post

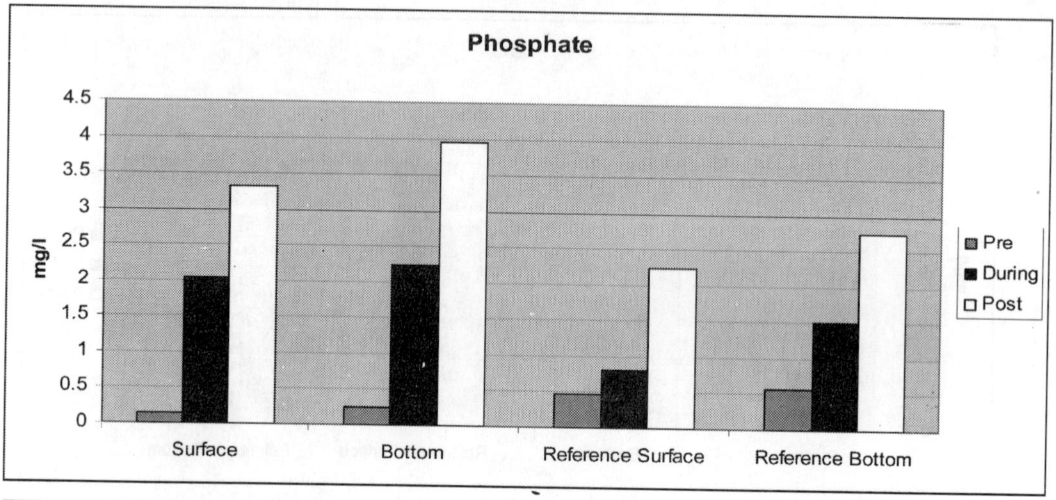

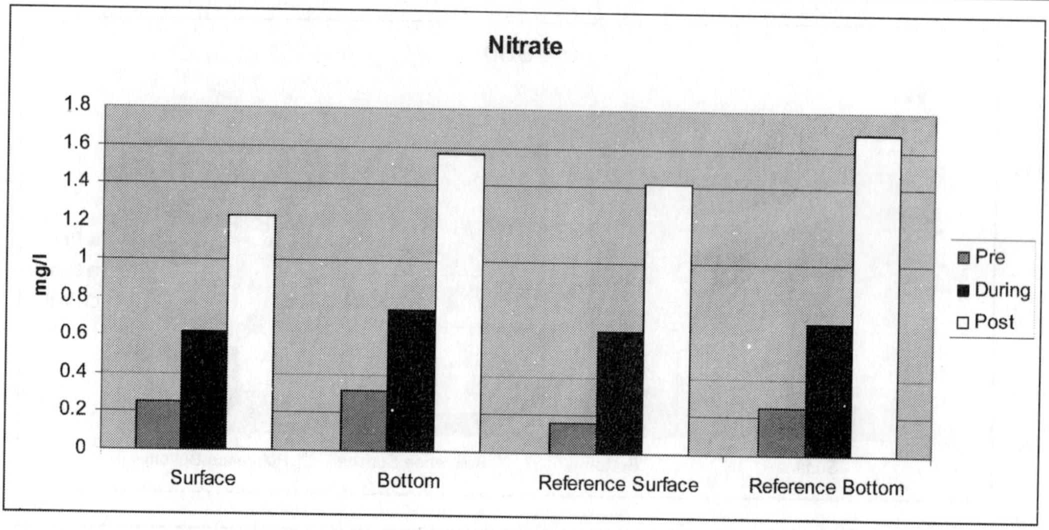

The addition of suspended particles during idol immersion increase the turbidity of water column.The value of turbidity, pre idol immersion was recorded as 8 FAU at the surface & 16 FAU at the bottom samples, which increase to 12 & 24 FAU at the surface & bottom respectively during idol immersion. After idol immersion the suspended particles settle down. Thereby reducing the value of turbidity to 8 FAU at the surface & 18 FAU at the bottom in the samples collected after the idol immersion. The plaster of paris used in making idols contains high concentration of calcium & immersion of idols in the water body increase the concentration of calcium in the water. Calcium when dissolves in water forms salts with carbonate & bicarbonate in water and this increase in the concentration of calcium is expressed as increase in the concentration of hardness of water. The concentration of calcium hardness was recorded as 56.6 mg/l in the surface sample and 65.4 mg/l in the bottom sample which increased to 108.5 mg/l and 115.5 mg/l respectively at surface &bottom. During idol immersion which further increased to 102 & 124.9 mg/l in surface & bottom samples respectively in post idol immersion the idol immersion activities increase the concentration of inorganic nutrients which are responsible for eutrophication of water body like nitrate & phosphate. The concentration of nitrate increase from 0.25 mg/l at the surface & 0.32 mg/l at the bottom to 0.6- 0.74 mg/l during idol immersion & 1.23 mg/l - 1.56 mg/l in post idol immersion in the surface & bottom samples respectively. Phosphorous is an important constituent of iorganic matter. On decomposition it is amongst the first inorganic species to release in the water. The concentration of phosphate was recorded in tune of 0.5 mg/l & 0.24mg/l. During the idol immersion high amount of organic matter in the form of poojan samagri, flowers , husk, bamboo etc. are added in the water. The concentration of organic matter is depicted directly by the concentration of BOD & COD in the water sample which recorded substantial increase during & post idol immersion. The concentration of BOD in pre idol immersion is recorded as in tune of 6 mg/l & 8 mg/l in the surface & bottom respectively which increase to 14 & 18 mg/l in the surface & bottom samples which further increase to 22 & 28 mg/l in post idol immersion. Similarly he concentration of COD was recorded in tune of 20 & 22 mg/l in pre idol immersion in the surface &bottom samples respectively which increases to 44 & 56 mg/l in during idol immersion & 68 &88 mg/l in post idol immersion in surface & bottom samples respectively. Thus it is evident, idol immersion activities in the water bodies deteriorate its water quality & the impact of this deterioration is visible even after a long time of idol immersion.

Recommendations

Number and size of idols should be reduced.

- Natural colours & paints should be used for decoration.
- Use of Plaster of paris should be minimized or avoided.
- Idol immersion site should be far from water intake points

References

- Aniruddhe mukherjee, 2002. “India idol immersion activities & management in water bodies”. India urban portal.
- Anju Vyas, D.D. Mishra, A. Bajpai, S.Dixit & N. Verma, “Environment Impact of Idol Immersion Activities in Urban Lake Bhopal, India.” Asian J. Exp. Sci, vol.20, no.2, 2006, 289-296.
- Dixit, A. and Tiwari, S., “Impact Assessment of Heavy Metal Pollution of Shahpura Lake, Bhopal, India.” Int.j.environ. Res., 2(1) 37-42, winter 2008.issn1735-6865.
- Shilpa, D. Joshi and M. Musaddiq, 2003. "Study on Distribution E-coli and Salmonella in Drinking Water"- Pollution Research.
- S. Radha, 2003. "Microbiological Analysis of Groundwater of Shri Padmavati Mahila University Campus A Report", Pollution Researches Journal.
- Suman mor, M. S. Bishnoi and N. R. Bishnoi, 2003. "assessment of groundwater quality of jind city", indian journal of environmental protection, jun.

- Tourbier, Jr. and R. W. Pierson, Jr. (ed.). 1976. Biological control of water pollution. University of Pennsylvania Press, inc., Philadelphia, pa. 340 pp.
- Tshobanoglous G and Culp, g.l., (1980). Wetland systems for wastewater treatment an engineering assessment. Us environmental protection agency, office of water programme operations, series water epa 430/9-80-007, Washington dc.
- V. Saravanan. Muna Ali and T.R. Sreekrishanan "water treatment of agro residue based pulp and paper mill effluents" "water and wastewater 'perspective of developing countries' edited by- Rema Devi & Naved Hasan December 2003.
- Wolverton B.C. and R. C. Mcdonald. 1979b. Upgrading facultative wastewater lagoons with vascular aquatic plants. J. Water pollut. Control fed. 51 305-313.
- Yount, J. L. 1964. Aquatic nutrient reduction and possible methods. Rep. 35th Ann. Meet., Fl. Anti-mosquito assoc. P. 83-85.

31

STUDY ON THE FISH SPECIES DIVERSITY OF THE RIVER NARMADA IN WESTERN ZONE

*Sunita Bakawale, R. R. Kanhere and Vinod Thakur**

Abstract: Fish diversity of river Narmada were studied during September 2006 to August 2007. After construction of Sardar Sarovar Dam on river Narmada a major part of the running water is being converted in to stagnant water. The fish fauna were studied during construction of dam on the river Narmada.The fish diversity is correlated with biological and various physico-chemical parameters that regulate the productivity and distribution of different species of the fishes. The fish population is abundant and majority of fishes are exploited for human consumption. In present study, attempts have been made to collect, classify and identify fish of river Narmada in its Western Zone. The study operated on a stretch of the river approximately 15 km from Piplud to Jangerwa from September 2006 to August 2007. The survey indicated that 51 species of fish were found in this zone of the river. The major fish abundance was noticed viz. major carps, minor carps and cat fishes. The several species of fish belonging order Clupiformes, Cypriniformes, Beloniformes, Opiocephaliformes, Mastacambelliformes, Siluriformes and Perciformes. In which maximum 37 species belonging to the order Cypriniformes. Some species of fishes like *Cirrihinus cirrihosa*, *Aspidoparia jaya*, *Colisa fasciatus*, *Labeo bata*, *Oreichthys cosuatis*, *Osteobrama cotio* etc. showed a declining trend in this stretch.The fish species diversity was decreasing. Diversity Index was lowest 0.53 in March and was highest 1 in October in the year 2006- 2007. The main reasons behind the decline of species are habitat destruction, introduction of exotic species, pollution and over fishing. An urgent need exists for studying the life history traits and demography of the most important threatened fishes, as lack of information on these aspects have significantly affected conservation efforts.

Keywords- Fish fauna, Diversity, River Narmada, Western Zone

Deptt. of Zoology, Govt. P.G. College Barwani (M.P.)
* Lecturer, School of Studies, Indore (M.P.)

Introduction

Several ways of measuring the species diversity of biological communities have been described (MacArthur 1965, Chao & Shen 2003, Magurran, 2005), and this diversity is considered by ecologists to be an indicator of community well-being (Magurran 1988). India has considerable ichthyofaunal diversity. Day (1875) described 1,418 species of fish under 342 genera, and a century later Jayaram (1981) listed 742 freshwater species under 233 genera, 64 families and 16 orders from the Indian region. Heda (2009) described 150 species belonging to 26 families. In spite of this rich diversity, literature concerning numerical studies of fish diversity is scarce.

The Narmada River is the fifth longest in India. It's already been dammed in several places but the main dam, the Sardar Sarovar is closet to the Sea. In the river Narmada carps should be in major quantity but in spite of that data's are reflecting that the No. of cat fish are increasing rather than the carps. Therefore fish species & Ecological condition of our area is being changing. No. of species are decreasing it mean's species diversity in the descending order. The first information regarding the ichthyofaunal assemblage of river Narmada is credited to Hora & Nair (1941), who recorded 40 species of fish. Kulkarni (1951) had observed Hilsa migration in Narmada which generally commenced in July when the first flood occurred. Karmachandani *et al.* (1967) have recorded 77 species. Rao *et al.* (1991) recorded 84 species belonging to 45 genera. Desai (1992), Bakawale and Kanhere (2006) and Verma & Kanhere (2007) have also studied the fish fauna of river Narmada.

Material & Methods

For the present study 14th stations were selected at the bank of river Narmada, Satmatra, Eklera, Kasrawad, Bagud, Piplud, Bhilkheda, Pendra, Nangaon, Pichhodi, Kathoda, Sondul, Jangarwa, Chikhalda & Koteshwar in the stretch of 15 ks. the period of investigation were one year from September 2006 to August 2007.

Fish species were collected with the help of local fishermen and the tribal people at various locations. The specimens were preserved in 5% formalin, morphometary of fishes with the helped of tools and identification of the fish was done with the help of literature Day (1878), Jayrarm (1981).

Species Diversity of fishes are measured by the following methods-

$$\text{Species Diversity} = \frac{\text{No. of Species}}{\sqrt{\text{Total Number of individuals}}} \times 100$$

Results & Discussion

Data collected during the study period, from at the sampling centers was utilized to estimate the fish diversity in the river Narmada. 51 Species of fish belonging to 7 orders and included under 15 families were collected in Narmada (Table-1). Perciformes contributed by 4 species, while ophiocephaliformes came thire in contribution and counted 3 species and rest species were from Clupiformes, Beloniformes, Siluriformes and Mastacembleformes (3,1,1, and 2 species respectively).

The monthly species diversity of the fishes were recorded as September 0.91, October 1, November 0.81, December 0.57, January 0.81, February 0.89, March 0.53, April 0.67, May 0.78, June 0.67, July 0.53, August 0.54 (Table- 2). The detailed monitoring and thorough comparisons of old collection and observations data with more recent ones showed that many species of fish in the river Narmada are declining and some have been disappeared. The present study reveals that at least 39 species, in comparison to the study of Rao *et al.* (1991), are at decline and can be considered as threatened species or endangered species. Verma & Kanhere 2007 have studied the threatened Icthyofauna of river Narmada in Western Zone.

The results indicate that reduction in the overall abundance of fish fauna in recovery compared to reference is a clear indication of the effect of habit destruction. Simultaneously, it was also revealed that the river has not recovered in the studied stretch. There is definitely some kind of disturbances in the river which is causing reduction in the abundance of fish fauna. The disturbances visibly seen were local fish harvest by traditional means of fish catch or unseen means like illegal use of electrical appliances or poisoning of the fish fauna using plants herb extraction by the local people. An urgent need exists for studying the life history traits and demography of the most important threatened fishes, as lack of information on these aspects have significantly affected conservation efforts.

Table 1 - List of fishes recorded in River Narmada during Sep. 2006 to Aug. 2007.

ORDER	FAMILY	GENERA
Cypriniformes	Cyprinidae	*Catla catla*
		Cirrihinus mrigala
		Labeo rohita
		Aspidopariajaya
		Cirrihinus cirrihosa
		Cirrihinus reba
		Crossochelius latius
		Discognathus lamta
		Labeo calbasu
		Labeofimbriatus
		Labeo bata
		Labeo gonius
		Nemacheilus botia
		Oreichthys cosuatis

		Osteobrama cotio
		Puntius chola
		Puntius sarana
		Puntius sophorae
		Tor putitora
		Tor tor
		Clarius batrachus
		Mystus aor
		Mystus seenghala
		Oxygaster bacaila
		Puntius ticto
		Ctenopharygdon idella
		Cyprinus carpio
		Hypothalmicthys moltrix
	Siluridae	*Ompok bimaculatus*
		Ompok pabo
		Wallago attu
	Bagridae	*Mystus bleekeri*
		Mystus cavasius
		Rita rita
		Rita pavimentata
	Schielbeidae	*Clupisoma garua*
	Heteropneustidae	*Heteropneustes fossilis*
Clupiformes	Clupeidae	*Gudusia chopra*
	Notopteridae	*Notopterus notopterus*
		Notopterus chitala
Beloniformes	Belonidae	*Xenthodon cancila*
Ophiocephaliformes	Ophiocephalidae	*Channa marulius*
		Channa punctatus
		Channa striatus
Perciformes	Centropomidae	*Chanda nama*
	Nandidae	*Nandus nandus*
	Gobioidae	*Glassogobius giuris*
	Anbantidae	*Colisa faciatus*
Mastacembaleformes	mastacembelidae	*Mastacembelus armatus*
		Mastacembelus pancalus
Siluriformes	Pangasiidae	*Pungasius pangasius*

Table 2. Species Diversity of Fishes in River Narmada 2006- 07

S.No.	Month	Species Diversity
1	September	0.91
2	October	1(high diversity)
3	November	0.81
4	December	0.57
5	January	0.81
6	February	0.89
7	March	0.53 (low diversity)
8	April	0.67
9	May	0.78
10	June	0.67
11	July	0.53 (low diversity)
12	August	0.54

References

- Bakawale, S. and R. R. Kanhere (2006). Fish fauna of river Narmada in West Nimar (M.P.). Research Hunt., 146-51.
- Day F. (1878). The fish of India. William Dawson's and Sons, London U. K. Reprint edition, today and Tomorrow Book Agency, Delhi.
- MacArthur, R. H. (1965). Patterns of Species Diversity. In Patric, R (ed.). Benchmark Papers in Ecology 13. Hutchinson Ross Publishing Company, Pennsylvania, 14pp.
- Magurran, A. E. (1988). Ecological Diversity and its Measurement. Princeton University Press, Princeton, NJ, 179pp.
- Magurran, A. E. (2005). Species abundance distributions pattern or process Functional Ecology 19 177-181.
- Chao, A. & T. J. Shen (2003).Non-parametric estimation of Shannon's index of diversity when there are unseen species in sample. Journal of Environmental and Ecological Statistics 10 429 - 443.
- Day, F. (1875). The Fishes of India Being A Natural History of the Fishes Known to Inhabit The Seas and Fresh Waters of India, Burma and Ceylon. Text and Atlas in 4 Parts. London, xx + 778195pp.
- Desai,V. R. (1992). Endangered, vulnerable and rare fishes of river systems in Madhya Pradesh. Proceedings of the National Seminar on endangered fishes of India. 25-26, April 1992 at N.B.F.G.R. Allahabad, p. 22.
- Jayaram, K. C. (1981). The Freshwater Fishes of India, Pakistan, Bangladesh, Burma and Sri Lanka. Zoological Survey of India, Calcutta, xxii+475pp.

- Heda, N. K. (2009). Freshwater Fishes of Central India A Field Guide. (2009). Vigyan Prasar, Department of Science and Technology, Government of India, Noida, 169pp.
- Hora, S. L. & K. K. Nair (1941). Fishes of Satpura range, Hoshangabad district, Central provinces, Rec. Indian Mus., 43(3) 361-373.
- Kulkarni, C.V. (1951). Hilsa fisheries in the Narmada River. J.Bombay Nat. Hist. Soc., 49(4) 614-623.
- Karamchandani, S. J.; V. R. Desai, M. D. Pisolkar and G. K. Bhatnagar (1967). Biological investigations on the fish and fisheries of Narmada River (1958-1966). Central Inland Fisheries Research Institute, Barrackpore, W.B., p.10.
- Rao, K.S.; S. N. Chaterjee & A. K. Singh (1991). Studies on the prepoundment fisheries potential of Narmada river in Western zone. J. Inland fish. Soc. of India, (I) 23 34-51. CIFRI, Barrackpore.
- Verma, D. and R.R. Kanhere (2007). Threatened Ichthyofauna of the River Narmada in Western Zone. Life Science Bulletin, 4 (1&2) 2007 (17-20.)

32

QUALITY OF WATER RESOURCES IN BUNDELKHAND REGION: A CASE STUDY OF MATATILA DAM, LALITPUR, UTTAR PRADESH, INDIA

Jitendra Kumar and Amit Pal

Abstract : Water as resource is under relentless pressure due to population growth, rapid urbanization, large scale industrialization and environmental concern. It is one of the most important commodities which man has exploited than any other resource for the substance of his life. In this context the status of water quality of Matatila dam was carried out for the study. Water samples were collected quarterly from ten different sites of Matatila dam during the month of November 2006 to October 2008. The ranges of Electrical Conductivity in study area are 265 - 297 μS/cm. Total Dissolved Solid (TDS) ranges from 159 - 214 mg/l. DO 5.8 - 10.4 mg/l; BOD 3.0 - 8.6 mg/l; COD 58 - 98 mg/l. In cationic abundance sodium is followed by calcium magnesium and potassium (Na>Ca>Mg>K) in dam water through out the year. Nitrate, Sulphate and Phosphate were found to below the permissible limit of WHO. The presence of heavy metals were followed by Mn>Pb>Fe>Cr>Cu>Zn>Al>Cd in the dam water throughout the year although Co, Ni and Hg were not detected.

Keywords: Cationic abundance, Heavy metals, TDS, Matatila dam.

Introduction

Water as resource is under relentless pressure due to population growth, rapid urbanization, large scale industrialization and environmental concern. It is one of the most important commodities which man has exploited than any other resource for the substance of his life (Mathur and Maheshwari, 2005). Water pollution has now reached a crisis point specifically in developing world. Almost every water body is polluted to an alarming level. Thus, estimation of quality of water is extremely important for proper assessment of the associated hazards (Warhate *et al.*, 2006).

Institute of Environment and Development Studies, Bundelkhand University, Jhansi (U.P.) 284128

In societies like our India with developing economics, the optimum development, efficient utilization and effective management of their water resources should be the dominant strategy for economic growth. But in recent years unscientific management and use of this resources for various purpose almost invariably has created undesirable problems in its wake, water logging and salinity in the case of agriculture use and environment pollution of various limits as a result of mining, industries and municipal use (Kumar *et al.*, 2008).

The water quality of urban Lakes/Pond/River has deteriorated sufficiently to cause serious disturbance to the biodiversity of Lake/Pond/River environments. Due to lack of proper planning and negligence of regulations, an appreciable amount of environmental degradation and ecological damage to Lake/Pond/River water.

Materials and Methods

Matatila Dam constructed during 1952-1964 on Betwa River in Lalitpur district, Uttar Pradesh. The catchments area is around 20 sq km with a maximum storage capacity of 1132 mcm. Length of the dam is 6.30 km and height is 33.53 meters. The dam generates about 45 MW of power.

Preservation and analysis of water samples were based on standard method proposed by American Public Health Association (APHA, 2005). pH, EC, TDS, Temp. and DO were analyzed on the spot by water quality analyzer kit (Elico, PE 138). NO_3, PO_4 & SO_4 were analyzed by UV Visible Spectrophotometer (Elico, SL 159). Na and K were analyzed by Flame Photometer (Systronic 130). The water quality was tested for its use for irrigation and interpreted in terms of Sodium Absorption Ratio (SAR) and % Na.

The water samples were taken in evaporating dishes and acidified to methyl orange with conc. HNO_3. Further 5mL conc. HNO_3 was added and evaporated to 10mL. Then it was transferred to a 125mL conical flask. 5 mL of conc. HNO_3 and 10mL $HClO_4$ (70 %) were added. Then heated gently, till white dense fumes of $HClO_4$ appear. The digested samples were cooled at room temperature, filtered through Whatman No. 41 and finally the volume was made upto 100mL with double distilled water. Then this solution was boiled to expel oxides of nitrogen and chlorine. The

Table- 1 Classification of water for irrigation use based on TDS, SAR and % Na

Classification	TDS (ppm)	SAR	Na %
Excellent	< 200	< 10	<20
Good	200-500	10-18	20-40
Fair	500-1500	18-26	40-60
Unsuitable	>1500	>26	>60

The Na % describing the Sodium hazard giving by (Warhate *et al.*, 2006)- % Na = 100 X Na^+ / (Na^+ + Ca^+ + Mg^+ + K^+)

solution was used for the determination of heavy metals. The metals were analyzed by Atomic Absorption Spectrophotometer (Perkin Elmer 200).

The tolerance limits for some parameters recommended by Wilcox (1965) are given below -

The Sodium or Alkali hazard in the use of water for irrigation is determined by absolute and relative concentration of cation and is expressed in terms of Sodium Absorption Ratio (SAR) and it can be estimated by formula (Singh, 2002)

$$SAR = Na^{+} / [(Ca^{+} + Mg^{+})/2]^{1/2}$$

Where, concentrations are expressed in milli equivalent per litre (meq/l.)

Results and Discussion

The average value of physico-chemical properties and heavy metal concentration of the Matatila Dam water during the month of November 2006 to October 2008 are shown in Table 2 and 3.

In temperature of Matatila dam water sample ranged between 16.0-30.7°C. The pH of water sample ranges between 7.09-7.75. The value of pH in all category of water is within permissible limit. Conductivity is a numerical expression of the ability of an aqueous solution to carry an electrical current. The ranges of Electrical Conductivity in present are 265 - 297 µS/cm. Total Dissolved Solid (TDS) ranges from 159 - 214 mg/l; DO 5.8 - 10.4 mg/l; BOD 3.0 - 8.6 mg/l; COD 58 - 98 mg/l. Nitrate is an important plant nutrient and causes eutrophication in receiving water bodies. NO_3 values (0.91-1.12 mg/l) and SO_4 values are (36.04-42.42 mg/l) also below the permissible limit of WHO guideline. PO_4 ranges from 0.05 -1.14 mg/l.

The cationic chemistry is dominated by sodium and Calcium. In cationic abundance sodium is followed by calcium magnesium and potassium (Na>Ca>Mg>K) in dam water through out the year (Table 2 and 3).

The tolerance limit for TDS, SAR and % Na of water may use for irrigation purpose as recommended by Wilcox (1965). The result has been found to be excellent for TDS with some exception, fair for SAR and unsuitable for % Na (Table 1, 2 and 3).

The effluents are usually treated by physicochemical treatment followed by biological treatment process. However, such treatment systems are not effective for removal of color, dissolved solids; trace metals etc. and the effluents are directly discharged into drains, public sewers, rivers etc. These effluents containing trace metals, when discharged on agricultural land for irrigation, increase the metal content of the soils and availability of metals to plants. Some of the metals are concern because of their toxicity to plant (Cu and Zn) while others (As, Cd, Pb etc.) are hazardous to human health (Singh and Chandel, 2006).

In the present study following metals were analyzed such as Al, Cd, Cr, Cu, Co, Fe, Pb, Mn, Ni, Zn and Hg. The metals data show Al, Cd, Cu & Zn found to be under the permissible limit

Table 2. Physico-chemical properties of Matatila Dam water during the month of November 06 - October 08

Sample No.	pH	EC	Temp	Total Hardness	Alkalinity	TDS	DO	BOD	COD	NO_3	PO_4	SO_4	Na	K	Ca	Mg	% Na	SAR
1.	7.201 ± 0.016	282.75 ± 1.431	23.56 ± 1.637	230.12 ± 3.283	113.5 ± 0.779	177.12 ± 2.392	8.70 ± 0.351	4.71 ± 0.162	77.37 ± 3.409	0.99 ± 0.012	0.34 ± 0.120	39.94 ± 0.441	145.15 ± 0.837	3.54 ± 0.068	65.04 ± 0.439	16.45 ± 0.601	63.06 ± 0.282	22.75 ± 0.175
2.	7.231 ± 0.013	275.75 ± 1.504	23.71 ± 1.632	224.12 ± 2.603	110.5 ± 1.356	182 ± 1.996	8.75 ± 0.308	4.62 ± 0.213	75.62 ± 3.292	0.98 ± 0.013	0.34 ± 0.121	39.76 ± 0.280	142.79 ± 1.785	2.97 ± 0.051	63.29 ± 0.591	16.05 ± 0.452	63.41 ± 0.125	22.66 ± 0.185
3.	7.66 ± 0.026	287 ± 1.086	23.57 ± 1.613	219.62 ± 2.844	112.37 ± 1.737	177.75 ± 1.381	8.6 ± 0.408	4.81 ± 0.139	78.62 ± 2.104	0.98 ± 0.012	0.47 ± 0.136	39.57 ± 0.397	146.84 ± 1.319	3.66 ± 0.119	62.18 ± 0.601	15.64 ± 0.463	64.31 ± 0.307	23.55 ± 0.220
4.	7.225 ± 0.021	276 ± 0.977	23.45 ± 1.645	217.25 ± 2.621	111.25 ± 1.287	186.87 ± 2.989	8.55 ± 0.426	5.55 ± 0.472	76.75 ± 4.206	0.99 ± 0.011	0.34 ± 0.121	40.04 ± 0.339	139.95 ± 1.023	3.59 ± 0.048	62.33 ± 0.395	14.96 ± 0.475	63.37 ± 0.136	22.51 ± 0.098
5.	7.309 ± 0.011	287.75 ± 1.252	23.63 ± 1.614	221.37 ± 2.774	112.37 ± 2.183	181 ± 3.394	8.6 ± 0.349	4.78 ± 0.253	74 ± 2.595	0.94 ± 0.006	0.47 ± 0.134	40.19 ± 0.236	148.59 ± 1.373	3.48 ± 0.058	63.46 ± 0.461	15.28 ± 0.537	64.37 ± 0.171	23.68 ± 0.165
6.	7.309 ± 0.014	290.5 ± 1.157	23.41 ± 1.607	215 ± 2.663	109.87 ± 2.024	192.62 ± 2.961	8.51 ± 0.402	4.83 ± 0.237	80.62 ± 1.737	0.94 ± 0.009	0.47 ± 0.135	39.83 ± 0.302	162.31 ± 1.112	3.01 ± 0.065	62.76 ± 0.393	14.15 ± 0.490	67.00 ± 0.250	26.18 ± 0.205
7.	7.216 ± 0.014	286.13 ± 2.169	23.33 ± 1.681	220.62 ± 2.563	110.37 ± 2.083	194.37 ± 1.793	8.65 ± 0.366	4.71 ± 0.175	75.87 ± 1.650	0.99 ± 0.017	0.21 ± 0.092	39.15 ± 0.279	145.01 ± 2.516	3.35 ± 0.050	63.29 ± 0.458	15.20 ± 0.536	63.88 ± 0.243	23.13 ± 0.316
8.	7.456 ± 0.021	288.88 ± 0.891	23.49 ± 1.622	221.75 ± 2.191	107.87 ± 1.859	192.62 ± 1.482	8.49 ± 0.409	4.82 ± 0.234	81.87 ± 2.029	0.98 ± 0.012	0.46 ± 0.135	40.02 ± 0.136	165.71 ± 1.755	3.71 ± 0.056	63.18 ± 0.540	15.55 ± 0.483	66.75 ± 0.346	26.42 ± 0.336
9.	7.350 ± 0.015	289.13 ± 1.661	23.54 ± 1.744	224.62 ± 2.982	112.12 ± 2.081	194.25 ± 2.492	8.55 ± 0.386	4.7 ± 0.171	75.62 ± 3.487	0.99 ± 0.014	0.33 ± 0.121	39.47 ± 0.243	156.90 ± 1.185	3.59 ± 0.067	63.87 ± 0.367	15.82 ± 0.808	65.31 ± 0.348	24.87 ± 0.280
10.	7.274 ± 0.016	286.13 ± 0.784	23.31 ± 1.700	217.25 ± 2.183	112.25 ± 2.528	194.37 ± 2.853	8.68 ± 0 384	4.73 ± 0.202	80.5 ± 2.507	0.98 ± 0.014	0.46 ± 0.136	39.45 ± 0.175	147.46 ± 1.018	3.35 ± 0.064	62.88 ± 0.470	14.63 ± 0.354	64.58 ± 0.136	23.68 ± 0.108

Values are mean ± SE (n=8); Units - Concentration in mg/l, except pH; Temperature (°C); EC (µS/cm); SAR (meq/l)

Table 3. Average Value of Metal Analysis of Matatila Dam water during the month of November 2006 to October 2008

Sample No.	Al	Cd	Cr	Cu	Co	Fe	Pb	Mn	Ni	Zn	Hg
1.	0.004 ± 0.0001	0.001 ± 0.00003	0.462 ± 0.148	0.008 ± 0.0002	ND	0.678 ± 0.012	0.918 ± 0.011	1.390 ± 0.022	ND	0.006 ± 0.0001	ND
2.	0.004 ± 0.0001	0.001 ± 0.00004	0.456 ± 0.148	0.008 ± 0.0001	ND	0.668 ± 0.013	0.903 ± 0.014	1.389 ± 0.033	ND	0.006 ± 0.0002	ND
3.	0.004 ± 0.0001	0.001 ± 0.00005	0.456 ± 0.146	0.008 ± 0.0002	ND	0.673 ± 0.011	0.912 ± 0.014	1.425 ± 0.026	ND	0.006 ± 0.0002	ND
4.	0.004 ± 0.0001	0.001 ± 0.00005	0.447 ± 0.141	0.008 ± 0.0002	ND	0.698 ± 0.008	0.939 ± 0.009	1.423 ± 0.022	ND	0.006 ± 0.0002	ND
5.	0.004 ± 0.0001	0.001 ± 0.00004	0.489 ± 0.156	0.008 ± 0.0001	ND	0.708 ± 0.012	0.943 ± 0.008	1.425 ± 0.016	ND	0.006 ± 0.0002	ND
6.	0.004 ± 0.0001	0.001 ± 0.00003	0.489 ± 0.156	0.008 ± 0.0001	ND	0.700 ± 0.013	0.947 ± 0.009	1.443 ± 0.018	ND	0.006 ± 0.0002	ND
7.	0.004 ± 0.0001	0.001 ± 0.00003	0.485 ± 0.155	0.008 ± 0.0002	ND	0.710 ± 0.015	0.942 ± 0.008	1.435 ± 0.014	ND	0.006 ± 0.0002	ND
8.	0.004 ± 0.0002	0.001 ± 0.00004	0.497 ± 0.160	0.008 ± 0.0002	ND	0.707 ± 0.016	0.953 ± 0.006	1.419 ± 0.015	ND	0.006 ± 0.0002	ND
9.	0.004 ± 0.0001	0.001 ± 0.00002	0.483 ± 0.155	0.008 ± 0.0001	ND	0.697 ± 0.017	0.959 ± 0.005	1.407 ± 0.014	ND	0.005 ± 0.0001	ND
10.	0.004 ± 0.0001	0.001 ± 0.00005	0.483 ± 0.152	0.008 ± 0.0001	ND	0.683 ± 0.010	0.958 ± 0.003	1.422 ± 0.015	ND	0.005 ± 0.0001	ND

Values are Mean ± SE (n=8); Unit- mg/l; ND= Not detected

prescribed by WHO (1993) in the dam water. Although Co, Ni and Hg were not detected in Matatila dam in any seasons however the concentrations of Cr, Fe Pb and Mn are far above the permissible limit of WHO through out the year.

Conclusion

Water quality has solicited a major global concern due to ever increasing human developmental activities that over exploit and pollute the water resources on surface and underground (Adhilakshmi *et al.*, 2004). Water pollution has, among other consequences, that of causing deep long term modification of biodiversity. Eutrophication is one of the most alarming long term alterations. In this context the status of water quality of Matatila dam was considered for the study.

It is concluded that water samples from Matatila dam during the November 2006 to October 2008 have been found to be excellent for TDS with some exception, fair for SAR and unsuitable for % Na that means it is suitable for some extent for agricultural purposes. The cationic chemistry is dominated by sodium and Calcium. Overall metal concentration are Mn>Pb>Fe>Cr>Cu>Zn>Al>Cd in the dam water. Co, Ni and Hg are totally absent in all seasons in the dam water.

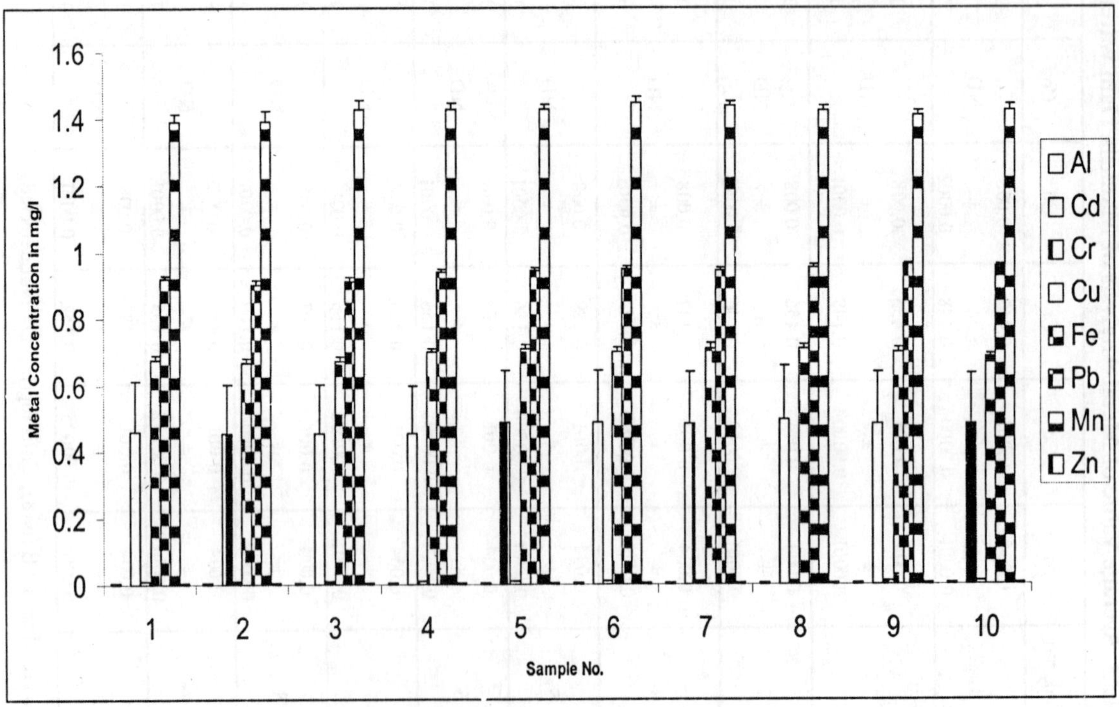

Fig. 1. Concentration of Heavy metals in water sample of Matatila Dam during the month of November 2006 to October 2008

References

- APHA (2005), Standard methods for examination of water and wastewater 21st ed. American Public Health Association, Washington.
- Adhilakshmi A., Mariappan P. and Sashirega Jasmine (2004), Assessment of water quality of Gedilam and Pennar Rivers at Cuddalore, Indian J. Environmental Protection, 24 (12), 888-893
- Kumar Jitendra, Singh Siddharth and Pal Amit (2008), Water quality of Turamdih and Jaduguda uranium mines and adjacent areas, East Singhbhum, Jharkhand, J. Ecophysiol. Occup. Hlth. 8 (2008) 7-14
- Mathur S.P. and Maheshwari (2005), Physico-chemical aspect of Pollution in Chambal River, Indian J. Environmental Protection.25 (10), 933-937.
- Singh, A. K. (2002). Quality Assessment of Surface and Subsurface Water of Damodar River Basin; Indian J. Environ Hlth, 44, 42-46.
- Singh, Virendra and Chandel, C.P. Singh (2006). Analytical Study of Heavy Metals of Industrial Effluents at Jaipur, Rajasthan (India), Journal of Environ. Science & Engg. 48 (2), 103-108.
- Warhate S.R., Yenkie M.K.N., Chaudhari M.D. and Pokale W.K. (2006), Impacts of Mining Activities on Water and Soil, Journal of Environ. Science & Engg. 48 (2), 81& 88.
- Wilcox, L.V. (1965). Effects of Industrial Wastes on Water for Irrigation Use, ASTM Tech Pub. No.273, 58.
- WHO (World Health Organization) (1993), Guidelines for drinking-water quality, vol. 1, Recommendations, Geneva, WHO, 130.

33

AN ENEMY OF AQUATIC ECOSYSTEM (WATER HYACINTH) CAN BE UTILIZED ECO-FRIENDLY

G. N. Jha, K. Borana, T.A. Qureshi*, Trapti Tiwari* and Vipin Kumar Sharma**

Abstract: The study was designed with the objective to utilize the world's most troublesome aquatic weed (*Eichhornia crassipes*) as an eco-friend or in economic manner in aquaculture. The weeds were collected from the Lower and Upper Lake of Bhopal and proximate composition were determined. Four isoproteic diets (Control, Exp-1, Exp-2 and Exp-3) containing 35% protein and different percentages of water hyacinth (0%, 10%. 20% and 30% respectively) were prepared and fed to the fingerlings of *Labeo rohita* for ninety days to see the utilization of this weed by the fish and to see the growth of the fish fed on different percentages of water hyacinth impregnated feeds. The FCR, SGR, AWG, Proximate composition and other parameters after the experiment showed that 20% water hyacinth added feed reduced the cost of fish production. There was no harm on fish health by the weed was observed. Therefore, water hyacinth can be utilized eco-friendly in aquaculture.

Keywords: *Eichhornia crassipes, Labeo rohita,* aquaculture, eco-friendly, isoproteic, proximate composition.

Introduction

The menace of aquatic weeds is reaching alarming problems in many parts of the world, but it is particularly severe in tropical countries, where abundant sunlight and favorable water temperature, increasing numbers of dams, barrage and irrigation channels foster aquatic plant growth. This problem is further aggravated because these unutilized weeds choke the water bodies, thus reducing the carrying capacity for aquaculture purposes. Such water bodies are often left unproductive with impeded light penetration and depletion of dissolved oxygen. Regrettably, there is hardly any simple or cost-effective way to control the infestation of these aquatic macrophytes in an environment-

*Directorate of Coldwater Fisheries Research (ICAR), Bhimtal-263 136 (U.K.)
Department of Zoology and Applied Aquaculture, Barkatullah University, Bhopal (M. P.);
E-mail.ghan_shark@yahoo.com

friendly manner. However, a perusal of the available literature shows that some of the aquatic weeds are highly nutritive and, therefore, one alternative solution to check the massive population of these weeds might be their utilization through incorporation as components of feedstuff for fish and prawn, in particular. In fact, significant effort has been directed toward evaluating the nutritive value of different non-conventional feed resources, including terrestrial and aquatic macrophytes, to formulate nutritionally balanced and cost-effective diets for fish and poultry (Edwards *et al.*, 1985; Patra & Ray, 1988; Ray & Das, 1995; Wee & Wang, 1987).

Bhopal is the capital of Madhya Pradesh state of India. Upper and Lower Lake are the life line of the city. The water of Upper Lake is very much clean than the Lower Lake so that drinking water is supplied from this Lake but the water from Lower lake is not utilized for drinking purposes due to pollution caused by high quantity of organic load. Organic water supports micro as well as macrophytes growth reduces light penetration resulting oxygen depletion and finally eutrophication. So that Lower Lake of Bhopal considered as eutrophic Lake. Both the major Lake of the city more or less sheltered by the water enemy '*Eichhornia crassipes*' commonly known as water hyacinth.

Water hyacinth belongs to Order-Commelinales, Family- Pontederiaceae, Genus- Eichhornia and Species- *E. crassipes*. It is a floating aquatic weed distributed worldwide and a major problem of aquatic ecosystem. It is widely used by the agriculturist, aquaculturist and other live stock culturist at some or at large scale as feeding or fertilizing them since long.

Several works have already been done worldwide in this aspect to see the utilization of this aquatic weed (water hyacinth) impregnated feeds by the fish and other animal. It has high quantity of organic fertilizers so that widely used by the agriculturist. In Kenya the experimental use of water hyacinth as an organic fertilizer and animal feed has been undertaken in places such as flower farms. In China the weed was widely used as animal food from the 1950s to the 1970s. As at that time, the economy in rural areas was very depressed and there was great shortage of food for animals. Boiled water hyacinth is used in Southeast Asia as a feed for pigs. The plants are chopped and sometimes mixed with other vegetable wastes, such as banana stems, and boiled slowly for a few hours until the ingredients turn into a paste, to which oil cake, rice bran and sometimes maize and salt are added.

There has been some use of *E. crassipes* for the removal of nutrients and heavy metal from sewage and sludge ponds (bioremediation) also been reported. World's most troublesome aquatic weed water hyacinth (*Eichhornia crassipes*) has been studied for its tendency to accumulate the heavy metal contaminants present in water bodies by Dixit in 2007. Still, there are scanty of information regarding water hyacinth utilization in the feed of Indian major carp.

Therefore this work was designed with the objectives- to prepare isoproteic feeds with 35% protein, impregnated with different proportion of water hyacinth that is locally available in abundance free of cost, to reduce down the cost of feed formulation, to see the water hyacinth impregnated feed utilization by one of the Indian major carp, '*Labeo rohita*' and over all to reduce down the cost of fish production.

Materials and Method

Weed collection

Water hyacinths were collected from Upper and Lower Lake of Bhopal by hand picking method with the help of hand net, rope and boat, brought to the experimentation site for further process.

Sample preparation

Collected weeds were washed properly disaggregated from roots. Some of the samples were grinded for proximate composition, minerals, etc. analysis in fresh and rest of the sample were left open for sun drying for feed preparation.

Feed preparation

Four isoproteic feeds with 35% protein (Control-without Eichhornia, Exp-1 with 10% Eichhornia, Exp-2 with 20% Eichhornia and Exp-3 with 30% Eichhornia) were prepared with Wheat bran, Rice bran, Soybean oil cake and Fish meal as basic ingredients while sodium alginate (3%) as binder. 2% commercial vitamin-mineral mixture were added in all the four formulated feeds. Eichhornia were added in replacing Wheat bran through Exp-1 to Exp-3. Ingredient quantity was selected as per Pearson-square formula.

Experimental set-up, feeding and management

Experiments were conducted in triplicate for 90 days with continuous flowing fiver tanks provided with inlet at one end and outlet at other end.

Fingerlings of nearly same length-weight of *Labeo rohita* were brought from a private fish farm, acclimatized in controlled condition and released into fiver tanks @ 10 numbers of fishes per tank.

Fishes were fed with above mentioned formulated feeds @ 5% of their body weight (feeding rate readjusted on fortnight basis depending on their utilization).

Regular siphoning was done to clean the tank and water parameter monitored regularly and kept under control.

Proximate composition analysis

Moisture, ash, ether extract (EE) and nitrogen-free extract (NFE) were determined by the methods of the Association of Official Analytical Chemists (AOAC, 1990). The crude protein content was determined by the Kjeldahl procedure, AOAC Method 920; the factor N X 6.25 was used to convert nitrogen into crude protein. The crude lipid was extracted using a Soxhlet apparatus in petroleum ether and quantity of lipid was determined after drying the extract in an oven.

Mineral analysis

The mineral content of the plant and fish (after feeding trial) namely, Na, K and Ca, were estimated after ashing by flame photometry. The total phosphorus content was determined as described by Umoren *et al.* (2005).

Growth analysis

For growth parameter analysis, length-weight was recorded fortnightly. FCR, SGR and AWG were calculated at the end of the experiment.

Table 1. Nutrient composition of water hyacinth

Composition	Upper Lake	Lower Lake
Moisture	90	90.5
Dry Matter	9.9	10.2
Crude Protein (% DW)	14.9	14.5
Crude Fat (% DW)	2.7	2.4
Ash	14	14.9
Na (%)	0.03	0.04
Ca (%)	2.65	2.7
K (%)	4.25	4.38
P (%)	0.45	0.4

Table 2. Length-weight of the observed fish

Tanks	Average Length (cm)		Average Weight (gm)	
	Initial	Final	Initial	Final
Control	5.8	14.2	2.7	12.2
Exp-1	5.5	12.9	2.6	12
Exp-2	5.7	13.8	2.4	11.9
Exp-3	5.4	12.5	2.5	11.5

Results and Discussion

In the present study, moisture and dry matter was found slightly high in the weeds of Lower Lake (90.5% and 10.2%, respectively) than the Upper Lake (90% and 9.9% respectively) while protein and lipid contents were observed slightly high in the weed collected from Upper Lake (14.9% and 2.7% respectively) than the Lower Lake (14.5% and 2.4% respectively) of Bhopal. In case of ash,

Na, Ca and K, it was in high concentrations in the weed of Lower Lake (14.9, 0.04, 2.7 and 4.38 respectively) than the Upper Lake (14, 0.03, 2.65 and 4.25 respectively) while P was slightly low in the weed of Lower Lake (0.4) than the Upper Lake (0.45) (table-1). Observations of the present study having nearly similar results with the results of Khadiga *et al.* (2009). He studied the proximate composition of water hyacinth. According to him water hyacinth contains, moisture-91%, dry matter 10.4%, ash 15.5%, lipids 2.9% and protein 14.2% of its dry weight. It having 2.25% N, 0.28% P, and 3.48% K of dry weight. The composition of an animal or plant can slightly vary with the sex, age, environment, etc. from where they are taken.

Paul (2004) conducted an experiment to test the potential of diets containing different amounts of water hyacinth meal as food for matrincha (*Brycon* sp.). Two diets with a similar protein content but different percentage of water hyacinth were tested and compared with a control ration. In 90 days, fish fed on a diet containing 9.5% of water hyacinth (34.9% protein) grew from 87.9 to 208.0g (1.2g/d). The food conversion factor was 1.7. Fish fed on a second diet (18.9% water hyacinth, 34.0% protein) grew during the same period from 84.5 to 190.3g (1.1g/d). The conversion factor was 1.8. Fish fed on a control ration (0% water hyacinth, 34.0% protein) grew from 89.3 to 190.2g (0.9g/d). The experiment indicated that matrincha grow well when fed diets containing water hyacinth and apparently were not adversely affected the addition of this material to the ration. In our study, it was observed that fish fed with 20% water hyacinth fortified feed (Exp-2) having satisfactory weight gain (395.83%), FCR (1.89) and SGR (10.55%) than the control (weight gain-351.85%, FCR-1.89 and SGR-10.55%), Exp-1 (weight gain-361.53%, FCR-1.91 and SGR-10.44%) and Exp-3 (weight gain-360%, FCR-2 and SGR-10), table-3.

Table-3 Efficiency of the formulated feeds

Tanks	Weight gain (%)	SGR (%)	FCR	Cost/kg feed (Rs)	Cost/kg Fish (Rs)
Control	351.85	10.55	1.89	19.86	37.53
Exp-1	361.53	10.44	1.91	18.86	36.02
Exp-2	395.83	10.55	1.89	17.86	33.75
Exp-3	360	10	2	16.86	33.72

Table-4 Composition of experimented fish

Compositions	Control	Exp-1	Exp-2	Exp-3
Moisture (%)	76.876	77.254	75.85	76.5
Dry Matter (%)	22.89	21.985	23.3	22.9
Ash (g/100g)	2.17	2.09	2.3	2.4
Crude Fat (g/100g)	2.23	2.35	2.48	2.3
Crude Protein (%)	12.35	12.4	12.62	12.39

Table-5 Mineral composition of experimented fish

Minerals (mg/100g)	Control	Exp-1	Exp-2	Exp-3
Na	275	282	284	262
K	740	860	978	990
Ca	285	280	286	295

The cost of feed decreases with increase in water hyacinth content and FCR value also increase accordingly, showed by this experiment (table-3). Lower the FCR shows higher the quality of the feed and also reduces the cost of fish production. In present study, it was observed that the Exp-2 (20% water hyacinth) and control (0% water hyacinth) having the lowest FCR (1.89), slightly increased in Exp-1 (1.91) with 10% water hyacinth while FCR was highest (2) in Exp-3 (having 30% water hyacinth). Cost of feed for control tank was highest (19.86/kg) than Exp-1(18.86/kg), Exp-2 (17.86/kg) and it was lowest for Exp-3 tank (16.86/kg). The cost of the produced fish was lowest (nearly similar) from Exp-3 (Rs. 33.72/kg fish) and Exp-2 (Rs. 33.75/kg fish), it was highest from control (Rs. 37.53/kg fish) and slightly low from Exp-1 (Rs. 36.02/kg fish). Lower the cost of production, higher the benefits from the produce.

Proximate composition analysis of all the experimental fish (control-Exp-3) did not showed very much significant variations. But, very slight variations in terms of protein and lipid were observed among 20% (Exp-2 12.62% and 2.48%, respectively) and 0% (control 12.35% and 2.23%, respectively) water hyacinth fortified feeds (table-4). While, in case of mineral composition, K was observed increasing (control 740mg/100g, Exp-1 860mg/100g, Exp-2 978mg/100g and Exp-3 990mg/100g) with increase in water hyacinth concentration (table-5), may be due to slightly high concentration of potassium (4.25-4.38%) in water hyacinth than the other (Na 0.03-0.04% and Ca 2.65-2.7%). Rest all (Na and Ca) were found near constant in the fish of all the tanks.

Conclusion

The present study has demonstrated that, *Eichhornia crassipes* can serve as an important source of protein, vitamins and minerals, suitable for incorporation in fish diet. It neither affected the growth nor the composition of the fish negatively. Incorporation of this weed in fish feed will certainly benefit the aquaculturist by reducing down the cost of fish production by the reduction in feed formulation cost and preventing from the use and harm of pesticides to control the weed as well. In this way, this aquatic enemy can be utilized eco-friendly.

References

- AOAC (1990). Methods of analysis (15th ed.). Washington, DC Association of Official Analytical Chemists.

- Edwards, P., Kamal, M., & Wee, K. L. (1985). Incorporation of composted and dried water hyacinth in pelleted feed for the tilapia *Oreochromis niloticus* (Peters). Aquaculture and Fisheries Management, 16, 233-248.
- Khadiga M. Sharshar and Amany M. Haroon (2009); Comparative investigations on some biological aspects in freshwater Crayfish (*Paracambarus clarkii*) fed on *Eichhornia crassipes*, *Eichhornia stagnina* L. and *Polygonum tomentosum* L. American-Eurasian J. Agric. & Environ. Sci., 5(4) 579-589, 2009.
- Dixit Savita; Tiwari Suchi (2007); Effective utilization of an aquatic weed in an eco-friendly treatment of polluted water bodies; J. Appl. Sci. Environ. Manage. September 2007 Vol. 11(3) 41 - 44.
- Patra, B. C., & Ray, A. K. (1988). A preliminary study on the utilization of the aquatic weed *Hydrilla Verticillata* (L.) Rayle as feed by the carp, *Labeo rohita* (Hamilton) growth and certain biochemical composition of flesh. Indian Biology, XX(I), 44-50.
- Paul Ulrich Saint; Ulrich Werder and Aylton Saturnino Teixeira (2004). Use of waterhyacinth in feeding trails with Matrincha (*Brycon* sp.) J. Aquat. Plant Manage. 19 18-22.
- Ray, A. K., & Das, I. (1995). Evaluation of dried aquatic weed, *Pistia stratiotes* meal as a feedstuff in pelleted feed for rohu, *Labeo rohita* fingerlings. Journal of Applied Aquaculture, 5, 35-44.
- Umoren, U. E., Essien, A. I., Ukorebi, B. A., & Essien, E. B. (2005). Chemical evaluation of the seeds of *Milletia obanensis*. Food Chemistry, 91, 195-201.
- Wee, K. L., & Wang, S. S. (1987). Nutritive value of Leucaena leaf meal in pelleted feed for Nile Tilapia. Aquaculture, 62, 97-108.

34

PHYSICO-CHEMICAL PROPERTIES OF WATER OF VISHAV STREAM IN KASHMIR VALLEY, J& K, INDIA

*Assadullah Sheikh, Amit Pal and *Ashok K. Pandit*

Abstract: The present investigation was undertaken from January to May, 2009 on Vishav stream of Kashmir valley. A stretch of about 40 km of the Vishav stream in the valley of Kashmir was studied within four assembly constituencies before it confluences with the river Jhelum at Sangam. The Vishav contributes major portion of discharge to the Jhelum and it is most live and perennial tributary of it. Due to the over-exploitation of fresh water resources and waste water disposal, the modification in the physical, chemical, biological characteristic of water have occurred, creating profound and deleterious effects on human health. The investigation deals with the general water chemistry at four sampling stations. The comparatively low DO, transparency and high nutrient level viz, Ca, N, P, Cl and CO_2 etc. all pointing towards the nutrient enrichment.

Keywords: Fresh water, nutrient enrichment, River Jhelum, Vishav stream

Introduction

Water, literally the source of life on earth, is a major constituent of hydrosphere, since without it life cannot exist. Essentially, all life depends upon the water, there can be no life without water and no water without life and is an elixir of life. This is evident from the Quranic verse i.e., "And we made from water every living thing" (2130)

وَجَعَلْنَا مِنَ الْمَاءِ كُلَّ شَيْءٍ حَيٍّ

Institute of Environment and Development Studies, Bundelkhand University, Jhansi (U.P) 284128, India
*Department of Environment Science, Unioversity of Kashmir, Srinagar -190006, J&K, India

Water is an indispensable commodity which should be easily accessible, adequate in quantity, free of contamination, safe and readily available throughout the year in order to sustain life (WHO, 1997). It is one of the most important commodities which man has exploited than any other resource for the sustenance of his life (Kumar *et al.*, 2008). Water pollution has now reached a crisis point specifically in developing world. Almost every water body is polluted to an alarming level. Thus estimation of quality of water is extremely important for proper assessment of the associated pollutants (Warhate *et al.*, 2006)

The valley of Kashmir is well known for its water resources. These water bodies are facing grave pollution problems. A number of indigenous and high quality species are diminishing. To cope up with these problems there is a need of application oriented limnological research so that we are able to better utilize this most important resources of ours. The given paper contains the work done on a stream named Vishav stream, situated to the south eastern corner of Kashmir Valley. The Vishav stream has remained untouched by limnogists mainly be far away from the capital city of Srinagar. Different physic-chemical parameters of this important and perennial stream have been investigated.

Vishav Stream is one of the most live and perennial tributaries of River Jhelum. It originates from Kounsernag at the foothill of the Pir Panjal range in the southeastern corner of Kashmir Valley, close to that of Jhelum. Vishav contributes major portion of the discharge towards the Jhelum. Infact, the Jhelum draws heavily on Vishav feeders in the initial stage. It carries huge discharge and is the only life line for the dense population in its catchment. Its major tributaries are Kandy Nallah (D.H.Pora) Vethvithro Nallah (Bonagam Verinag). The world famous waterfall Arabal on Vishav has created a deep picturesque channel in its basefoot and its stretch dashes into the plain of Kashmir in a style. From Arabal fall it flows in a boulder stage up to village Nowpora and in a subtle grade from Khudwani village it actually takes the river features and finally confluences with River Jhelum at Sangam in District Pulwama of J&K State. The maximum discharge is available in the month of July and August and decreases to lowest in the month of December and January.

Its water is used for drinking, irrigation, fishing, washing and navigation purposes and in return it receives the effluents carrying the organic and inorganic pollutants from its catchments. It has catchments area of 1230 Sq km. Due to increase in population and over exploitation of fresh water resources and waste water disposal, the direct or indirect modifications and alterations in creating the physical, chemical and biological characteristics of water have occurred profound and deleterious effects on human health and on the environment as a whole.

Materials and Methods

Study Area

In the present study an attempt was made to find out the physico-chemical characteristics of the Vishav Stream, at four sites viz.- Reshi Nagar (site-I), Nehama (site-II), Brazloo (site-III) and

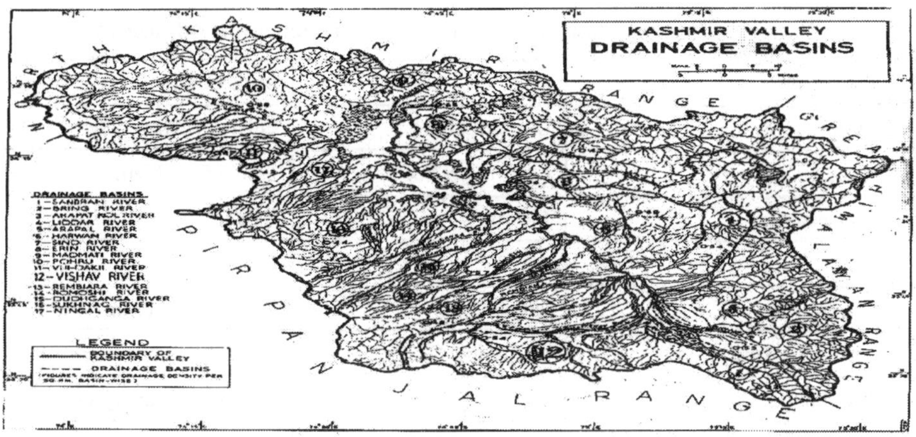

Fig. 1 Map of Kashmir Valley showing location of Vishav stream (as 12)

Khudwani (site IV), so as to have an insight into the level of pollution in the stream. In order to have reasonable and wholesome information of Vishav stream, a detailed study of the stream course was conducted on monthly basis from January to May 2009. The selection of sites was done keeping in mind different types of habitat features i.e., elevations, fast flowing areas, slow zones, as well as human interference.

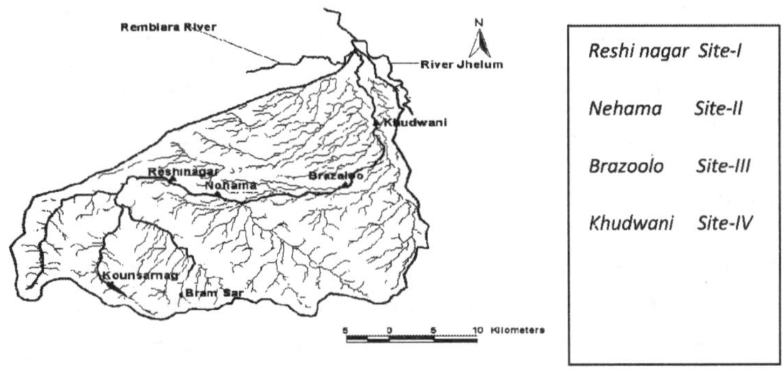

Fig. 2 Drainage system of Vishav stream showing selected study areas

The water samples were collected from selected sites in one liter polythene bottles. Separate samples were collected for dissolved oxygen in well stopper DO bottles of 30ml capacity. Initial fixation was done in the field and chemical analysis of water samples were carried in laboratory within 24hrs and analysis was done as per (APHA, 2005; Mackereth, 1963; Galterman and Clymo, 1969). Some parameters like Transparency, tempt, depth, velocity, were made on spot at samples sites, while rest of parameters were determined in the laboratory. The methodologies used in the study are given in a tabulated form (Table - 1).

Table - 1 Physico-chemical parameters and their methodologies

S.No.	Parameters	Unit	Methodology
1	Temperature	Degree Celsius	Celsius Thermometer
2	Velocity	Centimeters/Sec	Sacchi Disc Method
3	Transparency	Meters	Sacchi Disc Method
4	pH		Digital pH meter
5	Conductivity	μS/cm	Digital conductivity meter
6	Carbon Dioxide	mg/ litre	Complexometric titration
7	Alkalinity	mg/ litre	Titrimetry with H_2SO_4
8	Dissolved Oxygen	mg/ litre	Winkler's Method
9	Chloride Content	mg/ litre	Argentometric method
10	Total Hardness	mg/ litre	Tritrimetry with EDTA
11	Calcium Content	mg/ litre	Tritrimetry with EDTA
12	Magnesium Content	mg/ litre	Tritrimetry with EDTA
13	Total Phosphorous	μg/ litre	Stannous Chloride method
14	Orthophosphate Phosphorus	μg/ litre	Stannous Chloride method
15	Nitrate-Nitrogen	μg/ litre	Salicylate method
16	Ammonical-Nitrogen	μg/ litre	Phenate method
17	Nitrite-Nitrogen	μg/ litre	Adoni's method

Results and Discussion

The mean values for the various physico-chemical characteristics of water are depicted in Figures 3 to 15. The water temperature shows a close proximity with prevailing air temperature. A close relationship between air and water temperature was reported in the present study. The temperature of water which depends upon the exposure to the solar radiation, water depth, transparency and climate was recorded as minimum of 3.5°C during January at site-I and maximum 23.5°C during may at site-IV water temperature gradually increased downstream. The increase in temperature may be attributed to the decrease in altitude. Decrease in velocity also results in an increase water temperature (Imevbore, 1970).

Velocity shows a clear temporal and spatial fluctuation. The maximum velocity of 190cm/s was recorded at site-II during April while as minimum of 35cm/s was recorded at site -IV during January. The higher velocity may be due to melting of glaciers and rainfall while as lowest velocity was due to freezing of water at higher altitudes. The current velocity showed a decreasing trend towards lower elevations. The velocity of water is directly associated with volume of water (Hynes,

1970). The transparency at site-IV could attributed to the run off from immediate neighborhood being the contributor to the flow from higher reaches. The reduction in Sacchi transparency was due to high loading of dissolved organic matter, rainfall, incoming swage from catchments (Zutshi and Vass, 1972)

The pH of water fluctuated (Figure - 6) from near acidic 6.22 to slightly alkaline 8.69. pH is directly dependent upon the amount of CO_2 present (Juday *et al.*, 1980) and indirectly proportional to the photosynthetic activity (Pandit *et al.*, 2001). The stream showed an increased trend for pH from January to May towards downstream. The increase in pH may be associated with increase in DO, produced as a result of photosynthesis (Wetzel, 1975). The conductively values depicts the indication of total nutrient level ranged from minimum of 48μS cm-1 at site-I and maximum of 137 μS cm^{-1} at site-IV. The high value at site-VI may be due to nutrient enrichment (Quadri *et al.*, 1981; Bhat and Yousuf, 2004). Free Carbon dioxide ranged from 2-9 mg/l as against DO content. The maximum value of free CO_2 content was recorded at site-IV (Figure - 8). The high content of CO_2 at downstream is an indication of high degree of pollution (Cole, 1979). The increase in the CO_2 content may be due to respiratory activity of aquatic biota and decomposition. The total alkalinity (Figure - 9) ranges from 27-98 mg/l and was mainly due to bicarbonates and CO_2 content.

In consonance with the general fact that, the lotic systems containing relatively higher concentration of DO under natural conditions (Figure - 10). The present study revealed that DO content ranged from 5.8 mg/l at site-IV to 16.4 mg/l at site-I depicting the increase relationship with temperature as well as CO_2.The maximum mean value of chloride content 10.77 mg/l was recorded at site-IV against the minimum value of 6.27 mg/l at site-I (Figure - 11). The increase in chloride content may be due to feacal pollution and other wastes.

Total hardness was predominantly constituted by calcium and magnesium salts. The average value ranged from 70 mg/l at site-I and 101.30 mg/l at site-IV. There was an increase trend to downward stream increase in total hardness (Figure - 12) may also be due to soaps and detergents. Calcium and magnesium showed a similar trend to that of total hardness. The calcium hardness (Figure - 13) recorded maximum mean value of 45.93 mg/l at site-IV and 29.7 mg/l at site-I and magnesium content fluctuated from 15.94mg/l to 27mg/l. Increase in Ca^{++} and Mg^{++} downstream could be related to human interferences in the lower reaches of the stream. Calcium hardness may

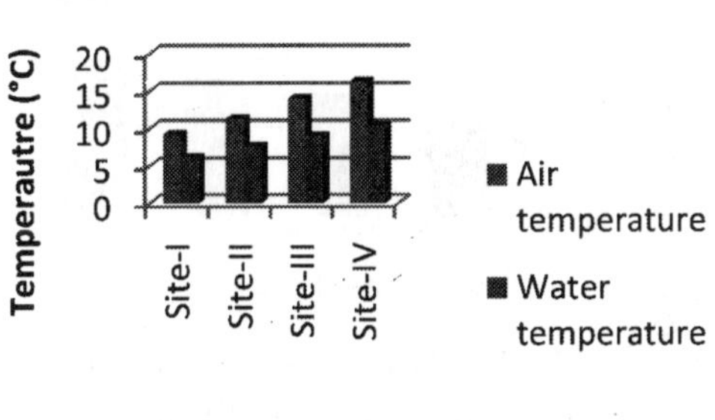

Fig 3: Mean values of temperature at four sites

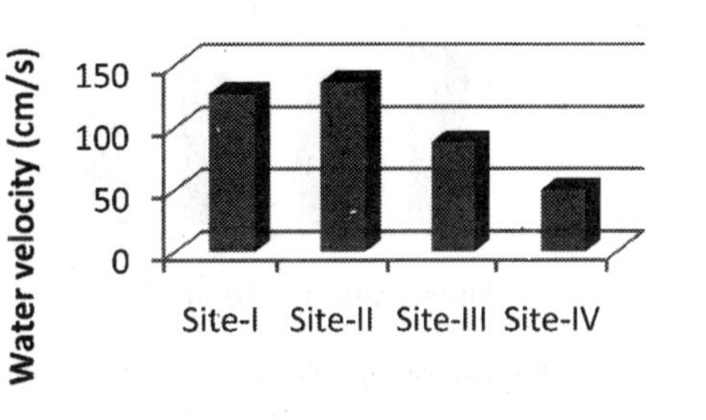

Fig 4: Mean value of water velocity at four sites

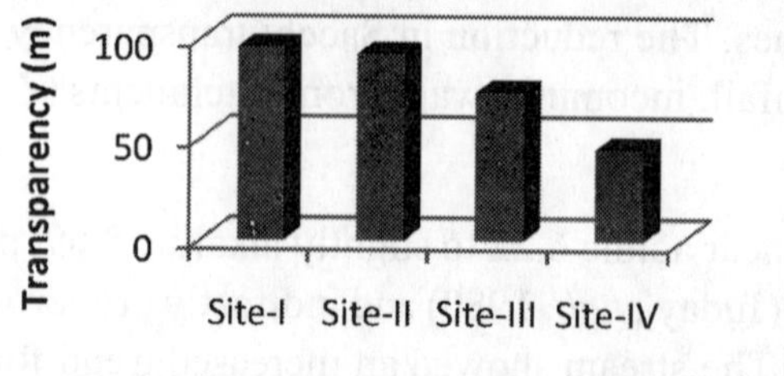

Fig 5: Mean value of transparency at...

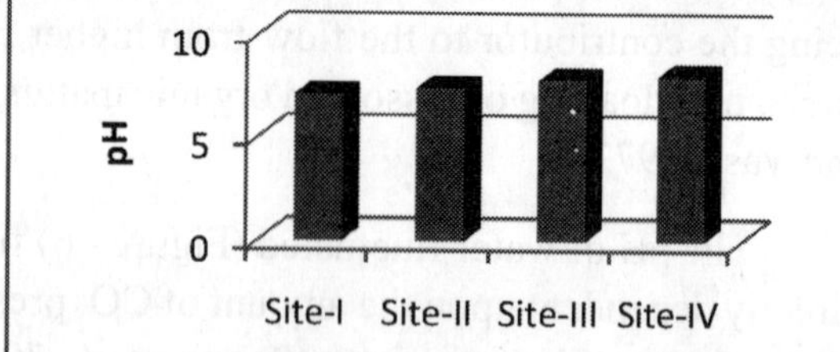

Fig 6: Mean value of pH at four sites

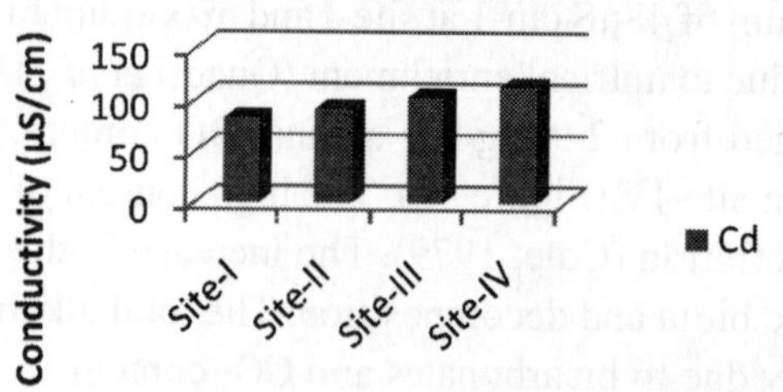

Fig 7: Mean value of Conductivity at...

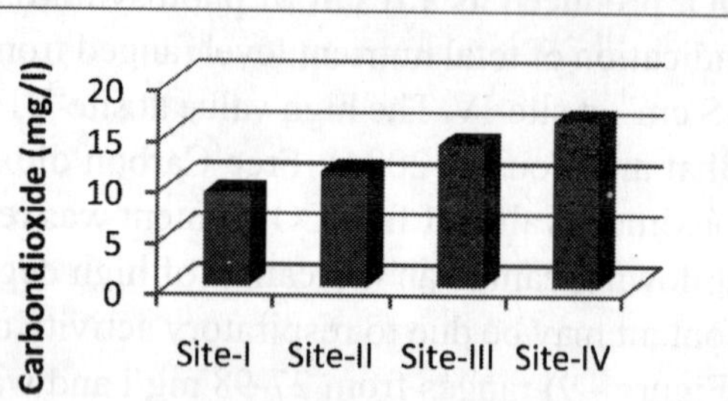

Fig 8: Mean value of free...

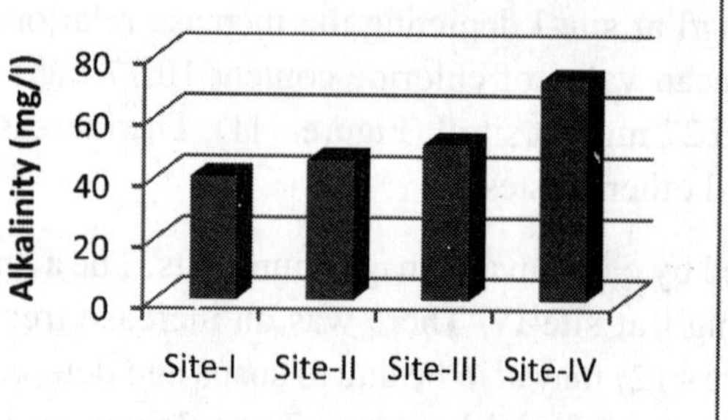

Fig 9: Mean value of Alkalinity at four sites

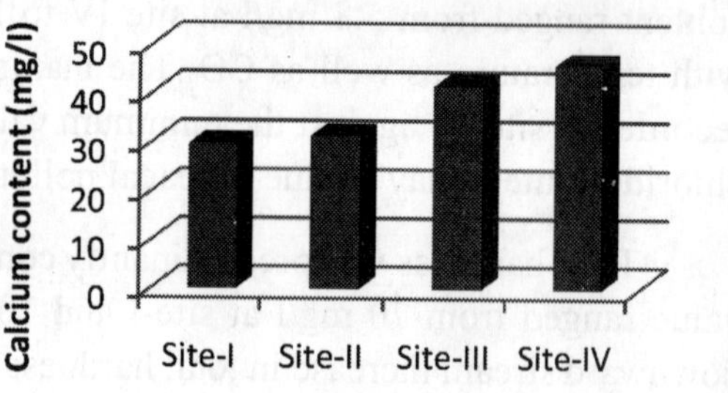

Fig 13: Mean value of calcium content at four sites

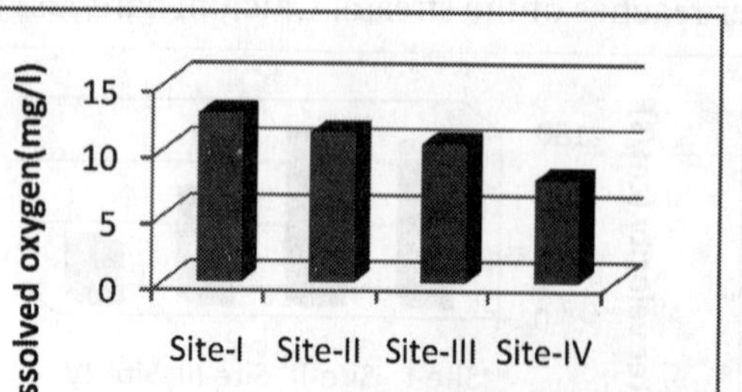

Fig 10: Mean value of dissolved oxygen at four sites

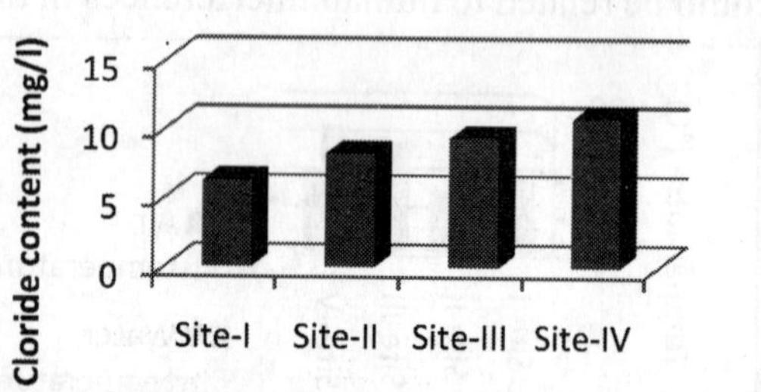

Fig 11: Mean value of chloride content at four sites

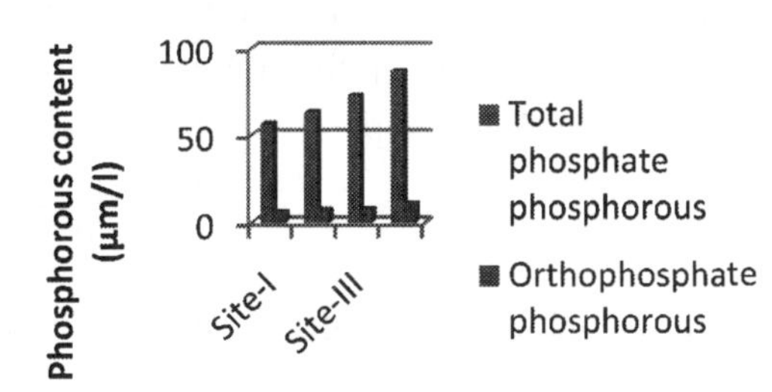

Fig 14: Mean values of phosphorous content at four sites

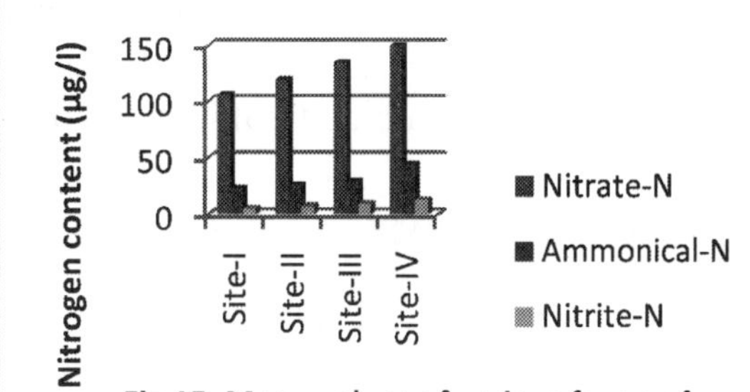

Fig 15: Mean values of various forms of nitrogen at four sites

also be due to decomposition of remains of plants and animals (Cole, 1970). The total phosphate phosphorus depicted clear spatial variations. The highest concentration of total phosphorus (Figure - 14) was recorded at site-IV with an average value of 86.8µg/l and the lowest concentration at site-I with an average value of 56.6µg/l. The orthophosphate phosphorus was low and ranged between an average value of 6µg/l at site-I and 10.94 µg/l at site-IV. The major sources of phosphorus in water include domestic sewage, agricultural effluents. The concentration of phosphorus showed an increased trend to downward stream may be due to increased human interference and death and decomposition of living biota.

Nitrate was the most dominant form of nitrogen. The concentration of the nitrogen compound was in the progression of NO_3-N>NH_3-N> NO_2-N (Bhat, 2004). In general site-IV depicted the high concentration of nitrate-N (Figure - 15). The higher value at site-IV may be due to application of nitrogen fertilizers and sewage contamination. It may also be due to decomposition of organic matter. The NH_3-N was the second most dominant form nitrogen after nitrate. The lowest concentration of NH_3-N was recorded at site-I which recorded an average value of 23.08 µg/l and the highest concentration of NH_3-N was recorded at site-IV with mean value of 44.3µg/l. The high value of NH_3-N at site-IV may be due to the consequences of decomposition of organic matters (Kaul, 1977). The nitrite nitrogen concentration was relatively low as compared to NO_3-N and NH_3-N, owing to the fact that NO_2-N gets oxidized to NH_3-N. The highest mean value of NO_2-N was recorded at site-IV (13.2 µg/l) and lowest average value was recorded at site-I. The highest value at site-IV may associated with either oxidation of ammonia, reduction of nitrates and due to the disposed of untreated sewage. An increased trend in concentration of NO_3-N, NH_3-N and NO_2-N was observed from January to May towards the lower elevation of stream with increase in temperature.

Conclusion

The Vishav stream has most ecological importance, a backbone for agriculture and water supply schemes for adjacent areas. The said water body is undergoing ecological modification. The present investigation revealed that the deterioration of stream water quality at lower reaches as low DO, transparency and high concentration of Ca, N, P, Cl, CO_2 etc. all depicting the nutrient enrichment.

The Site-IV (Khudwani site) is subjected to more anthropogenic pressure and therefore experience more pollution problems as compared to other sites. The main sources of pollution are untreated sewage disposal and agricultural wastes from the area of catchments.

Acknowledgements

The authors are thankful to Director, Centre of Research for Development (CORD), The University of Kashmir, Srinagar for providing laboratory facilities.

References

- Agarwal, D.K., Gour, S.D., Tiwari, I.C., Narayan Swami, N. and Narwash, S.M. (1976). Physico-chemical characteristic of Ganga water in Varanasi. Ind. J. Env. Hlth., 18 201-206.
- APHA. (2005). Standard methods for examination of water and wastewater 21st ed. American Public Health Association, Washington D.C.
- Bhat, F.A. and Yousuf, A.R. (2004). Limnological features of some lotic systems of Kashmir in Bio-resources, pp. 57-70.
- Cole, G.R. (1979). A text Book of Limnology 2nd ed. The Mosley Co. London, New York.
- Hynes, H.B.N. (1979). The ecology of running waters. Liverpool University Press Liverpool, England.
- Imevbore, A.M.A. (1970). The chemistry of River Niger in Kainji reservoir area. Arch Hydrobiol; 67(3) 412-431.
- Kaul, V. (1977). Limnological Survey of Kashmir lakes with references to tropic status and conservation. Int. J. Ecol. Environ. Science, 3 29-44.
- Kumar Jitendra, Singh Siddharth and Pal Amit (2008). Water quality of Turamdih and Jaduguda uranium mines and adjacent areas, East Singhbhum, Jharkhand, J. Ecophysiol. Occup. Hlth. 8 7-14
- Mackereth, F.J. H. Haron, H. and Tulling J.F. (1978). Water analysis, Fresh Water Biol. Assoc. Sci. Publ. no. 36.
- Pandit, A. K., Rathore, S. A. and Bhat, S. A. (2001). Limnological features of freshwater of Uri, Kashmir. J. Res. Dev., 1 22-29.
- Qadri, M.Y., Naqash, S.A., Shah, G.M. and Yousuf, A.R., (1981). Limnology of trout streams of Kashmir, J. Indian Inst. Sci., 63 137-141.
- Rai, U.N. and Pal, Amit (2001). Algae as indicator and mitigator of toxic metals in aquatic environment, Applied Botany Abstracts, 21 (1); 60-71.
- Saxena, K.L., Chakras, R.N., Khan A.Q., Chattopadhya, S.N. and Chandra, H. (1966). Pollution Studies of River Ganga near Kanpur. Environ. Health 8(4) 270-285.
- Warhate S.R., Yenkie M.K.N., Chaudhari M.D. and Pokale W.K. (2006), Impacts of Mining Activities on Water and Soil, Journal of Environ. Science and Engg. 48 (2) 81 - 88.
- Wetzel, R.G. (1975). Limnology. Standard publishers, Philadelphia.
- Yousuf, A.R. and Shah, G.M. (1988). Comparative Limnology of some fresh habitats of Kashmir. Geobios, 7 58-61.
- Zutshi, D.P. and Vass, K.K. (1972). Limnology of high altitude Kashmir lakes. Verh. Int. Verein. Limnol., 18 599-604.

35

UTILIZATION OF A WETLAND ECOSYSTEM BY WADERS THROUGHOUT THE YEAR

Chandni Patel, Nirjara Gandhi and Geeta Padate.

Abstract: Wetland, though an important ecosystem with reference to its productivity, is probably an unpredictable system with reference to availability of prey base." This is especially true for wetlands associated with irrigation reservoir. The water level declines in such reservoirs as the water is used up for irrigation. However, there are some reservoirs which are inundated with water from larger reservoirs/dams making them more complex ecosystems. One such reservoir present near Vadodara is surveyed for Wader populations. Waders- the long-legged birds feed in the shallow, marshy areas of the wetland. In the present study, Wader density is compared with Total water bird density of Wadhwana Irrigation reservoir, a Nationally Important Wetland near Vadodara, Central Gujarat. The seasonal data collected by transect method and analyzed with the Shannon-Weiner diversity index, evenness and density indicates that waders form major group with 42% in summer and 88% in monsoon while only about a quarter in post-monsoon and winter. Utilization of this habitat by waders in various seasons is discussed.

Keywords: Ecosystem, Long legged birds, Utilization

Introduction

Wetland ecosystems are the shallow parts of aquatic bodies that support diversity of flora and fauna. The balance of these ecosystems is based on the food web structure and interactions among the communities i.e. the biodiversity supported by it. This interdependency of the communities in an ecosystem is affected by the variability in factors like the local climate, availability of water, nutrients and substrate. The effects are more prominent in aquatic ecosystem especially the wetland which is a dynamic ecosystem.

Division of Avian Biology, Department of Zoology, Faculty of Science, The M. S. University of Baroda, Vadodara-390002. Gujarat, India

Wetland ecosystems accounts for about 6% of the global land area and are among the most threatened of all environmental resources (Turner, 1993). Wetlands have the highest biodiversity and biological productivity levels in the world (Whittaker and Likens, 1973; Gibbs, 1993). India is blessed with water resources in the form of numerous rivers and streams. By virtue of its geographical position, varied terrain and climatic zones it supports a rich diversity of inland and coastal wetlands. The peninsular India has very few natural wetlands however; a number of man-made water storage reservoirs are constructed almost in every village, supplying water for human needs. The shallow regions of these reservoirs form wetland habitats. These wetlands provide feeding and breeding grounds for a variety of fauna. Gujarat has 36% of the total wetland area of the country with 2,092 km^2 area under inland wetlands. Being located in the arid and semi-arid zones of India, Gujarat has a large number of manmade reservoirs which covers 1,394 km^2 areas as inland wetlands. Gujarat falls in the migratory route of birds. Migratory birds from Europe and Asia pass through this part of the country and enjoy the varied habitats. The wetlands in Gujarat provide a huge prey base to diversity of these migratory species. They are full of migratory birds during winter and in some wetlands huge congregations of birds amounting to several thousands are also observed. Further, several wetlands in Central Gujarat are inundated with Narmada water from Navagam dam. Water is transferred to various smaller and medium sized irrigation reservoirs. This has influenced the duck population in the area (Padate *et al.*, 2008). Due to such developments the condition of the reservoir is significantly altered (Deshkar, 2009). These changes can be easily monitored by studying the changes in the biodiversity patterns. Birds, that represent wide array of feeding strategies, are one of the ideal bio-indicators and useful models for studying the status of environmental fitness (Newton, 1995; Urfi *et al.*, 2005). Various groups of water birds that are found in and around wetland are waders, divers, swimmers, etc. In a wetland ecosystem which comprises of shallow waters, waders can be considered as health indicators (Custer and Osborn, 1977; Kushlan, 1993; Hoffmann *et al.*,1994). The present study concentrates mainly on this group. Waders are the long-legged birds with long neck and/or beak preferring shallow marshy lands and wade on soft marshy base. These wading birds respond behaviourally by moving to local concentrations of food availability and move away from unsuitable areas. Hence, their presence and absence is used to assess the transient conditions of wetlands (Erwin, 1985; Dugan *et al.*, 1988). In this study, we examined the use of the wetland by waders during different seasons in a Nationally Important Wetland, inundated frequently by water from Sardar Sarovar on Narmada River. We also analyzed the differences between the wader population and other water bird population. The population parameters such as species richness, relative density and diversity of birds, the frequently used parameters as indicators of habitat quality (Nilsson and Nilsson, 1978; Weller 1978; Sampath and Krishnamoorthy, 1990; Nagarajan and Thiyagesan, 1996) are assessed in the present study.

Study Area

The study area is an irrigation reservoir, Wadhwana Irrigation Reservoir (WIR) (22°10′N, 73°30′′E) situated about 50 kms. south- east of Vadodara city in Central Gujarat (Fig. 1). This irrigation reservoir was constructed about 100 years ago in the year 1909-1910 by His Highness Shrimant Maharaja Sir Sayajirao Gaekwad III of erstwhile State of Baroda at the Wadhwana village, Taluka Dabhoi of Vadodara District, with a view to make the farmers independent of rain water dependency. The dam is mainly an earthen dam of 8.2 kms. With the periphery of 11.2 kms. Earlier, water of only Jojwa dam on Orsang River was brought to this reservoir. However, after construction of Navagam dam on Narmada River, Narmada water is also diverted towards this reservoir. The full capacity of the reservoir is 5 billion cubic feet. It irrigates about 8815 hectare land of 25 villages surrounding the dam. Inlet and outlet canals are present with several sub canals distributing water throughout the agricultural area in the fields. In recent years, the reservoir has been receiving the

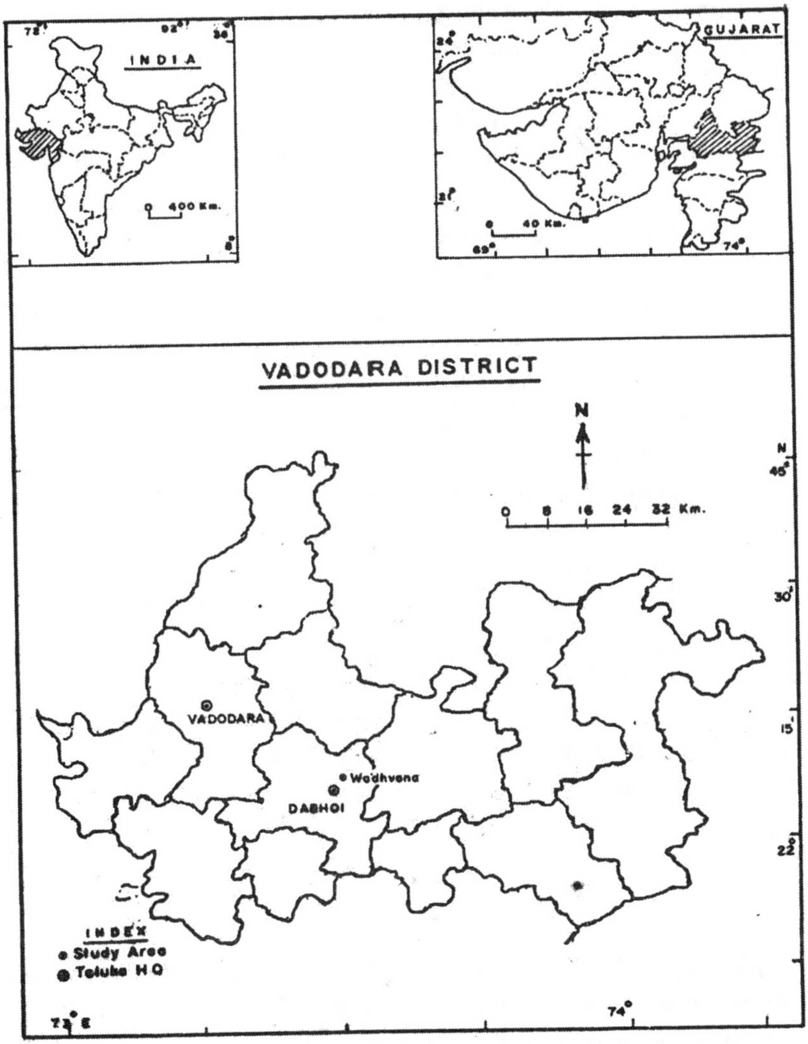

Fig. 1. Map showing geographical location of Wadhwana irrigation reservoir in Vadodara district, Gujarat, India.

water from the famous Sardar Sarovar on Narmada River. Because of Narmada inundation around the turn of Century, it has become a perennial water reservoir. Cattle grazing are moderate but as the reservoir is large, it does not influence the water significantly. However, no major source of pollution is observed at the reservoir. On the basis of the waterfowls assemblage supported by this wetland, it was declared as a wetland of National Importance in 2005.

Materials and Methods

The irrigation reservoir was visited twice in a month. The census of birds was conducted while walking on the dam during morning hours, half an hour after sunrise, which is known to be the best time for the observation of birds. The water birds present on both the sides of transect, in water as well as on the land were counted while walking on the edge of the wetland. The length and the width of transect were 3.2 kms. and 0.5 km. respectively. Direct count was carried out with the help of binoculars having the magnification of 10 × 50. The study was carried out from October 2008 to November 2009. The birds were identified on the basis of field guides by Ali and Ripley, (1983); Kazmierczak, (2006) and Grimmett *et al.*, (2001). The birds are categorized into two groups. Group I- Waders and Group II- other Water birds so as to compare their densities and in turn the utilization of the wetland by them. The Diversity indices like Species Richness, Species diversity - Shannon-Wiener index (H′) and Equitability (E) (Krebs, 1985) as well as Density (Rodgers, 1991) were calculated for each group as well as for each visit. The birds observed in the reservoir as well as those present in the agricultural fields and observed to move in and out of the reservoir within transect width were also counted avoiding double counts. The density is calculated as per km^2 for the transect method (Rodgers, 1991). Total number of species observed per visit is considered as Species richness. Shannon-Wiener Diversity Index is calculated as $H' = -\pi \ln \pi$ (for maximum number of birds) where pi is total sample belonging to the ith proportion of a species, calculated as proportion of the total number of individuals of all the species and ln is the natural log. Evenness/equitability is calculated as $E = H'/H$ max, where, H is information content of sample (bits/individuals) = index of species diversity (Krebs, 1985; Javed and Kaul, 2002). For the statistical analysis the data for 3 months is pooled according to the seasons as Post-monsoon September, October, November; Winter December, January, February; Summer March, April, May and Monsoon June, July, August. The values are presented as Mean ± Standard Error. The seasonal variations are statistically analysed with ANOVA while t-test is performed between the two groups of birds using Prism software.

Results and Discussion

In a wetland, water conditions like depth influences the size of birds present whereas the availability of prey and hydroperiod influence the number of aquatic birds. With major changes in seasonal water coverage and, therefore in prey availability, wading birds show long-distance shifts in foraging locations. The study was thus conducted for four seasons i.e. Post-monsoon, winter, summer and monsoon. Total 60 species of water bird were observed over the year. Of these 38 species were waders. The mean density for both, the waders (Group I) and the other water birds (Group II) was maximum during winter (491±130 and 2120±1270 birds/km^2 respectively) because of the presence of migratory birds in far-reaching numbers (Fig. 2). The statistical differences in the seasonal density with ANOVA was significantly significant ($P<0.01$, $F(3,19)$ 7.02) for waders and non-significant ($P>0.05$, $F(3,19)$ 2.57) for other water birds. When t-test is used for comparison of density between the two groups of birds, significant differences ($P<0.05$) were observed during summer while non-significant differences ($P> 0.05$) were observed during post-monsoon, winter and monsoon. When the mean species richness is considered it was highest during summer for waders (Group I) (16.7±1.05) while it was highest during winter (29±1.3) for other water birds (Group II) (Fig. 2). The seasonal differences in species richness with ANOVA showed significant differences ($P<0.05$, $F(3,19)$ 3.76) for waders and significantly significant differences ($P<0.01$, $F(3,19)$ 5.14) for other water birds. However, when season-wise comparison was carried out with t-test for species richness, highly significant differences ($P<0.001$) and significantly significant

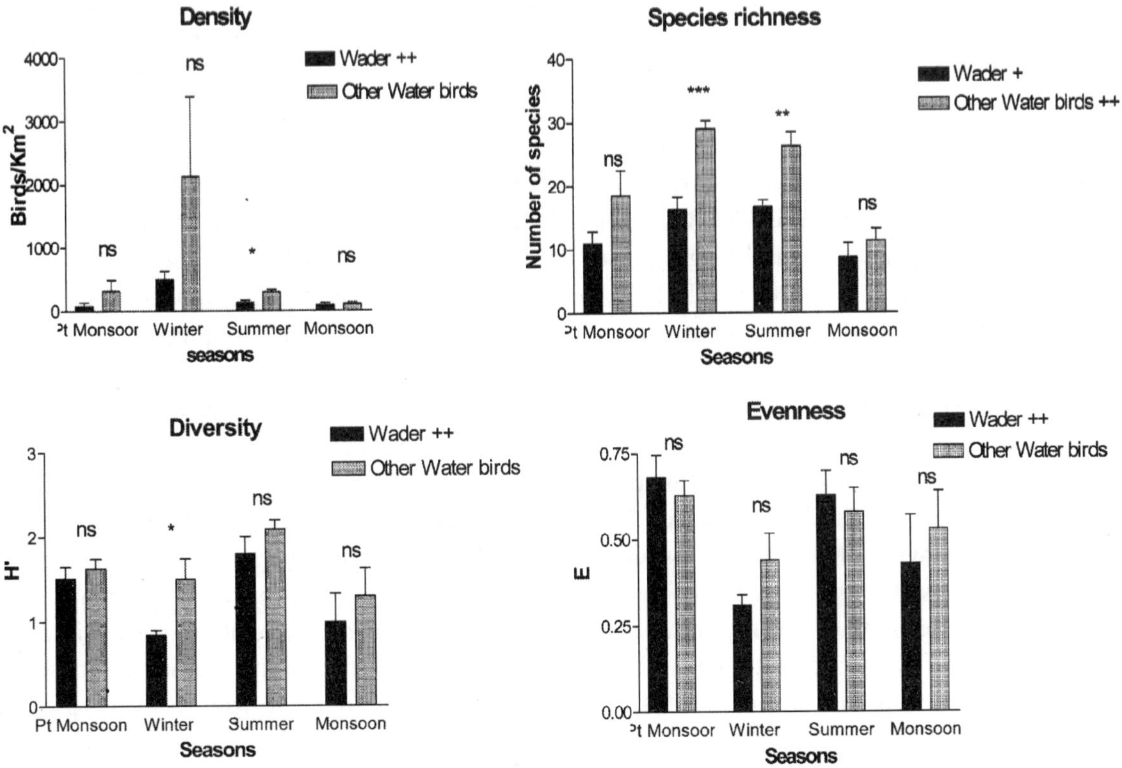

Fig. 2. Density, Species richness, Diversity index (H') and Evenness (E) of waders and other water birds at Wadhwana irrigation reservoir.

ns P>0.05, */+ P<0.05, **/++ P<0.01, ***/+++ P<0.001
*For t-test and + For ANOVA

differences ($P<0.01$) were noted during winter and summer respectively, while non-significant differences ($P>0.05$) were noted during post-monsoon and monsoon. This wetland supports large number of migratory birds during winter and many water-dependent resident species in summer. However, during monsoon and post-monsoon when water is available everywhere, which increases productivity too, majority of species are busy in nesting and nurturing activities hence their presence at wetland is lacking with absence of migratory species which are yet to arrive (Padate *et al.*, 2008).

The mean Shannon-Weiner diversity index was higher for both the groups during Summer (1.8±0.2 and 2.09±0.107 for group I and group II respectively) (Fig. 2) when good number of water-dependent species are present. However, the seasonal fluctuations were significantly significant ($P<0.01$, $F(3,19)$ 5.08) for waders and non-significant ($P>0.05$, $F(3,19)$ 3.121) for other water birds reflecting fluctuating water levels. When t-test is applied for individual seasonal for comparison of the two groups, significant differences ($P<0.05$) were observed during winter and non-significant ($P>0.05$) during the other three seasons. This again reflects that as water level fluctuates seasonally there are significantly significant ($P<0.01$) variations in wader diversity whereas one or other species of Group II (other water birds) is always utilizing the wetland. When evenness is considered it was low for both the groups during winter (0.31 + 0.03, 0.44 + 0.07 wader and other water birds respectively). Higher species richness at the same time indicates that the conditions at wetland favour one particular species of birds that utilizes the wetland in large number bringing down the evenness index. The ANOVA applied for seasonal distribution of waders with reference to evenness showed significantly significant differences ($P<0.01$, $F(3,19)$ 4.93) whereas that for other water birds was insignificant ($P>0.05$, $F(3,19)$ 1.34). However, the differences between the two groups for each season were non-significant (t-test). The Ruff (Philomachus pugnax) dominated the wader population (Group I) while the Common Coot (Fulica atra) dominated among the other water birds (Group II). The mean Evenness was maximum during Post-monsoon for Group I (0.68±0.066) and Group II (0.68±0.042) (Fig. 2). Several factors such as physicochemical factors, food resources, vegetation cover and interspersion, as well as habitat and landscape configuration are responsible for the distribution of birds (Pescador and Peris, 2009). All these factors are influenced by seasonal changes. Thus, seasonal wader and other water bird numbers do reflect the annual carrying capacity that is a function of the availability of resources such as food and space. The differential seasonal trend is seen in the abundance of both the groups, with relatively high wintering bird numbers reflecting habitat quality. There also occurs difference in the species composition among the two groups themselves. That is, one or the other species of each group is always present at the wetland. Differences in species composition between seasons are related to seasonal species replacement due to the latitude/altitude where the wetland is located (Pescador and Peris, 2009).

Aquatic birds constitute one of the most characteristic components of wetlands (Weller, 1999). However, these species are rarely distributed in a uniform way because none of them occur at one place throughout their annual cycles. They may occur either during their breeding, moulting, or wintering periods. Certain lakes and ponds can be important areas of concentration during the annual migration of certain species. The abundance of these birds is associated with local environment characteristics as well as specific demands of each species. Water bird assemblage on a local scale is known to depend upon habitat characteristics, availability, distribution, density of food and the availability of suitable sites for reproduction or resting (Wiens, 1989). Further, surface area is also considered as an important factor related to the gradient of species composition and the abundance of aquatic birds. (Uresk and Severson, 1988; Celada and Bogliani, 1993; Guadagnin *et al.*, 2005). In present study, it has been noted that Wadhwana is an important site for both the waders as well as other water birds. Water nutrients are associated with increased plant production in the form of productivity, and as productivity increases so does usage by birds (Guadagnin *et al.*, 2005). According to Pescador and Peris, (2009) though the carrying capacity of the water body cannot be measured directly, it can be measured indirectly by the existing aquatic bird density and the area-abundance relationships only appear useful for wintering water birds, which require less specialized habitats, except during breeding. The significantly lower diversity (H′) of waders during winter and non-significantly higher diversity (H′) during other seasons compared to other water birds with higher mean species richness and density indicate that the fluctuating water levels of irrigation dam inundated with waters from Orsang and Narmada rivers is less preferred habitat for waders. Different species of waders (Total 38 species) do visit the reservoir where mean density is mainly influenced by birds like cattle egrets during other seasons and by ruffs during winters. The ruffs get shelter in the grasses that grow in the very shallow marshy areas. About 15,000 or more ruffs are reported roosting at night at wadhwana and some of them do spend day at wadhwana influencing density of waders.

Acknowledgements

The authors are thankful to the authorities for providing all the necessary facilities for this study. The 1st and 2nd authors are also thankful to UGC for Research fellowship under the RFSMS scheme.

References

- Ali, S. and Ripley, S. D. 1983. Handbook of the Birds of India and Pakistan (Compact Edition) University Press Bombay. Bombay, India.
- Celada, C. and Bogliani G. 1993. Breeding bird communities in fragmented wetlands. Bollettino di Zoologia, 60 73-80.
- Custer, T. W. and Osborn, R. G., 1977. Wading birds as biological indicators 1975 Colony survey. U.S. Fish Wildl. Serv. Spec. Sci. Rep. Wildl., 2061-28.
- Deshkar, S. L. 2008. Avifaunal diversity and ecology of wetlands in semi-arid zone of Central Gujarat with reference to their conservation and categorization. Ph.D thesis submitted to The Maharaja Sayajirao University of Baroda, Vadodara, India.
- Dugan, P., Hafner, H. and Boy, V., 1988. Habitat switches and foraging success in the little egret (Egretta garzetta). In Acta 19 Congr. Int. Ornithologici, H. Ouellet (Ed.), Ottawa, Canada, pp. 1868-1877.

- Erwin, R. M., 1985. Foraging habitats of nesting wading birds Spatial use and social influences. Auk, 100 960-970.
- Gibbs, J. P. 1993. The importance of small wetlands for the persistence of local populations of wetland-associated animals. Wetlands, 13 25-31.
- Grimmett, R., Inskip S. and Inskipp, R. 2001. Pocket Guide to the Birds of Indian Sub Continent. Oxford University Press. New Delhi.
- Guadagnin, D.L., Schmitz-Peter, A., Carvalho, L.F. and Maltchik, L. 2005. Spatial and temporal patterns of waterbird assemblages in fragmented wetlands of Southern Brazil. Journal of Waterbird Society. 28 268-272.
- Hoffman, W., Bancroft, G. T., Sawicki, R. J. 1994. Foraging habitat of wading birds in the water conservation areas of the Everglades. In Everglades the ecosystem and its restoration. Steven Davis and John Ogden (Eds.) pp. 585-614.
- Javed, S. and Kaul, R. 2002. The Field Methods for Bird Surveys. Bombay Natural History Society, Mumbai.
- Kazmierczak, K. and Ber van Perlo. 2006. A field guide to the Birds of India. Pica Press (an imprint of Helm International Ltd.) East Sussex.
- Krebs, C. J. 1985. Ecology the experimental analysis of distribution and abundance. Third edition, Harper and Row Publishers, New York.
- Kushlan, J. A. 1993. Responses of Wading Birds to Seasonally Fluctuating Water Levels Strategies and Their Limits. Colonial Waterbirds 9(2) 155-162.
- Murphy, S. M., Kessel, B. and Vining Source, J. L. 1984. Waterfowl Populations and Limnologic characteristics of Taiga Ponds. Journal of Wildlife Management, 48 1156-1163.
- Nagarajan, R. and Thiyagesan, K. 1996 Shore birds and substrate quality of the Pichavaram wetlands, south India. Ibis, 138 710-721.
- Newton, I. 1995. The Contribution of Some Recent Research on Birds to Ecological Understanding. The Journal of Animal Ecology, 64 675-696.
- Nilsson, S. G. and Nilsson, I. N. 1978. Breeding bird community densities and species richness in lakes. Oikos, 31 214-221.
- Padate, G., Deshkar, S. and Sapna S. 2008. Influence of Narmada water inundation on the duck populations of Wadhwana irrigation reservoir. In proceedings of Taal (2007) The 12th World Lake Conference, Sengupta M. and Dalwani R. (Eds.) pp. 131-136.
- Pescador M. and Peris S. 2009. Seasonal and water mass size effects on the abundance and diversity of waterbirds in a Patagonian National park. Waterbirds, 32(1) 25-35.
- Rodgers, W. A. 1991. Techniques for wildlife census in India A field manual. Wildlife Institute of India, Dehradun, India.
- Sampath, K. and Krishnamoorthy, K. 1990. Bird fauna and limnology of the Koliveli tank, Tamil Nadu. India. In Vijayan. V.S. (ed) Proc. Seminar on Wetland Ecology and Management 47-48 Bombay Natural History Society, Bombay.
- Turner, K. 1993. Economics and Wetland Management. Ambio, 20 (2) 59-63.
- Uresk, D.W. and Severson, K. 1988. Waterfowl and shorebird use of surface-mined and Livestock water impoundments on the Northern Great Plains. Great Basin Naturalist, 48 353-357.

- Urfi, A. J., Sen, M., Kalam A. and Meganathan, T. 2005. Counting birds in India Methodologies and Trends. Current Science 89 1997-2003.
- Weller, M.W. 1978. Management of Freshwater marshes for wildlife. In Good, R.E., Whingham, D.F. and Simpson, R.L. (eds) Freshwater Wetlands Ecological process and management potential. New York Academic Press. As Cited in Murphy, S. M, *et al.*, 1984. Loc. Cit.
- Weller, M.W. 1999. Wetland birds habitat resources and conservation implications. Cambridge University Press, Cambridge. 288 pp.
- Whittaker, R. H. and Likens, G. E. 1973. Primary production the biosphere and man. Hum. Ecology, 1 357-369.
- Wiens, J.A. 1989. The ecology of bird communities. Cambridge University Press, Cambridge. 288 pp.

36

BUTTERFLY FAUNA AROUND TWO IRRIGATION RESERVOIRS WITH DIFFERENT ANTHROPOGENIC PRESSURES

Nirjara Gandhi, Chandni Patel and Geeta Padate

Abstract: In arid and semi arid regions of India several irrigation reservoirs are built to store rain water to suffice the needs of agricultural land. In recent times anthropogenic pressures are increasing around some of them leading to deteriorate their conditions. Shallow regions of these reservoirs as well as area surrounding the reservoir are the niches that support large number of organisms including Insects of the various insects, Lepidopterans form a major part. The beautifully coloured Lepidopterans are easy to monitor. In the present study Lepidopteran fauna around two water bodies; one with anthropogenic pressures while other without anthropogenic pressures; in semi arid zone of Gujarat are considered. The study was carried out from March '09 to November '09 covering three seasons. The Species richness, Shannon-Weiner diversity index and Evenness were calculated. The percentage occurrence of the butterfly families present was also calculated. In all, 25 species belonging to 5 families were reported around the water body with anthropogenic pressures while 29 species belonging to 5 different families were observed around the water body without anthropogenic pressures with 18 species common to both. However, their occurrence varied according to the pressures faced. The most represented family at both the Irrigation Reservoir was Pieridae. The seasonal changes in diversity and occurrence are discussed.

Keywords: Semi arid regions, Niches, Diversity

Introduction

Under the concept of conservation of biodiversity, rather than conserving an individual species, conservation of ecosystem has taken momentum. The conservation of Biodiversity which started with Rio De Janeiro Earth Summit on Biodiversity in 1992 at Brazil has reached from National to

Department of Zoology, Faculty of Science, The Maharaja Sayajiroa University of Baroda, Vadodara- 390002. Gujarat. India.

State level where the identification of biodiversity up to village level is under consideration. The ministry of Environment and Forest, Government of India has also given emphasis on a comprehensive piece of legislation that aims to protect and conserve fragile ecosystem like wetlands. Wetland is the second most productive ecosystem in the world next to the tropical rain forests (Ramachandra *et al.*, 2002). Wetlands are the transitional zone between the land and water, saturated with water. These are highly productive shallow areas which support varieties of submergent and emergent vegetation and a large variety of fauna. This ecosystem being the most vulnerable, identification and regular monitoring of various species present around it is also important of the various species supported by wetlands, generally waterbirds are considered as the most important component. However, there are several other organisms which are directly or indirectly dependent on the wetland ecosystem. Largest group of organisms other than the birds found around a wetland ecosystem are the insects.

Insects form one of the major components of any ecosystem. Due to their ability to adapt to any and every habitat present on earth, insects have conquered planet Earth. Many orders of Class Insecta form an intimate parts of the wetland ecosystems. There are also large group of insects which have one or the other aquatic stage in their life-cycle. It has been observed that of the various insect orders found around a waterbody, Lepidopterans forms the major volume, especially on the vegetation present around the wetland. Lepidopterans, being medium to large sized and beautifully coloured, are easy to monitor. Although, easy monitoring is possible in case of Lepidopterans very less attention is paid on the butterfly fauna around wetlands. Very sparse information is available regarding the butterfly fauna of wetlands (Levente, 2005; Shapiro, 2009) in comparison to the immense data documented regarding the butterflies of forests (Murray, 2000; Sundufu and Dumbuya, 2008; Wadatkar and Kasambe, 2008; Jalil *et al.*, 2008 ; Sharmila, 2009; Fermon *et al.*, 2009; Peixoto and Benson, 2009). Further, reports on influence of anthropogenic pressures on butterfly fauna are lacking.

In the present study Lepidopteran fauna around two water bodies one under anthropogenic pressures; Timbi Irrigation Reservoir (TIR), while other without anthropogenic pressures; Jawla Irrigation Reservoir (JIR) in semi arid zone of Gujarat are considered. Earlier studies document avifauna, plankton and molluscan diversity of these reservoirs (Deshkar, 2008; Rathod, 2009). On the basis of these reports, these reservoirs were considered as "in good condition". However, recently due to the increasing anthropogenic pressures, the conditions at TIR have started changing. The butterfly fauna of both the reservoirs was monitored for the period of 9 months and the differences in their occurrence and diversity are considered in detail.

Study area

Arid and semi arid regions of India depend primarily on the rain water for their water requirements. For the storage of rain water many irrigation reservoirs were built decades ago. The rain water is stored to suffice the agricultural as well as domestic demands. Two such reservoirs Timbi Irrigation reservoir (TIR) and Jawla Irrigation reservoir (JIR) are present in Vadodara district of Central Gujarat. Both the reservoirs are under different anthropogenic pressures.

Jawla Irrigation Reservoir (JIR) (Plate 1)

Jawla Irrigation Reservoir (22°34' 20"N and 73°19' 24"E) is located about 40 kms North of Vadodara city and 1.5 km away from the Savli town in the out skirts of Jawla Village. The reservoir spreads in 0.77 km^2 area. The water from this reservoir is supplied to agricultural fields of villages surrounding the reservoir. It is also bounded by an earthen dam, which has a periphery of approximately 3 kms. It is a seasonal reservoir which solely depends on the rain water. The wetland formed is still in a better condition as it is away from the human habitation and urban pressures. The village has a different water tank for all the domestic needs. Hence the Jawla Irrigation reservoir faces comparatively little anthropogenic pressures.

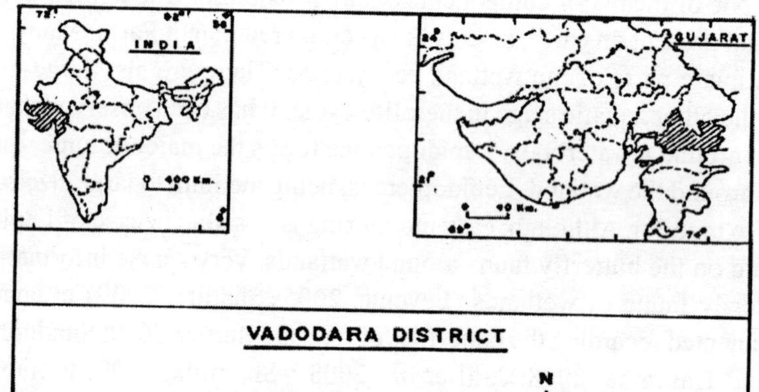

Plate 1. Map showing the location of both irrigation reservoirs.

Timbi Irrigation Reservoir (TIR) (Plate 1)

Timbi irrigation reservoir (22°18' N 73°16' E) is located about 15 kms East of the Vadodara city. This Irrigation reservoir was constructed in the year 1947-48 by the grants provided by His Highness Shrimant Maharaja Sayajirao Gaekwad III of the erstwhile state of Baroda at village Shripor Timbi in the Waghodia Taluka of Vadodara District. The reservoir spreads in 0.47 km^2 area. The water from this reservoir is supplied to agricultural fields of eight villages surrounding Shripor Timbi village. It has an earthen dam, which has a periphery of approximately 3 kms. It has become almost perennial wetland due to the inundation from the Sardar Sarovar of Navagam dam on river Narmada. This wetland is facing increasing anthropogenic pressures. The water of the wetland is used for the domestic purpose such as washing the clothes and utensils, taking bath and grazing animals. The sanitation activity around the wetland has also increased recently. Many educational institutes and farm houses are coming up in the area leading to increase in the anthropogenic activities.

Materials and Methods

The study was conducted from March 09 to November 09. Timbi Irrigation Reservoir was visited 16 times while Jawla Irrigation reservoir was visited 12 times in the period of 9 months divided as Summer/Monsoon/Post-monsoon. The observations were made between 800a.m to 10 a.m +1 hour during each visit.

The butterflies were recorded using Transect method. The butterflies present on both the side of the earthen dam were observed while walking on feet. The butterflies appearing on either side of the earthen dam were counted as they were encountered while walking on the fixed transect. The butterflies were identified on the basis of standard books by Gay *et al.*, (1992), Haribal (1992), Kunte (2000), Parasharya and Jani (2007) and Kehmikar (2008). The butterflies observed were divided according to the families. The transect length of Timbi Irrigation reservoir was 1.33 km. while that of Jawla Irrigation reservoir was 0.89km.

The diversity indices, Species Richness- No. of species; Species diversity- Shannon Wiener index (H/) and Equitability (E) of the butterflies were calculated for each visit. The percentage occurrence of the butterflies was also calculated. Total No. of species observed per visit is considered as species richness. To estimate diversity- Shannon-Wiener Diversity Index is calculated as H′= -? π ln π (for maximum number of butterflies) where π is total sample belonging to the ith proportion of species, calculated as proportion of the total number of individuals of all the species and ln is the natural log. Evenness/equitability is calculated as E = H′′ H max (Krebs, 1985; Javed and Kaul, 2002). The Jaccard's index of Similarity was calculated as C/(A+B)–C. Where C is the number of species common at both reservoirs, A the number of species at JIR and B the number of species at TIR. The Percentage occurrence of butterflies was calculated in two ways. First, occurrence in comparison to other flying insects and second occurrence of a particular family i.e. depending on species encounter of various species belonging to a particular family during every visit. The data was subjected to t- test for comparison between two sites and ANOVA for the seasonal differences at each site. The statistical analysis is performed using Prism 3.

Results

In all 29 different species of butterflies were observed at Jawla Irrigation reservoir while 25 at Timbi Irrigation Reservoir (Table 1). These butterflies mainly represented 5 families i.e. Papilionidae, Pieridae, Lycaenidae, Danidae and Nymphalidae. Family Papilionidae, the swallowtails is represented mainly by a single species Lime butterfly at JIR while 3 species Common Rose, Crimson Rose and Tailed Jay at TIR. This is one of the two uncommon families. Family Pieridae, the whites and the yellows/sulphurs has 14 species at JIR and 11 at TIR that includes Jezbels, Orange tips, Emigrants, Grass yellows, etc. This is the richest family at both the reservoirs. Family Lycaenidae, the blues that includes the Grass blues and Gram blues, Cupids and Pierrots are represented by 7 species at JIR and only 3 species at TIR. Gram blue is a very common species observed at both the reservoirs. The family Danidae is represented by a single species Plain Tiger at JIR and 2 species Common Crow and Plain Tiger at TIR. This is the second uncommon family of the study area. However, Plain Tiger is the most common species found throughout the study period around both the reservoirs. Nymphalidae, the Brush-footed/four-footed butterflies, that includes the Pansies, Leopard, Eggfly, Browns, etc. is represented by 6 species each at JIR and TIR. Peacock pansy was common species at JIR.

Jaccard's similarity index (Table 2) showed that about 50% of the species are common at both the reservoirs. When the seasonal differences in Jaccard's similarity index are considered it is found that maximum number of species (45%) are common during Post-monsoon while only 11.76% during Summer and about 25% of species during Monsoon.

The mean butterfly species richness of JIR is 6.75 + 3.79 while that of TIR was 5.43 + 3.34 (Table 3, Fig.1.a). When the seasonal differences in the species richness are considered the overall butterfly diversity of JIR is higher than TIR across the seasons studied (Table 4, Fig.1.b).The difference in seasonal species richness with ANOVA are significant (P′ 0.05, F(2,9) 5.037) for JIR

Table 1 Species Richness of butterflies with respect to family classification

	Jawla	Timbi
Papilionidae	1	3
Pieridae	14	11
Lycaenidae	7	3
Danidae	1	2
Nymphalidae	6	6

Table 2 Jaccard's similarity Index

Total	50
Summer	11.76
Monsoon	25
Post-monsoon	45

Table 3 Mean Species richness, Diversity, Evenness and % Occurrence at Jawla Irrigation Reservoir and Timbi Irrigation Reservoir

	Species Richness	H′	E	% Occurrence
Jawla	6.75 ± 3.39	1.35 ± 0.67	0.66 ± 0.30	26.05%
Timbi	5.43 ± 3.34	1.08 ± 0.45	0.66 ± 0.20	28.48%

Table 4 Seasonal Species richness, Diversity, Evenness and % Occurrence at Jawla Irrigation Reservoir and Timbi Irrigation Reservoir

	Species Richness	**H′**	**E**	**% Occurrence**
Jawla				
Summer	4.33 ± 3.55	1.00 ± 0.83	0.56 ± 0.44	21.66%
Monsoon	7.66 ± 1.15	1.56 ± 0.07	0.75 ± 0.01	29.11%
Post-monsoon	10.66± 2.08	1.83 ± 0.09	0.78 ± 0.04	29.62%
Timbi				
Summer	3 ± 1.67	0.82 ± 0.45	0.65 ± 0.32	18.75%
Monsoon	4 ± 1.66	0.89 ± 0.42	0.65 ± 0.10	23.88%
Post-monsoon	8.83 ± 2.56	1.46 ± 0.10	0.68 ± 0.08	34.64%

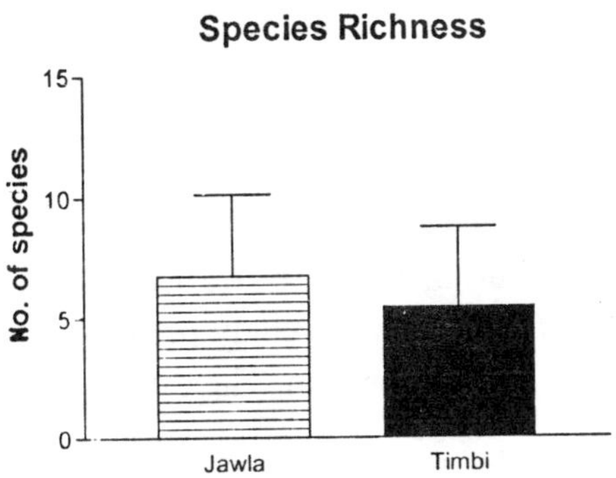

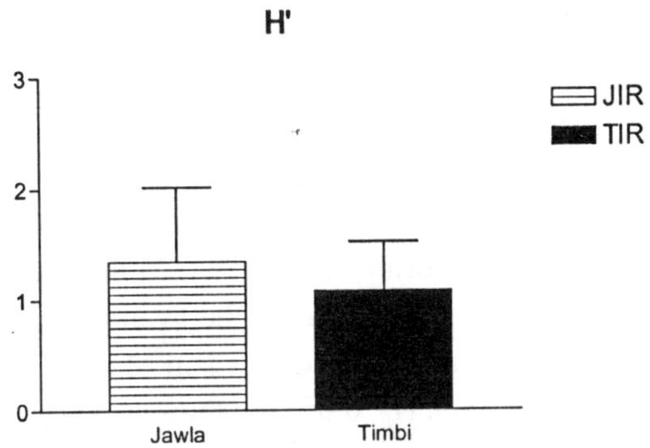

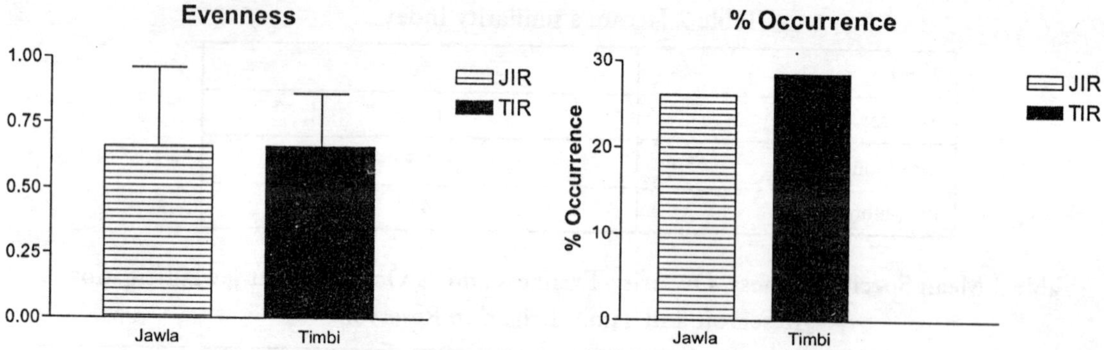

Fig. 1.a The comparison between the total species richness, Diversity index H', Evenness and % occurrence of butterflies at Timbi irrigation reservoir and Jawla irrigation Reservoir

and highly significant (P′ 0.001, F(2,13) 13.41) for TIR. The species richness is highest during post monsoon compared to other seasons studied with 10.66 + 2.08 at JIR and 8.83 + 2.56 at TIR with non-significant differences between the two reserviours when t-test is used (P′ 0.05), while lowest during summer with 4.33 + 3.55 at JIR and 3 + 1.67 at TIR and non-significant differences (P′ 0.05). During monsoon the species richness at JIR is comparatively higher 7.66 + 1.15 than that of TIR (4 + 1.63) with significant differences (P′ 0.05).

The mean Shannon-Weiner diversity index (H') was 1.35 + 0.67 at JIR while 1.8 + 0.45 at TIR (Table 3, Fig.1.a). However, when the seasonal differences are considered they are higher at JIR compared to that at TIR across the three seasons studied (Table 4, Fig.1.b). The seasonal variations show non- significant differences (ANOVA- P′ 0.05, F(2,9) 2.045) at JIR and significant differences (P′ 0.05, F(2,13) 5.680) at TIR. H' is maximum during post monsoon with 1.83 + 0.09 at JIR and 1.46 + 0.1 at TIR with significantly significant differences (t-test P′ 0.01) and minimum during summer with 1.00 + 0.83 and 0.82 + 0.45 respectively and non significant differences (P′ 0.05). During Monsoon it is 1.56 + 0.07 at JIR and 0.89 + 0.42 at TIR and has significant differences (P′ 0.05).

The mean Evenness was 0.66 ± 0.30 at JIR and 0.66 ± 0.11 at TIR (Table 3, Fig 1a). The seasonal differences in evenness is having same scenario as that of species richness and diversity index with higher values for JIR except summer (Table 4, Fig 1.b). When ANOVA is applied, the seasonal differences are non-significant for both the reservoirs (P′ 0.05) with F values F(2,9) 0.5659 and F(2,13) 0.1573 for JIR and TIR respectively. At JIR, the evenness varied non-significantly with 0.56 + 0.44 during summer and higher values of 0.75 + 0.01 during monsoon and 0.78 + 0.04 during post monsoon. At TIR, the evenness is almost constant throughout the study period with 0.65 + 0.32 in summer, 0.65 + 0.1 in Monsoon and 0.68 + 0.08 in post monsoon. The differences are non-significant (P′ 0.05) for the seasonal comparisons.

The total percentage occurrence of butterfly species over all the other flying insect species is 26.05% at JIR and 28.48% at TIR (Table 3, Fig 1.a). When seasonal % occurrence is considered for TIR it is observed that there is variation across the seasons with 18.75% in summer increasing to 23.88% in Monsoon and reaching to highest level of 34.64% in post-monsoon. At JIR the % occurrence showed fewer variations with 21.66% in Summer, followed by 29.11% in Monsoon that almost remained constant in Post -monsoon (29.62%) (Table 4, Fig.1.b).

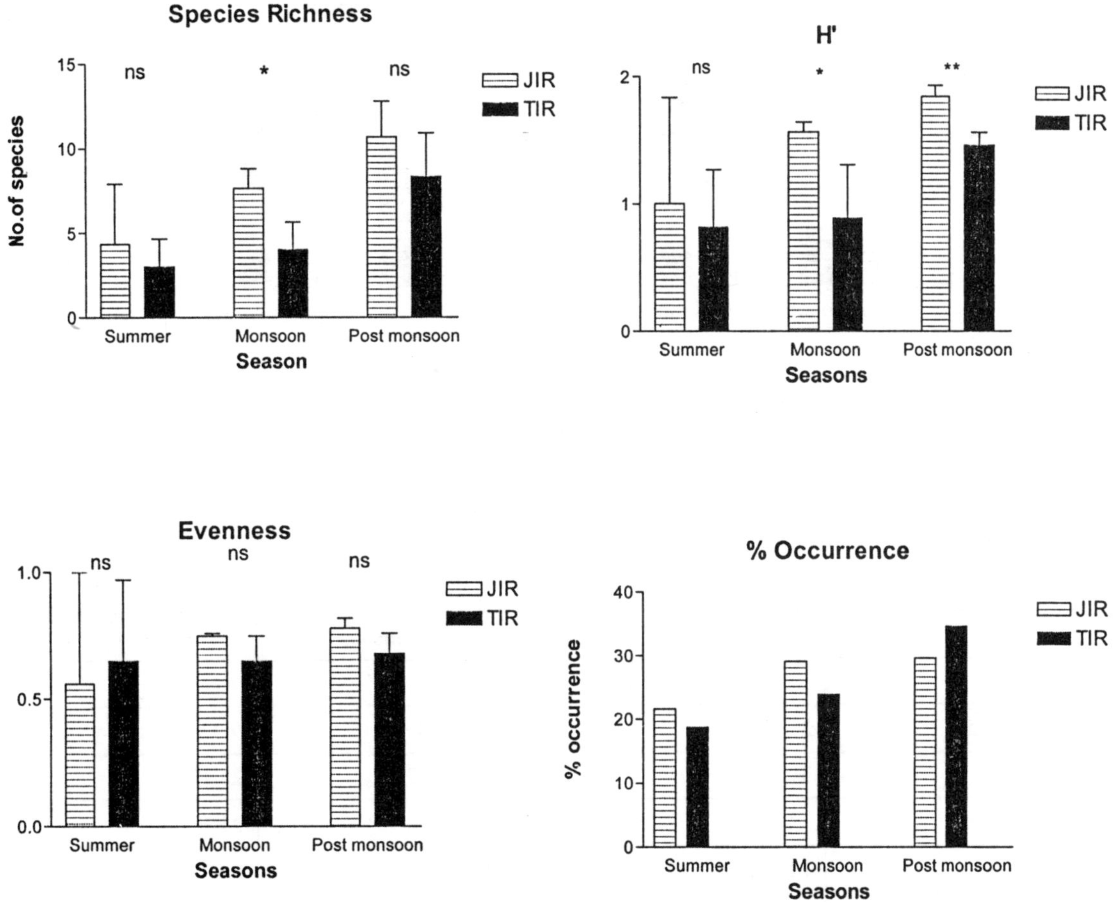

Fig. 1.b The seasonal comparison between the species richness, Diversity index H', Evenness and % occurrence at Timbi irrigation reservoir and Jawla irrigation Reservoir.

When family wise % occurrence is considered for complete study period (Table 5, Fig. 2), Lycaenidae is the most common family with 31.25% at JIR and Pieridae with 31.11% at TIR while Papilionidae is the most uncommon family around both the water reservoirs with 3.75% and 6.67% at JIR and TIR respectively. Pieridae also has 30% occurrence at JIR. The % occurrence of Nymphalidae is 23.75% at JIR and 26.67% at TIR. While Danidae represented by Plain tiger, the most common species at both the reservoirs is less represented family with respect to species number. Its occurrence is 11.25% at JIR and 16.67% at TIR.

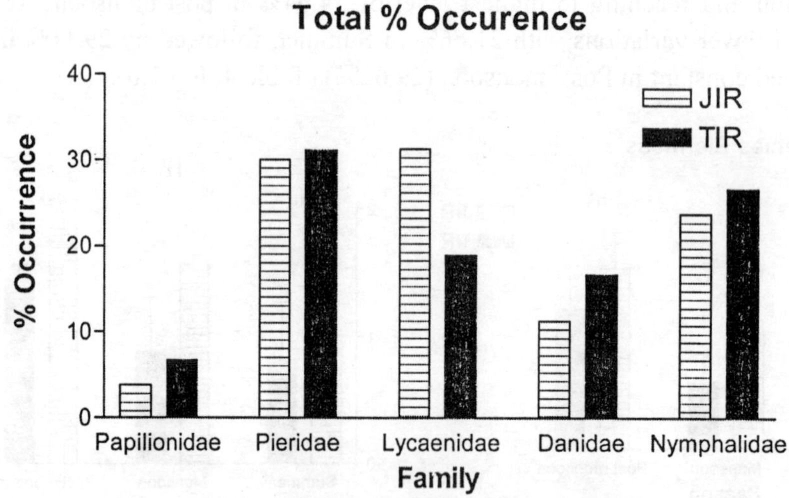

Fig. 2. Total % occurrence of butterflies compared to other flying insects at Jawla Irrigation Reservoir and Timbi Irrigation Reservoir

Table 5. Percentage occurrence of Family at Jawla Irrigation Reservoir and Timbi Irrigation Reservoir.

	Total	Summer	Monsoon	Post monsoon
Jawla				
Papilionidae	3.75%	0	0	9.37%
Pieridae	30%	26.92%	39.13%	28.12%
Lycaenidae	31.25%	38.46%	30.43%	25%
Danidae	11.25%	8.33%	13.04%	9.37%
Nymphalidae	23.75%	23.07%	17.39%	28.12%
Timbi				
Papilionidae	6.67%	11.11%	0	7.54%
Pieridae	31.11%	11.11%	18.75%	35.84%
Lycaenidae	18.89%	16.67%	43.75%	20.75%
Danidae	16.67%	33.33%	18.75%	11.32%
Nymphalidae	26.67%	27.78%	18.75%	24.52%

The seasonal occurrence of families (Fig. 3, Table 5) shows that at JIR Papilionidae accounts only 3.75% of total occurrence and is found only during Post monsoon with 9.37% of the total butterfly population of the season while at TIR where it has 6.67% total occurrence, it is absent in monsoon and has 11.11% occurrence in summer and 7.54% occurrence in Post monsoon. Pieridae is one of the common families at JIR. Though it has 30% total occurrence, its occurrence is low during summer (26.92 %), increases in monsoon (39.13%) and declines in post monsoon. While at TIR it increases gradually from 11.11% of summer to 18.75% of monsoon to 35.84% of post monsoon with total occurrence of 31.11%. The next family Lycaenidae with its 31.25% total occurrence at JIR, decreases from 38.46% of summer to 30.43% in monsoon and 25% in post monsoon. While at TIR with total 18.89% it fluctuates over the seasons with 16.67% of summer to 43.75% of monsoon to 20.75% of post monsoon. The next family Danidae with comparatively lower representation exhibited 11.25% total occurrence at JIR with seasonal fluctuations of 8.33% in summer, 13.04% in monsoon and 9.37% in post monsoon. Its total occurrence at TIR is higher with 16.67%. It declined from 33.33% of summer to 18.75% of monsoon to 11.32% of post monsoon. The % occurrence of family Nymphalidae with total % occurrence of 23.75% at JIR and 26.67% at TIR fluctuated parallely around both the reservoirs. At JIR, it is 23.07% in summer that declines to 17.39% during monsoon and increases again in post monsoon (28.12%) and at TIR it is 27.78% during summer that declines to 18.75% during monsoon and increases to 24.52% in post monsoon.

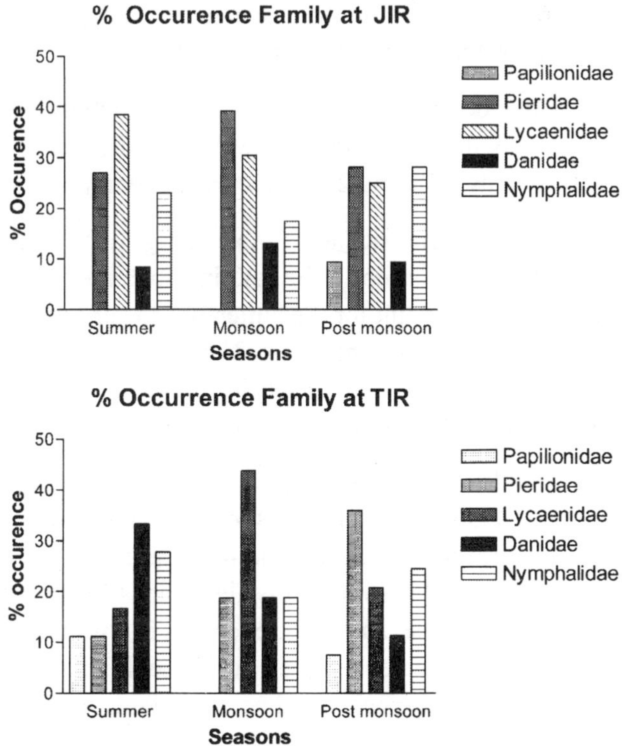

Fig. 3 The seasonal % occurrence of families at Jawla Irrigation Reservoir and Timbi Irrigation Reservoir

Discussion

Both the Irrigation reservoirs of present study are located in the monsoon type of climatic areas, hence they depend mainly on the rain water. One of the reservoir, TIR is inundated with the Narmada water and hence has perennial type of ecosystem, while JIR is not inundated and frequently dries off in the hot summers. Another difference between these two reservoirs is influences from the human activities. JIR is located a little away from the human habitation and hence faces less human pressures while around TIR human activities are increasing in the form of development of Institutes as well as farm houses. JIR is mainly surrounded by Agricultural and rural matrix while TIR is getting surrounded by sub-urban matrix. This is reflected by more of human activities like usage of water for domestic needs like washing of clothes, bathing, as sanitation site, cattle grazing, etc. Fishing activity is also going on all throughout the year. Thus, TIR can be considered as reservoir under the Anthropogenic pressures while JIR as a reservoir in its natural form without much of disturbance to its habitants.

In the present study it was observed that around the reservoir with less of anthropogenic pressures (JIR) more variety of butterflies are present in comparison to the one with anthropogenic pressures (TIR) except for the % occurrence. This indicates that different species of butterflies do visit wetland under anthropogenic pressures but do not stay there for longer duration. The native species mainly prefer natural habitats while those species that adapt to human presence thrive in the areas with anthropogenic influences (Kark *et al.*, 2007).

The Jaccard's similarity index (Table 2) suggests that the habitats are quite similar because of sharing of 50% (18) species by both the reservoirs. However, the difference is greater during summer and monsoon respectively as only 11% and 25% species are common.

The Seasonal variations in species richness (Table 4, Fig. 1.b) however are same for both the water reservoirs suggesting that Post monsoon is the best time for butterflies to explore and reproduce. During summer the heat makes it difficult for the butterflies to survive, during monsoon the rains while in winter the cold makes the conditions worse. Number of species found at JIR in the harsh conditions of summers resembles the number of species found at TIR during monsoon. This leads us to summarize that the human disturbances play a major role in the species distribution.

The species diversity index is increasing from Summer to Post monsoon at both the sites (Table 4, Fig. 1.b). But the diversity index at TIR is low as compared to JIR. The values suggest that the diversity at JIR and TIR are comparable between seasons. The evenness shows less of variation in the seasonal analysis at TIR. Evenness is higher at TIR as compared to JIR only during summer because of less species richness. At JIR the evenness is low during summer. At JIR there is higher variation in the numbers of species reported during each visit of summer while at TIR all throughout the summer the number of species is almost constant.

When compared to other flying insects the percentage occurrence shows that butterflies dominated among all the flying insects at both the reservoirs mainly during post monsoon (Table 4,

Fig. 1.b). When the seasonal data is considered it is observed that at TIR post monsoon was the most dominant period for the butterflies while their dominance was very low in summer and monsoon. At JIR butterfly occurrence fluctuated less with lower occurrence during summer and higher occurrence during monsoon and post monsoon. The differences in the seasonal occurrence of butterflies indicate that during unfavourable conditions of summer and monsoon JIR is a comparatively favourable site whereas during favourable post monsoon it is TIR. The farmhouses and the Institutional garden with flowering plants around TIR attract more butterflies compared to the agricultural matrix of JIR. At JIR the common species are observed all throughout the year and the % occurrence do not show much variations, while at TIR the common species that are not seen during summer and monsoon are seen during post monsoon and hence the increase in percentage occurrence.

Percentage occurrence of families show that Pieridae, Lycaenidae and Nymphalidae are the abundant families consisting about 85% of the butterfly population observed while Papilonidae and Danidae are less abundant with only 15 % of the population. However, when the occurrence of single species is considered it was found that Plain tiger of Danidae out numbers all the butterflies all throughout the year. Plain tiger is most common butterfly appearing during all the visits. Due to the appearance of the less common species with the decrease in the temperatures and arrival of the rain the percentage occurrence of Plain Tiger is affected.

The higher species richness as well as percentage occurrence of butterflies can be correlated to the occurrence of families Pieridae and Lycaenidae at JIR as compared to TIR (Table 1, 5). The differences in total % occurrence of the two families are negligible at JIR and more at TIR. Among the two, Pieridae is more common at JIR during monsoon and summer while at TIR only during post monsoon. This indicates that during favourable season they wander around extensively at TIR. An opposite trend is seen with family Lycaenidae that is known to prefer grasses. Grasses are more common during monsoon at TIR and during summer and Post monsoon at JIR. The green grasses developed during rains attract this family to the TIR as the agricultural matrix has less grass. The total representation of family Papilionidae and Danidae is marginally higher at TIR (Table 1, 5). Further, the seasonal data of Papilionidae suggests that it is better distributed at TIR while less distributed at JIR. It is represented by 3 species at TIR during 2 seasons as compared to single in the peak season post monsoon at JIR. This family with Danidae is probably better adapted to human presence as can be evinced by seasonal data of Danidae. Danidae is represented more at TIR compared to JIR all throughout the study period. The main representative of this family Plain tiger is the most common species and probably adapted up to some extent to the urban conditions. It is one of the most common species of gardens of Vadodara city too (Personal observation). This is also a common species of summer at TIR probably attracted to the floral plants. Family Nymphalidae has an almost same scenario for the total % abundance at both the reservoirs. At TIR Blue pansy and Peacock pansy increased the abundance while at JIR it was only Peacock pansy which were in higher numbers.

Conclusions

The differences in butterfly populations around two reservoirs are mainly due to the surrounding matrix that is reflecting the anthropogenic activities.

Acknowledgement

The authors are thankful to the Head, Department of Zoology for providing necessary facilities to carry out this work. 1st and 2nd authors are also thankful to UGC for providing RFSMS scholarship.

References

- Deshkar, S.L. 2008. Avifaunal diversity and Ecology of wetlands in semi-arid zone of Central Gujarat with reference to their conservation and categorization. Ph.D. Thesis submitted to The M.S. University of Baroda. Vadodara.
- Fermon, H., Schulze, C.H., Waltert, M. and Muhlenberg, M. 2001. The butterfly fauna of the Noyau Central, Lama Forest (Republic of Benin), with notes on its ecological composition and geographic distribution. African Entomology 9(2) 1-9.
- Gay, T., Kehmikar, I. and Punetha, J.C. 1992. Common butterflies of India. WWF India-Oxford University Press, Bombay.
- Haribal, M. 1994. Butterflies of Sikkim. Himalaya and their natural History, Sikkim Nature Conservation Foundation, Gangtok.
- Jalil, M. F., Mahsol, H.F., Wahid, N. and Ahmad, A. H. 2008. A preliminary survey on the butterfly fauna of Sungai Imbak Forest Reserve, a remote area at the centre of Sabah, Malaysia. Journal of Tropical Biology and Conservation 4 (1) 115 - 120.
- Javed, S. and Kaul, R. 2002. The Field Methods for Birds Surveys. Bombay Natural History Society, Mumbai.
- Kark, S., Iwaniuk, A., Schalimtzek, A. and Banker, E. 2007. Living in the city can anyone become an "Urban exploiter"? Journal of Biogography 34 638 651.
- Kehmikar. I. 2008. The book of Indian butterflies. Bombay Natural History Society, Oxford University Press, Mumbai, Delhi, Calcutta, Chennia.
- Krebs, C. J. 1985. Ecology the experimental analysis of distribution and occurrence. Third edition, Harper and Row Publishers, New York.
- Kunte, K. J. 2000. India- A lifescape - Butterflies of Peninsular India. Indian Academy of Sciences, Bangalore and Universities Press, Hyderabad.
- Levente, A. 2005. Biomonitoring of butterfly fauna in the Darva region (Lepidoptera Diurna). Natura somogygiensis 7 63-73.
- Murray, D. L. 2000. A Survey of the Butterfly Fauna of Jatun Sacha, Ecuador (Lepidoptera Hesperioidea and Papilionoidea). Journal of Research on the Lepidoptera 3542-60
- Parasharya, B.M. and Jani, J.J. 2007. Butterflies of Gujarat. Anand Agricultural University, Anand, Gujarat, India.

- Peixoto, P. and Benson, W.W. 2009. Daily activity patterns of two co-occuring tropical satyrine butterflies. Journal of Insect Science 954.
- Ramachandra, T., Kiran, A., Ahylaya, N. and Deepa, R.S. 2002. Status of wetlands of Bangalore. Technical Report 86. Available at www.wgbis.ces.iisc.ernet.in/energy/TR86/welcome.html
- Rathod, J.Y. 2009. Avifauna of Urban area Its significance and implications under various human disturbances. Ph. D. Thesis submitted to The M.S. University of Baroda, Vadodara.
- Shapiro, A. M. 2009. Revisiting the pre-European butterfly fauna of the Sacramento Valley, California. Journal of Research on the Lepidoptera 41 31-39.
- Sharmila, E. J. 2009. Butterflies of Hawa valley in a reserve forest - Tamil Nadu. Current Biotica 3 (2)135-142.
- Sundufu, A. and Dumbuya, R. 2008. Habitat preferences of butterflies in the Bumbuna forest, Northern Sierra Leona. Journal of Insect Science 864.
- Wadatkar, J.S. and Kasambe, R. 2008. Butterflies of Melghat Tiger Reserve, Maharastra with notes on their occurrence, status and larval host plants. The Ecoscan 2(2) 165-171.

37

TOXIC EFFECTS OF HEAVY METALS

Ranjeeta Choudhary, Preeti Gupta and Monika Vishwakarma**

Abstract: Toxic heavy metals change the biological structures and systems into inflexible & irreversible conformations leading to deformity in the body or death. Almost all metals are toxic at higher concentration and some are lethal even at very low concentrations, although heavy function metals within limits are essential for aquatic organisms, plants as well as human to survive. This chapter deals with the study of harmful effects on biological system, testing and treatment of selective heavy metals.

Keywords: Heavy metals, Toxic effect

Introduction

There are over 50 heavy metals, out of which only 17 are considered to be very toxic. These occur in very small quantities in the earth's crust and hence are called trace metals. Metals with densities above 5000 kg/m^3 are known as "heavy metals", thus metals like mercury (Hg), lead (Pb), copper (Cu), zinc (Zn) nickel (Ni), etc are generally known as "toxic heavy metals". Heavy metals like mercury, lead, arsenic, cadmium, selenium, copper, zinc, nickel, and chromium, should be given particular attention in terms of water pollution and runoff/discharge effects. Toxicity levels depend on the type of metal, its biological role and type of organisms that are exposed to it. These metals are a cause of environmental pollution (heavy-metal pollution) from a number of sources, including lead in petrol, industrial effluents, and leaching of metal ions from the soil into lakes and rivers by acid rain. Heavy metal toxicity can be fatal for mental functions, energy, nervous system, kidneys, lungs and other organ functions.

Some of the heavy metals are considered to be essential for plant & animal nutrition & they serve some useful biological functions viz. zinc, copper, manganese, selenium, chromium, and molybdenum. Under supply of these essential trace heavy metals leads to deficiency, optimum

Sant Hirdaram Girls College, Bhopal, NIIST, Bhopal; Email: ranjeetachoudhary@gmail.com supply helps in healthy growth & over supply leads to toxicity & even eventual death of the organism. Living organisms require varying amounts of "heavy metals". Excessive levels can be damaging to the organism. Other heavy metals such as mercury, plutonium, and lead are toxic metals that have no known vital or beneficial effect on organisms, and their accumulation over time in the bodies of animals can cause serious illness or death depending upon the dose. Usually people inhale or ingest small amounts of metals which build up in the body slowly interfering with function and slowly causing illness which may ultimately end in death.

Some well-documented environmental disasters associated with heavy metals

Chemetco's illegal construction and use of a secret discharge pipe which CEO, Denis L. Feron and others used to deliberately pump hundreds of tons of heavy metal waste into the Mississippi River from 1986 to 1996. Feron is now an international fugitive and one of the Environmental Protection Agency's Most Wanted environmental criminals.

The Minamata disease caused by mercury pollution

Heavy metals occur naturally with large variations in concentration. Heavy metal pollution can arise from many sources but most commonly arises from the purification of metals, preparation of nuclear fuels, Electroplating, etc. Through precipitation of their compounds or by ion exchange into soils and mud, heavy metal pollutants can localize and lay dormant. Toxic heavy metals can also be present in industrial, municipal, and urban runoff.

The toxicity of heavy metals depends on a number of factors. Specific symptomatology varies according to the metal in question, the total dose absorbed, and whether the exposure was acute or chronic. The age of the person can also influence toxicity. For example, young children are more susceptible to the effects of lead exposure because they absorb several times the percent ingested compared with adults and because their brains are more plastic and even brief exposures may influence developmental processes. The route of exposure is also important.

Study of Effect of Selective Heavy Metals

Mercury (Hg)

There are two kinds of mercury. The simple one atom Hg is called "Inorganic Mercury". The other type of mercury is called "Organic Mercury". The Organic mercury is 100 times more dangerous since it can easily penetrate cell walls, is easily absorbed in fatty tissues, and is easily absorbed into nerve and brain cells.

The Organic mercury is a mercury atom attached to - CH_3, a "methyl group". The mercury from fish is Organic mercury and it's chemical symbol is CH_3HgCH_3. Another kind of Organic mercury is one that is produced in the body by mixing Thiolmethyl Transferase (an enzyme, abbreviated THT), H_2S (gas fermented in gut), a methyl donor (someone who donates CH_3) to get CH_3SH; and then the CH_3SH mixes with Inorganic Mercury (Hg) to get CH_3SHgSH_3C, which is a form of Inorganic mercury. The THT enzyme steps is actually needed to detoxify the poisonous H_2S gas to get it out of the body. When the Hg picks up the CH_3 to become Organic, it is said to have been "methylated". In this example, the H_2S gut gas "methylated" the mercury. Mercury in the body is usually "methylated" in the colon, since that is where it picks up the chemistry to do it.

The human body contains about 13 mg Hg, about 70% of which is present in fat & muscle tissue. Hg containing insecticides & fungicides cause Hg toxication.The retension of absorbed Hg in soft tissues is high and follows the distribution.

Kidney > liver > intestinal & colon walls > brain > heart > lungs > respiratory mucosa > muscle > skin.

Sources: Dental amalgam, tuna fish, swordfish, felt, algaecides, floor waxes, adhesives, fabric softeners, chlorine production, contact lens solution, preparation H, diuretics, mercurochrome, merthiolate, childhood vaccines.

Symptoms and diseases: Adrenal gland dysfunction, anorexia, birth defects, brain damage, depression, dermatitis, dizziness, fatigue, hearing loss, hyperactivity, insomnia, kidney damage, memory loss, migraines, mood swings, nervousness, pain in limbs, skin rashes, schizophrenia, thyroid dysfunction, peripheral vision loss.

How Mercury affects health

Kidneys- mercury accumulates in the kidneys causing kidney damage.

Energy- mercury compounds inhibit the enzyme ATPase, which impairs energy production in all body cells.

Nervous system- Degeneration of nerve fibers and reduced motor conduction speed.

Elemental mercury is relatively inert in the gastrointestinal tract and also poorly absorbed through intact skin, yet inhaled or injected elemental mercury may have disastrous effects. Exposure to metals may occur through the diet, from medications, from the environment, or in the course of work or play. Where heavy metal toxicity is suspected, time taken to perform a thorough dietary, occupational, and recreational history is time well spent, since identification and removal of the source of exposure is frequently the only therapy required.

Manganese (Mn)

Mn is an essential trace element but is toxic at higher concentrations. The human body contains 12 mg Mn, about 45% of which is present in bones and rest is in soft tissues including brain. It is essential for mammals. This heavy metal is frequently associated with iron deposits. This metal is known to block Calcium channels, and with chronic intoxication results in CNS dopamine depletion. This latter condition duplicates almost all the symptomology of Parkinson's Disease, While the concentration in the surface water does not exceed official public health limits, it is in the high range of acceptable levels.

Sources Mn is used in dry battery cells, ferro-manganese & other iron alloys, glass, ceramics, electric coils, etc. These industrial processes are the major sources of Mn pollution. Most of the Mn in the environment is due to burning of fossil fuels. Use of Mn bearing fertilizers also contributes to air & water pollution.

Symptoms and diseases: Manganese psychosis which is an irreversibly brain disease characterized by uncontrollable laughter, euphoria and impulsiveness. It causes chlorosis and necrosis in leaves of plants. It causes growth retardation, fever,muscular fatigue & eye blindness.

Arsenic

Sources coal combustion, paints, rat poisoning, beer, pesticides, table salt, seafood from coastal waters (oysters, shrimp, muscles), fungicides, drinking water, wood preservatives.

Symptoms and diseases enzyme inhibitor, anorexia, diarrhea, nausea, vomiting, chronic anemia, drowsiness, dermatitis, stomatitis, liver dysfunction, hair loss, headache, vertigo, fever, stupor, herpes, jaundice, fluid loss, throat constriction, spasms, respiratory tract infection, garlicky odor to breath or stool, keritosis, pallor, goiter.

How Arsenic affects health

Skin-Arsenic is a carcinogen that accumulates in hair, nails, and skin causing depigmentation, rashes on palms and soles of feet.

Musculoskeletal system- Arsenic poisoning will retard the growth of limbs, resulting in deformities and birth defects.

Kidneys- Build up of arsenic in the kidneys will quickly cause kidney damage.

Nervous system- A decrease in mental ability due to damage of brain cells.

Digestive system- Arsenic will cause serious abdominal cramps accompanied by diarrhea and anemia.

Arsenic can produce all three types of toxcity at difference dosages, acute, sub-acute, and chronic. One sign of acute exposure is edema of the eyelids, and gastrointestinal irritation, and both central and peripheral neuropathies frequently occur. During chronic intoxication "garlic breath", skin sensitivity, dermatitis, and keratitus frequently occur. All types of arsenic exposure can cause kidney and liver damage, and most severe exposure leads to erythrocyte hemolysis.

Cadmium

Sources: Tap water, fungicides, marijuana, processed meat, rubber, seafood (cod, haddock, oyster, tuna), sewage, tobacco, colas (especially from vending machines), tools, welding material, evaporated milk, airborne industrial contaminants, batteries, instant coffee, incineration of tires/ rubber/plastic, refined grains, soft water, galvanized pipes, dental alloys, candy, ceramics, electroplating fertilizers, paints, motor oil and motor exhaust.

Symptoms and diseases: Alopecia, anemia, arthritis, cancer, lung disease, cerebral hemorrhage, cirrhosis of the liver, enlarged heart, diabetes, emphysema, hypoglycemia, hypertension, impotence, infertility, kidney disease, learning disorders, migraines, inflammation, renal disease, osteoporosis, schizophrenia, strokes, vascular disease, high cholesterol, growth is impaired, cardiovascular disease.

How Cadmium affects health

Energy- Cadmium inhibits essential enzymes in the Krebs energy cycle.

Nervous system- Cadmium directly damages nerve cells. It inhibits the release of acetylcholine and activates cholinesterase, resulting in a tendency for hyperactivity of the nervous system.

Bones and joints: by altering calcium and phosphorus metabolism, a toxic level of cadmium can contribute to arthritis, osteoporosis, and neuromuscular diseases.

Cardiovascular system- Cadmium replaces zinc in the arteries, which contributes to arteries being brittle and inflexible.

Excretory system- Cadmium accumulates in the kidneys, resulting in high blood pressure and kidney disease.

Dental- Cadmium toxicity can alter calcium and vitamin D activity, resulting in cavities and tooth deformities.

Lead

Sources: Ash, auto exhaust, cigarette smoke, coal combustion, colored inks, pesticides, rainwater, food cans with lead solder sealing, toothpaste, wine, manufacturing batteries, cosmetics, hair dyes, lead pipes, liver, glazed ceramics, pencils, lead-based paint, industrial emissions.

Symptoms and diseases: Abdominal pain, ADD, adrenal insufficiency, allergies, anemia, anxiety, arthritis, blindness, cardiovascular disease, autism, colic, constipation, convulsions, depression, dyslexia, epilepsy, fatigue, gout, hallucinations, headaches, hostility, hyperactivity, hypertension, hypothyroidism, impotence, liver dysfunction, hyperkinesis, mental retardation, mood swings, menstrual problems, muscular dystrophy, multiple sclerosis, nephritis, nightmares, nausea, numbness, Parkinson's disease, poor concentration, psychosis, renal dysfunction, restlessness, schizophrenia, seizures, stillbirths, SIDS, tooth decay, vertigo, weight loss.

How Lead affects Health

Bones- Instead of calcium, lead is incorporated into bone.

Brain- Lead can inhibit copper-dependent enzymes needed for neurotransmitters, causing hyperactivity.

Energy- Fatigue is triggered by increasing rate of destruction of red blood cells. Also, lead inhibits copper and iron-dependent enzymes in the Krebs cycle.

Kidneys- Gout can occur from lead toxicity raising uric acid levels and impairing kidney functions.

Minerals- Lead inhibits calcium, zinc, manganese, copper, and iron causing deficiencies.

Thyroid gland- Lead can inactivate the thyroid hormone thyroxin because it interferes with the iodine uptake to the thyroid gland.

In most individuals there is a "lead balance", that is one excretes as much as they take in, and the tissue levels are below the concentrations which result in pathological changes. However, an increase in the rate of intake will result in accumulation or a "positive lead balance". Since lead is chemically very similar to calcium, it is handled by the body as if it were calcium. Thus the first place to which it is transported is to the plasma and the membrane sites in soft tissues. It is then distributed to the other sites where calcium plays an important role, most notably in the teeth of developing children and in bone at all ages.

Copper

Sources: Copper cookware, copper pipes, dental alloys, fungicides, ice makers, industrial emissions, swimming pools, shellfish, perch, bluefish, lobster, walnuts, almonds, soybeans, wheat germ, yeast, beer, chocolate, corn oil, gelatin, liver, lamb, mushrooms, avocado, birth control pills.

Symptoms and diseases: Acne, allergies, alopecia, insomnia, nausea, spaciness, tooth decay, strokes, PMS, yeast infections, urinary tract infections, mood swings, kidney disorders, depression, cystic fibrosis, arthritis, anxiety, anorexia, multiple sclerosis, inflammation, pancreatic dysfunction, vitamin deficiencies, paranoia, migraines, libido decreased, nervousness, osteoporosis, senility, stuttering, phobias, diabetes, autism, estrogen dominance.

How Copper affects health

Nervous system: Copper buildup can result in a tendency for hyperactivity in autistic children. It can cause stuttering, insomnia, and hypertension.

Skin: An excess of copper can cause oily skin, loss of skin tone (due to its ability to block vitamin C), and can cause a dark pigmentation of the skin, usually around the face.

Hair and Nails: Copper can cause nails to be brittle and thin. It can contribute to hair loss, especially in women.

Treatment of acute and chronic effect of heavy metals

Metal	Acute	Chronic	Toxic Concentration	Treatment
Arsenic	Nausea, vomiting, "rice-water" diarrhea, encephalopathy, MODS, LoQTS, painful neuropathy	Diabetes, hypopigmentation/ hyperkeratosis, cancer lung, bladder, skin, encephalopathy	24-h urine = 50 µg/L urine, or 100 µg/g creatinine	BAL (acute, symptomatic) Succimer DMPS (Europe)
Cadmium	Pneumonitis (oxide fumes)	Proteinuria, lung cancer, osteomalacia	Proteinuria and/or = 15 µg/ g creatinine	*
Copper	Blue vomitus, GI irritation/ hemorrhage, hemolysis, MODS (ingested); MFF (inhaled)	Vineyard sprayer's lung (inhaled); Wilson disease (hepatic and basal ganglia degeneration)	Normal excretion 25 µg/24 h (urine)	BAL D-Penicillamine Succimer
Lead	Nausea, vomiting, encephalopathy (headache, seizures, ataxia, obtundation)	Encephalopathy, anemia, abdominal pain, nephropathy, foot-drop/ wrist-drop	Pediatric symptoms or [Pb] = 45 µ/dL (blood); Adult symptoms or [Pb] = 70 µ/dL^{-1}	BAL $CaNa_2$ EDTA Succimer
Manganese	MFF (inhaled)	Parkinson-like syndrome, respiratory, neuropsychiatric	No clear reference standard	*
Mercury	Elemental (inhaled) fever, vomiting, diarrhea, ALI: Inorganic salts (ingestion) caustic gastroenteritis	Nausea, metallic taste, gingivo-stomatitis, tremor, neurasthenia, nephrotic syndrome; hypersensitivity (Pink disease)	Background exposure "normal" limits 10 µg/L (whole blood); 20 µg/L (24-h urine)	BAL Succimer DMPS (Europe)

Testing of Heavy Metals

When heavy metal toxicity is suspected, a good screening test is a hair analysis. Scientists have been evaluating hair specimens for types of metals and concentration as early as 1931. A hair analysis acts as a timeline for exposure to various elements. It is a safe and non-invasive way to collect tissue and help a doctor to treat the underlying problems more efficiently.

At Caring Medical, the hair analysis measures for 12 toxic elements including mercury, lead, arsenic, aluminum, and cadmium. It measures for 33 nutrient elements including calcium, zinc, manganese, and cobalt. Additional elements measured in the hair analysis are iron, potassium, sodium, phosphorus, and titanium.

The hair is generally taken from the nape of the neck. That is, if the hair is long enough. About an inch, in length, closest to the skin is needed. (Approximately a heaping teaspoon) For bald men or women with short hair, this isn't always the route taken. Hair samples can also be taken from the under arm or the pubic area. Besides length, the problem of hair treatments can arise. Hair chemicals, including dyes, can contain lead that will attach to the hair and may contaminate the sample. For the most accurate results it is best to refrain from chemically treating hair for at least 2 months.

Hair analysis is a very useful tool in detecting what causes chronic symptoms but is unable to detect very recent heavy metal exposures. For recent, acute exposures to metals such as mercury or lead a blood test is suggested. A 7ml tube of blood is needed for the test. In the blood various heavy metals can be tested including mercury and lead.

The most sensitive test though for heavy metal load according to various natural medicine specialists is the provocative urine test. The person is infused with a chelating agent such as EDTA or DMPS and then the urine is collected for 24 hours. The chelating agent after being infused intravenously will go into the tissues of the body and the bind the various heavy metals including cadmium, lead, and mercury. It is then excreted out of the urine where the heavy metal levels are measured. This is often the only way a person can tell what types of heavy metal exposure they have had. For children, such as those with autism, chemet (DMSA), an oral (taken by mouth) chelating agent is given for 3 days. The child's urine is then collected for 6 hours to look for heavy metal poisoning. As a side note, oral chelation therapy has been a very effective treatment for autism.

Treatment of Heavy Metal Poisoning

The best treatment for heavy metal poisoning is chelation. Chelation is a treatment where a synthetic amino acid called EDTA (ethylene diamine tetraacetic acid) is administered intravenously. It was originally used in the 1940's to treat solders with lead poisoning and approved by FDA due to good results. When EDTA enters the bloodstream it binds chelates with metal atoms and flushes out with the urine. This is usually within the first 24 hours of the chelation treatment. EDTA chelation is used to treat heavy metal poisonings viz. lead and cadmium. Lead poisoning can also be treated orally using the chelating agent DMSA (dimercaptosuccinic acid). Mercury poisoning is treated with DMPS (dimercaptopropane sodium sulfonate) which is administered intravenously. DMSA (chemet) can also be used for mercury poisoning. The EDTA chelation IV lasts for approximately 3 hours. It is generally given weekly for a period of time based on the individual case. Kidney and liver functions are monitored every few treatments. This is because the chelation must pass out the urine with all the metals that it has collected throughout the blood stream. DMPS IV chelation therapy takes about 20 minutes.

In general after six to ten chelations, the urine is again collected to see if the chelation is still removing heavy metals. If not, the chelation is stopped. If heavy metals such as lead or mercury are still present then the chelation is given another six to ten treatments and the urine test repeated at that time.

Discussion

The maximum permissible limit for drinking water as recommended by World Health Organization (WHO) for Hg is 0.001 ppm, manganese is 0.50 ppm, copper is 1.50 ppm, cadmium is 0.01 ppm, lead is 0.10 ppm, arsenic 0.05 ppm, and for is zinc 15 ppm. Concentration of heavy metals should be below the maximum permissible limit otherwise they may have disastrous effects. Many of the elements that can be considered heavy metals have no known benefit for human physiology. Lead, mercury, and cadmium are prime examples of such "toxic metals." Yet, other metals are essential to human biochemical processes.

Nutritional status is extremely important in protecting from metal poisoning. Deficiency in minerals and trace elements may result in absorption of heavy metals by the body. Cell membrane breaks down and rebuilds every two weeks but does not release the heavy metals if essential fats are not ingested or if bad fats are ingested. The liver, which performs detoxification 100% of the time, cannot perform this important task without all the essential nutrients. Chronic occupational exposure to metal dusts has also been linked to the development of pneumoconiosis, neuropathies, hepatorenal degeneration and a variety of cancers. These syndromes develop slowly over time and may be difficult to recognize clinically.

Conclusion

From the above study, it is concluded that heavy metal on acute exposure or on chronic exposure affect different body organs. Excessive levels of heavy metals can be fatal. Unlike organic pollutants, heavy metals do not decay and thus pose a different kind of challenge for remediation. Therefore the water quality must be checked for the concentration of heavy metals and if it is above the permissible limit, it must be pretreated before using for domestic purpose.

References

- Environmental Chemistry by V.P. Kudesia, 658-673, 1st, 2000.
- Environmental Chemistry by A.K. De, 80-105, 4th, 2000.
- Centers for Disease Control, National Center for Environmental Health Accessed Oct. 2006.
- Lead Poisoning, National Safety Council Accessed Oct. 2006.
- Environmental Chemistry and pollution control by S S Dara, 177-216, 7th, 2004.
- K.Schwartz, Clinical Chemistry & Chemical Toxicology of metals, 3, Elsevier 1977.
- S.D. Lee, Biochemical Effects of Environmental pollutants, Ann Arbor, Mich, 1977.
- Engineering Chemistry by B.K. Sharma, E-90, 3rd, 2001.
- Journal of Environmental Science and Technology, 88-94, Vol-2, 2009.

38

A NOTE ON—INTEGRATED MANAGEMENT OF WATER RESOURCES WITH REFERENCE TO BIODIVERSITY AND LIVELIHOOD

H.S. Sidhu

Abstract: The Fisheries in the Madhya Pradesh state promise to contribute a vital share in rural economy, employment generation, up liftment of socially and economically Schedule Caste and Schedule Tribe and backward class people, those are earning their livelihood from fisheries activities and provide fish to the food basket. The objectives of integrated management of water resources with special reference to livelihood includes:

- To bring additional water area under fish culture.
- To attain self sufficiency in fish seed production
- To promote fishermen welfare programmes.
- To train fishermen and fish farmers specially to increase productivity of the pond and reservoirs
- To promote fishermen cooperatives

Madhya Pradesh is one of the land locked state blessed with vast potential aquatic resources in the form 2.88 lakh ha. of reservoirs, 0.56 lakh ha. of village ponds and tanks, the total 3.44 lakh ha. water area available out of which 3.36 lakh ha. of water area brought under fish culture which is 97% of total water area available. Fish fauna comprises of 172 species as per stated in the report of Zoological Survey of India.

Freshwater resources are one of the most precious commodities for the living beings and particularly to human beings. Aquatic resources of water should be assured on the basis of biotic and biotic parameters in order to provide the complete spectrum of information for the conservation assessment of fish diversity and better fish-management and sustainable utilization.

Stringent measures to curb the destruction of fishery are to be adopted with immediate effect. Rivers are natural home of Indian fishes. Development of riverine fisheries should be given top priority in

Director of Fisheries, Govt. of M.P., Bhopal.

the five-year plans The Government has also identified deep pools where the fishing is banned under Fisheries Rules 1972. However, it needs more effective Centre.

In view of the above, mass awareness campaign to save fish from destruction by pesticides, herbicides, dynamites and electric current should be organized. There is need to bring awareness amongst people associated with this trade to understand the importance of biodiversity of the lotic waters. Dissemination of information to fishers about the importance to fishers about the importance of fishes in the river, ecology and organization of Matsya Bachao Ghosthies in the villages situated at the banks of big pools is needed. Mahsheer is the important sport fish, which needs conservation and propagation in the State, its natural breeding centers need to be protected to save it for the posterity.

Under the devolution of power of the Panchyats the State Government has transferred rights of management of reservoirs up to 10 ha. to the Gram Panchayats 10 ha. to 100 ha. to the Jan pad Panchayats and 100 ha. to 1000 ha. to the Zila Panchayats and 1000 ha. to 2000 ha. of water area will be given to Department of Fisheries / Fisheries Fedreation Ltd. The reservoirs beyond 2000 ha. have been transferred to the M.P. Fisheries Co-operatives Federation Ltd. an apex body of the primary fisheries co-operatives. The Panchayat Raj bodies lease out these waters to the local fisher co-operatives on long term lease basis.

Keywords: Land locked state, Spectrum of information, Fisheries federation

Introduction

The Fisheries in the Madhya Pradesh state promise to contribute a vital share in rural economy, employment generation, up liftment of socially and economically Schedule Caste and Schedule Tribe and backward class people, those are earning their livelihood from fisheries activities and provide fish to the food basket.

Objectives

- To bring additional water area under fish culture.
- To attain self sufficiency in fish seed production.
- To promote fishermen welfare programmes.
- To traind the fishermen and fish farmers.
- To promote fishermen cooperatives.
- To increase productivity of the pond and reservoirs by ensuring optimum.
- credit facilities and training etc.
- Management of Lentic and Lotic aquatic resources.

Madhya Pradesh is one of the land locked state blessed with vast potential aquatic resources in the form 2.88 lakh ha. of reservoirs, 0.56 lakh ha. of village ponds and tanks, the total 3.44 lakh ha. water area available out of which 3.36 lakh ha. of water area brought under fish culture which is 97% of total water area available.

The state is drained directly by 3 main river basins viz. Narmada, Tapti & Mahi and indirectly bay to more i.e. Ganga and Godavari and a network of 17088 kms of rivers and their tributaries drain the State due to tropical status of these river system. The rivers and revolutes remain turbulent in rainy season and dry in summers. Some of the important rivers like Narmada,Tapti, Mahi, Pench and the tributaries of Ganga-Yamuna like Chambal, Betwa, Ken, Sone & Sindh have their origin in the State.

The Narmada river is one of the major and important river and has been a vital source of live hood for people residing along its banks. Besides, it has also been a shelter for a large number of aquatic organisms either plant origin or animal origin. The river receives water from numbers of small river, rivulets, streams & nallas coming from high altitudes and forms an entire river system known as Narmada river system. Narmada is unique , as it is the only river, which flows to the west. A number of untouched natural treasures are assimilated in this river. Among them, one is Mahseer.

The fishing in rivers in free to the local fishermen. A sizable population of fishermen lives adjoining the important rivers and earns their livelihood from capture fisheries.

Policy

Under the devolution of power of the Panchyats the State Government has transferred rights of management of reservoirs up to 10 ha. to the Gram Panchayats 10 ha. to 100 ha. to the Jan pad Panchayats and 100 ha. to 1000 ha. to the Zila Panchayats and 1000 ha. to 2000 ha. of water area will be give to Department of Fisheries / Fisheries Federation Ltd. The reservoirs beyond 2000 ha. have been transferred to the M.P. Fisheries Co-operatives Federation Ltd. an apex body of the primary fisheries co-operatives. The Panchayat Raj bodies lease out these waters to the local fisher co-operatives on long term lease basis.

Culture practice

The water area available in the state in the form of ponds and tanks i.e. 0.56 lakh hac. dependent on rainfall. Village ponds and tanks at present most of the tank and ponds are used for carp culture by following extensive and modified extensive culture practices. About 70 to 80% of village ponds are seasonal in nature therefore the growing period is very less so that carp culture is not economical, the fish formers cannot earn their livelihood. The present fish productivity of ponds in the state is 1512 kg/hac./yr which is less than the National productivity is 2228 kg/hac/yr. Therefore, it is necessary to implement the improve carp culture practices to enhance the fish production and profitability. Simultaneously introduction of the Giant fresh water prawn Magur culture and short cycle culture help in further enhance of profitability of aquaculture in Madhya Pradesh.

Reservoirs are larger water bodies, where mainly the capture-cum-culture technique is applied. With an estimated 2.88 lakh hac. of water area, the reservoir constitute about 83.73% of the total water area of the state, while contribution to the state's in total fish production is 17.78%. The average productivity of reservoir in the state is 52.29kg/hac./yr which is higher than the national fish productivity level.

Strategies

- All water bodies will be stocked with quality fish seed.
- To accelerate production of fish seed, seasonal village ponds will be use for seed rearing.
- Fish seed production and fish production in private sector will be encouraged.
- Pen and cage culture will be encouraged for raising fish fingerlings.
- Aquarium fish breeding and seed raising will be encouraged.
- Prawn culture will be introduced in private sector.

All tanks and reservoirs of Forest Department and Local bodies will be developed by the Fisheries Department.

Welfare Schemes such as 100% insurance cover through National Welfare Fund for Fishermen and Janshree Insurance Policy of LIC, Housing and community facilities, financial assistance during close season under CSS "Saving-cum-Relief " to fishermen will be promoted and women SHG would be encouraged.

Renovation of seasonal ponds would be undertaken through NREGA Programme resulting in the availability of approximately 6,000 ha, additional water area would be available for pisciulcutre during 11th Five Year Plan.

Additional fish seed demand due to increase in water area in the 11th Five year plan will be meet by developing fish seed production infra-structure in the vicinity of NVDA reservoirs.

Fish culture is a biological activity hence timely availability of inputs is of critical important and directly impact to fish production. Accordingly and timely availability of credit at reasonable terms is of vital importance. For that fishermen credit card programme is also undertaken in the state with the help of commercial and cooperative banks.

Fish seed Production

Fish seed is the main input for the fisheries development out of total 3.43 lacs ha. of water area available, 3.35 lacs ha. is under culture. An additional 0.65 lacs ha. area will be available due to rennovation of rural ponds, const, of new ones and through the complitation of Narmada Valley Development projects. The total water area which will be under culture by the end of 11th five year plan would be 4.00 lacs ha. raising from the present status of 3.43 lacs ha.This requires 6100 lacs of St.fry production against the present production level of 5235 lacs St.fry. This envisages a growth rate of 17% per annum.

To achieve this target, it is approved to pool up the existing resources of the production; create new facilities for production and rearing both in public and private sector. All the established technologies for enhancing production and rearing of fish seed are to be implemented in the State, so as to make the best use of infrastructure available with the Department and Fisheries Federation.

To accelerate the fish seed production and provision of vocational opportunities for educated employed, private pisciculturists, fisherwomen groups and their cooperatives fisheries federation, Govt. fish farms, self help groups and educated unemployed for seed production. With these objectives an additional water area 0.65 lakh ha. in the form of reservoirs/village ponds and 170 ha. additional rearing space will be developed by the Fisheries Department in the next three years.

Fish diversity in Madhya Pradesh

172 species of fishes have been reported by Zoological survey of India in the state. However, due to abstraction of water for irrigation and ecological transformation and anthro-pogenic factors, availability of certain fishes in the catches have dwindled. Fishes like *Bagarius bagarius* silondia *silondia, Eutropiichtchys vacha, Rita pavimentata, Rita rita, Anguilla bengalensis, Chitala chitala, Pangasius pangasius, Tortor, Barilius bala* and *Labeo fimbriatus* have been observed reduced in recent catches. Therefore, there is an urgent need for river ranching in selected deep pools to restore the endemic fauna as well as to make it available to the fisherman inhibiting the banks of these rivers for their livelihood through capture fisheries.

Future prospective

Habitat destruction water quality deterioration and alternation of hydrological regimes have wide ranging adverse effect on productivity, fish community structure and fish yield of open water abstraction have already affected our Riverine habitats, feeding grounds and migratory paths of aquatic organisms, disturbing the natural feeding grounds and life history traits of fish. In coming decades, more and more water course alterations will take place and these will impose further constraints on the open water aquatic resources, particularly for fisheries and aquatic biodiversity.

In view of the above the definite policies/guidelines is needed for dealing with such changing eco-patterns, restoration of aquatic ecosystems, rehabilitation and conservation of depleted fish stocks, and all management technology for sustainable fish production form Inland waters, and measures and methods for increasing fishing efficiency and socio-economic upliftment of fishing communities. The strategy of fisheries development will evolve management methods to bridge the gap between potential and actual production from open water fisheries resources. The required guidelines for restoration of aquatic environment, rehabilitation and conservation of fisheries would have to be evolved by pursuing appropriate research and extension strategy.

Future research in Inland Fisheries would have to be undertaken with emphasis on restoration of water quality and conservation of fisheries resources in a manner that would sustain productivity with a view to achieve these objectives and able to formulate definite guidelines for management of each type of open water Inland resources of the State, related responsibilities of leadership in the areas of fisheries resources management and conservation, restoration and rehabilitation and socio-economic development of the fisheries community.

Causes responsible for deterioration of fish diversity

- Habitat distortion due to construction of dams (reservoirs)
- Submergence of fish breeding grounds.
- Water pollution
- Indiscriminate fishing
- Over exploitation of water for agricultural purposes
- Lack of advanced fisheries knowledge among fishery community

Remedial measures to conserve fish diversity

- Conservation of fish ecology.
- Stocking of endemic varieties seeds in deep pools of rivers.
- Strictly follow up of "Closed Season" rules to conserve fish breeders.
- To promote the selective fishing.
- Check the entry of harmful exotic fishes.
- Cryo-preservation of endangered and rare fish species.
- Declaration of fish sanctuaries.
- Active people participation to protect fish fauna especially endangered species.
- Training to fishing community for rational and sustainable fishing.

Steps taken by the Government of Madhya Pradesh for fish diversity conservation

Declaration of "Closed Season" during the breeding period i.e., 16th June to 15th August every year "The Madhya Pradesh Fisheries Act, 1948", to provide protection, conservations and development of fishes has been enforced in the State since 1948 and Amendment 1981.

Protection, Conservation and Development of the fishes in Madhya Pradesh in the provision of **Prohibit or regulate all or any of the following matters that is to say**

The gear uses of fixed engine.

The constructions, temporary/ permanent, a weirs dams and bunds.

The dimensions, size of mesh and kind of nets to be used and the mode of using them.

The use of more than one method of catching fish at one time.

Prohibit the destruction of, or any attempt to destroy, fishes by explosives, gun, bow, arrows or the like in Inland waters.

Prohibit the destruction of or any attempt to destroy fishes by using a chemical or any other substance likely to cause pollution of water.

Prohibit the throwing into any water of any solid or liquid which may be harmful to fishes in such water.

Prescribe the seasons during which the killing, catching or sale of fishes of any prescribed species shall be prohibited; and

Prescribe a minimum size or weight below which no fish of any prescribed species shall be sold.

If any person contravenes any of the provisions of this Act or rules made there under, he shall on conviction be punishable with imprisonment for a term which may extend to one year or with fine which may extend to five thousand rupees or with both.

To conserve the indigenous biomass 37 deep pools has been declared as sanctuaries.

Department of Fisheries, M.P. made a survey to identify rivers deep pools to stock fish seed. There are 890 main deep pools available in the State.

State Govt. of M.P. has taken up an innovative programme for river ranching, fish seed stocked in selected deep pools of river.

Department of Fisheries, M.P. has also made an effort to collect the seed of Mahsheer from Narmada River near Hoshangabad, which comes under the endangered species and seed rearing in different fish seed farms for the raising of Mahaseer breeder.

To conserve the fish diversity there is a need to motivate fish farmers-fisher folk to adopt diversified fishery for example promotion of cat fish culture, which includes *Clarias batrachus*, *Heteropneustes fossilis*, feather fishes like *Chitala chitala*, *Macrobrachium rosenbergii* and *Macrobrachium malcolmsonii* and fin fishes like *Mahseer* and *Labeo bata*.

People's awareness and sensitization

Mass awareness campaign to save fish from destruction by pesticides, herbicides, dynamites and electric current should be organized. There is need to bring awareness amongst people associated with this trade to understand the importance of biodiversity of the lotic waters.

Dissemination of information to fishers about the importance to fishers about the importance of fishes in the river, ecology and organization of Matsya Bachao Ghosthies in the villages situated at the banks of big pools is needed. Mahsheer is the important sport fish, which needs conservation and propagation in the State, its natural breeding centers need to be protected to save it for the posterity.

Stringent measures to curb the destruction of fishery are to be adopted with immediate effect. Rivers are natural home of Indian fishes. Development of riverine fisheries should be given top priority in the five- year plans. The Government has also identified deep pools where the fishing is banned under Fisheries Rules 1972. However, it needs more effective Centre.

Freshwater resources are one of the most precious commodities for the living beings and particularly to human beings. Aquatic resources of water should be assured on the basis of biotic and biotic parameters in order to provide the complete spectrum of information for the conservation assessment of fish diversity and better fish-management and sustainable utilization.

Rapid industrial growth in the post independence of Indian has resulted in deforestation, further shrinkage of forest cover, embankments of rivers, unscrupulous exploitation of river water for irrigation, siltation of river beds and reservoir, industrial and domestic water pollution, unattendent fish catches by the fisherman, destructive fishing, damming, habitat fragmentation, loss of habitat, predation, diseases etc, has drastically decreased the fish diversity in the natural water as a result, several fish species has declined and some of them have become endangered.

39

IMPACTS OF MINING AND ALLIED ACTIVITIES ON WATER QUALITY OF RANIGANJ COALFIELD AND ADJACENT AREAS, WEST BENGAL, INDIA

Nitin P. Varma, A. K. Singh and Amit Pal*

Abstract; Physico-chemical characteristics of mine water of the upper catchments of the Raniganj Coalfield (W.B.) and adjacent areas were studied to evaluate the major ion chemistry of water. Water samples have been collected from underground and open cast mines of Ranigang mining areas. Mine water have been analyzed for pH, EC, TDS, F, Cl, HCO_3, and metals viz. Cu, Cr, Mn, Zn etc. Water chemistry of the area reflects continental weathering, aided by mining and other anthropogenic impacts. The assessment of the mine water quality showed that the water is alkaline in nature. Anion chemistry is dominated by HCO_3, SO_4 and Cl and cation chemistry showed that the 34% of the samples were Na>Ca>Mg>K. Metals like Zn, Cu, Fe, Cr were more than the permissible limit prescribed by WHO as well as BIS. Study showed that the water of the study area may not suitable for direct uses in drinking or domestic purposes and need proper treatment before its utilization.

Keywords: Anthropogenic, Ion chemistry, Raniganj coalfield, Water quality.

Introduction

Population growth, urban expansion and industrial development have persistently raised the demand for water supply and consequently increased the exploitation of groundwater. This has not been followed by an equal expansion in water supply and sanitary drainage, thus resulting in water scarcity and poor environmental conditions in many parts of the country. Overexploitation of groundwater and disposal of domestic wastes and industrial effluents poses threats to the groundwater resources of the area, both in terms of quantity and quality. Surface and sub-surface water resources is at risk from contamination due to rapid and unplanned urbanization, industrialization and

Institute of Environment & Development Studies, Bundelkhand University, Jhansi - 284 128, India
*Central Institute of Mining & Fuel Research, Dhanbad - 826 001, India

indiscriminate disposal of domestic, industrial, agricultural and mining wastes at many part of the country (Singh, 2002; Singh *et al.*, 1996).

Mining is one of the major activities causing water pollution and threatens the quality and quantity of surface and ground water resources in many part of the country. Dumping of overburden (OB) and spoils (sometimes on the surface water body itself) or spreading of OB through rolling and washing may cause the chemical pollution. In mining areas, large volume of water is pumped out daily from ground. Being underground water, mine water may be expected to be comparatively less polluted. However, in practice this water is also polluted due to mining as well as human activities. The term "mine water" encompassing all natural water emanating from mine site including mine pit, waste rock piles and tailing dam leachates. As there is no proper water management plan in most mines, major parts of this water is discharged into the open channels. There, it pollutes the channel and becomes unused. Not only this, the mining and related activities also damage the aquifer and make the area as water scare. Mine water can vary greatly in the concentration of contaminants present, and in some cases it may even meet the drinking water specification. Mining by its nature consumes, diverts and can seriously pollute water resources. The origin and impacts of mine water arise at several stages of the mining cycle.

Surface mine development and underground mine working below piezometric level invariably change the hydraulic gradient, thus affecting ground and surface water flows regime vis-à-vis their water quality and quantity. The presence of water in mining sites create a range of operational and stability problems and requires a drainage in a way to avoid slope stability problem, oxidation of metallic sulfides and corrosion of mining machinery and equipment. The quality of the drainage water depends on a series of geological, hydrological and mining conditions, which vary significantly from mine to mine. Mining by its nature consumes, diverts and can seriously pollute water resources. The origin and impacts of mining on water resources arise at several stages of the mining cycle the mining processes itself and/or at mineral processing operation sage. Water pollution in mining areas is mainly due to

The dewatering which is undertaken to make mining possible.

Seepage of contaminated leachates from waste rock piles and tailing dams.

Flooding of workings after extraction has ceased.

Discharge of untreated waters after flooding is completed.

Mine water pumped out from under ground and opencast mines.

Leaches/wash off from overburden and solid waste deposits.

Wastewater from coal based and related industries.

Municipal waste from human settlements

Mining needs excavation, spoils and OB dumping and creates mined out faces and dumps slopes. These provide potential surface for erosion which if not properly protected will be washed by rain water and redistributed surface runoff to result "siltation". These cause reduction in infiltration rate and volume of surface water bodies. This invites flood hazard and turbidity of the water and may even cause drying up of surface water bodies or diversion of streams. The spreading of OB and spoils on surrounding land damage greenery and cause water pollution (if any component in the spreading is soluble in water). Damage to greenery is also caused by vacating the land for mining, rehabilitation, direct excavation, OB & spoils dumping and damages. The greenery damage will cause loss of all the positive impacts and hence damage to the surface and ground water potentialities. This will also cause the decreased infiltration, increased run-off, increased erosion and evaporation of soil moisture, increased barren earth surface, its drying up, loss of ground water through capillary rise and lowering of water table. This, through a chain effect, results decrease in precipitation.

Further the huge pumping out of water from underground mines cause lowering of water pressure in the aquifer which if caused in coastal regions invites intrusion of sea water into the aquifer. This is known as "salt water intrusion" and hence ground water pollution. The interface between fresh and salt water is a zone of mixing. Acid mine drainage has been recognized as one of the largest environmental problem facing the metal mining sector. Acid mine drainage and associated heavy metal contamination is caused by natural oxidation taking place when minerals are exposed to air and water. Sulfide oxidizing bacteria also play an important role. Mine drainage is found when pyrite, an iron sulphide is exposed and reacts with air and water to form sulfuric acid and dissolved iron. Some or all of this iron can precipitate to form the red, orange, or yellow sediments in the bottom of streams containing mine drainage (Fig. 1.2). The acid run off further dissolves heavy metals such as copper, lead, mercury into ground or surface water. The rate and degree by which acid mine drainage proceeds can be increased by the action of certain bacteria, acidic metal laden drainage from abandoned coal mines can have substantial effects on aquatic resources. Problems that can be associated wit mine drainage including contaminated drinking waste disrupted growth and reproduction of aquatic plants and animals and corroding effects of the acid on parts of infrastructures such as bridges. As with any environmental damage, there are costs associated with trying to come up with a solution to the problem. The Raniganj coalfield in West Bengal and partly in Jharkhand is the easternmost of the Damodar Valley coalfields and is the next only in importance to the Jharia coalfield. It is situated about 185 Km North-West of Calcutta. Raniganj coalfield is the birth place of coal mining in our country. In 1774, first mining operation in the country was started in this coalfield by summer and Healy. Since then, a number of underground and opencast coal mining are in operation in this important coalfield spreading over Burdwan, Birbhum, Bankura and Purulia districts in west Bengal and Dhanbad district in Jharkhand.

In Raniganj coalfield area, coal deposits are extracted by both opencast and underground mining methods since long times. In the mining process huge quantity of water are generated and discharge on surface or in natural water bodies without any productive use and many times without any prier treatment. This discharge of mine water has a visible detrimental effect on the surface water quality and aquatic lives of the region and many times it also affects the underground water resources of the region. This discharge mine water as such may not be usable and contain unacceptable levels of heavy metals, toxic anions, organic and biological contaminants. However,

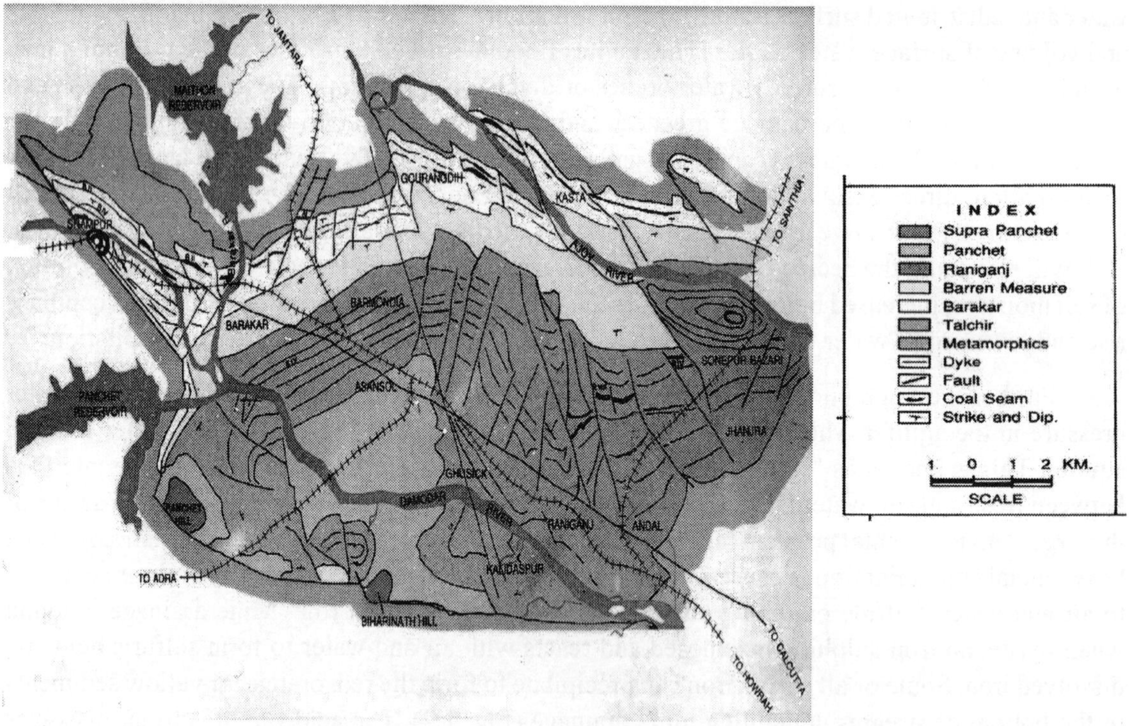

Fig. 1 Map of the Raniganj Coalfield area

many times concentrations of the contaminants are not very high and it may possible to utilize the mine water for human need after treatment (Kumar *et al.*, 2008). However, for proper utilization of available water resources of the mining areas for different uses and its management, a baseline water quality data and continuous monitoring of water quality of the mining regions is pre-requisite. In the present study, an attempt has been made to carry out quantitative and qualitative assessment of mine water discharges from the coal mines of the Raniganj Coalfield area. The hydrochemical data of water of these areas has been evaluated in terms of its suitability for domestic, industrial and irrigation uses.

Material and Methods

The present study have been carried out at Raniganj coalfield areas which lies in West Bengal and partly in Jharkhand states (Figure - 1) at the easternmost part of the Damodar valley coalfields. It is the next only in importance to the Jharia coalfield. Raniganj coalfield is situated about 185 Km North-West of Kolkata. Asansol is the most important town located in the central part of the coalfield. Asansol and Raniganj are the major railway stations in this region. The Raniganj coalfield is the birth place of coal mining in the India. In 1774, first mining operation in the country was started in this coalfield by summer and Healy. In 1820, first coal company M/S Alexander & Co. was established. In 1835, first Indian enterprise i.e. M/S Carr & Tagore Co. was formed. In 1843, the first joint stock coal company i.e. M/S Bengal Coal Co. was formed. In the 1850's this coalfield were the major coal producer, constituting over 90% of the total country's coal production. Presently, in this area coal are produced by underground as well as opencast mining methods by the Eastern Coalfield Ltd. (ECL), a subsidiary of Coal India Ltd. In addition, a small part of the coalfield is also operated by BCCL (Bharat Coking Coal Ltd.), IISCO and some private companies.

The Raniganj coalfield area covers about 1530 km^2, spreading over Burdwan, and Birbhum, Bankura and Purulia districts in West Bengal and Dhanbad district in Jharkhand. The major part of Raniganj coalfield is in Burdwan district and smaller portion in Birbhum, Purulia, Bankura and Dhanbad districts. The Kolkata-Delhi national highway (NH-2) as well as Hawrah-Delhi railway line of Eastern Railway pass through this coalfield area. A network of roads and railway branch link the surrounding areas. The topography of the Raniganj Coalfield is characterized by gently undulating area. A few linear hills mark the presence of intrusives. The range of elevation varies from 65 to 75 m above seas level. The highest elevations are marked by Panchet hills (643m) and Biharinath hills (451m). The drainage pattern in the study area is mainly dendritic in nature. The major part of Raniganj coalfield lies between two rivers - The Damodar and Ajay River, flowing almost parallel to each in the region. The average distance between two rivers is around 30 km. The drainage pattern of the area is mainly controlled by the Damodar River. The Damodar River traverses in the southern part of the coalfield flowing due east. The Ajay River is the other important river of this coalfield flowing in northern part of the coalfield. Other major river of the area is Barakar River, flowing in the western part of the coalfield and joined the Damodar River at Barakar. Apart from these, a number of small streams and jores are found throughout the area.

For detail geochemical analysis a systematic sampling of mine water discharges from different opencast as well as under ground from the entire regions of the Raniganj coalfield area have been carried out on seasonal basis. One liter water sample have been collected in high density polyethylene bottle from different parts of the area. The water sample also collected from the surface water resources as well as tube well and dug well from different part of the Raniganj coalfields area including mining, industrial and urban sites for surface and ground water quality assessment. The suspended sediments will be separated by 0.45 lim Millipore membrane filter using vacuum pump before storage for further chemical analysis. Filtered solution have been analysed (Maiti, 2001) for major cations (Ca, Mg, Na, K, NH_4) and major anions (SO_4, Cl, PO_4, NO_3, F) by Ion Chromatograph (Dionax-120) and heavy metals by AAS (Varian). The HCO_3 and dissolved silica (H_4SiO_4) in water will be determined by potentiometric titration and molybdosilicate method respectively (APHA, 2005).

Results and Discussion

To assess the water quality of mine water of the Raniganj coalfields, 50 samples were collected from underground and open cast mines in the months of March-April 2008. The mine water samples

Table - 1 Physico-Chemical Characteristics of Mine Water of Raniganj Coalfield Area

Sam. Code	Areas	Sample Site	Water Type	pH	EC	TDS	F	Cl	HCO_3	SO_4	NO_3	Silica	Ca	Mg	Na	K
1	Nirsa	Chapapur colliery	U/G	8.0	485	446	0.7	17.1	261.1	31.8	1.2	13.8	42.4	28.7	44.5	4.4
2	Nirsa	Chapapur colliery	D/W	8.1	1326	1261	0.9	27.0	326.1	565.7	0.6	7.6	123.6	134.7	59.4	15.6
3	Nirsa	Banjana Colliery	U/G	8.2	907	887	2.1	61.8	268.8	313.8	0.5	12.8	92.7	89.6	41.4	3.9
4	Nirsa	Banjana Colliery	D/W	8.2	927	904	2.2	36.9	349.0	262.9	3.4	7.8	80.6	109.9	41.7	9.4
5	Nirsa	Khudia Colliery	U/G	8.0	1169	1162	1.1	143.3	249.7	452.0	2.5	10.7	124.8	127.2	42.3	8.6
6	Nirsa	Khudia Colliery	U/G	8.1	1160	1120	1.3	110.1	238.2	456.2	0.8	9.7	121.0	134.4	39.8	8.9
7	Nirsa	Khudia Colliery	D/W	8.4	1131	1101	0.9	106.0	253.5	446.7	1.0	9.7	107.1	129.7	38.5	7.8
8	Nirsa	Khudia OCP	O/W	8.2	888	834	1.3	27.2	284.1	302.2	0.5	5.6	73.9	111.4	21.4	6.8
9	Nirsa	Khudia OCP	O/W	8.4	879	813	1.3	27.8	268.8	307.8	0.3	4.3	66.7	108.8	20.4	6.7
10	Mugma	Lakhi Mata Colliery	U/G	8.2	879	564	0.8	10.6	154.1	247.4	7.7	8.1	62.9	47.2	18.8	6.0
11	Mugma	Lakhi Mata Colliery	D/W	8.0	690	571	0.8	51.2	196.2	158.0	13.3	13.5	69.2	39.6	23.4	5.4
12	Mugma	Shyam Pur B Colliery	U/G	8.1	733	592	0.8	50.8	200.0	155.3	13.0	13.0	73.9	46.4	30.5	8.5
13	Mugma	Shyam Pur B Colliery	D/W	8.1	631	529	0.8	56.2	116.0	210.0	6.5	14.1	58.8	37.4	22.6	7.0
14	Mugma	Mandman Colliery	U/G	8.0	1030	771	0.4	22.9	203.8	312.8	0.5	13.1	104.5	52.2	55.3	5.9
15	Mugma	Mandman Colliery	D/W	7.9	1003	888	0.7	25.4	284.1	360.2	0.9	8.2	99.7	76.5	25.0	7.3
16	Mugma	Rajapura OCP	I/W	8.1	766	646	0.8	28.0	180.9	258.2	5.7	10.1	45.1	87.3	24.1	6.0
17	Mugma	Rajapura OCP	O/W	8.0	755	576	0.8	28.3	161.8	216.7	6.0	9.3	42.7	83.5	22.0	4.4
18	Salanpur	Mohanpur OCP	I/W	8.0	1227	1107	1.0	18.2	314.6	462.6	0.9	11.7	134.8	123.9	32.4	6.5
19	Salanpur	Mohanpur OCP	O/W	8.0	1034	982	0.9	20.4	288.7	388.9	5.0	13.5	112.8	109.9	33.3	8.4
20	Salanpur	Mohanpur OCP	S/W	8.1	920	895	0.9	20.7	245.8	366.8	0.8	10.9	103.0	102.4	34.8	8.5
21	Salanpur	Sangramgarh OCP	O/W	8.1	940	925	0.9	26.5	234.4	410.7	7.0	10.6	83.6	105.5	37.1	8.8
22	Salanpur	Sangramgarh OCP	I/W	8.0	997	976	1.4	28.3	276.9	387.9	7.9	11.3	113.6	102.8	38.0	7.9
23	Salanpur	Dabour Colliery	U/G	7.9	972	948	1.2	34.5	214.8	415.9	0.9	12.3	121.7	96.3	42.2	7.7
24	Salanpur	Dabour Colliery	D/W	8.0	964	938	1.4	34.6	222.9	399.9	0.4	12.3	122.9	95.5	41.0	7.4
25	Salanpur	Benjomeharg OCP	I/W	7.6	982	967	0.5	13.7	89.2	585.9	4.4	7.5	132.9	103.6	20.5	8.6
26	Salanpur	Benjomeharg OCP	O/W	7.6	966	948	0.9	15.3	96.8	556.9	5.1	8.1	132.1	104.5	19.6	8.9

Contd...

27	Salanpur	Gorandi OCP	I/W	7.9	415	341	0.5	23.4	123.6	100.7	5.3	6.4	37.5	21.9	16.7	5.2
28	Salanpur	Gorandi OCP	O/W	7.9	412	336	0.6	24.0	119.7	100.5	4.8	6.3	36.8	22.0	16.1	4.9
29	Salanpur	Debu(Old) OCP	I/W	7.8	409	337	0.6	22.7	112.1	115.1	0.4	7.6	36.3	22.1	15.7	4.1
30	Sodepur	Methani Colliery	U/G	8.2	1123	1087	0.7	153.8	394.9	209.2	7.4	16.3	24.6	102.7	167.0	10.0
31	Sodepur	Methani Colliery	D/W	8.2	1125	1077	0.8	153.8	391.1	206.8	10.7	15.7	18.9	110.0	159.6	9.9
32	Sodepur	Patmohana Colliery	U/G	8.5	882	807	0.8	47.9	398.7	135.6	3.2	13.2	37.5	29.4	136.7	4.4
33	Sodepur	Patmohana Colliery	D/W	8.5	865	770	0.5	41.2	387.2	117.7	3.2	13.3	34.7	28.4	139.0	4.6
34	Sodepur	Bejdih Colliery	U/G	7.5	1358	954	0.5	205.7	383.4	115.8	6.1	22.9	52.5	61.3	103.9	2.2
35	Sodepur	Dubeshwari Colliery	D/W	8.3	537	409	0.7	26.6	238.2	24.6	4.4	NA	25.9	11.0	75.1	2.8
36	Sodepur	Chinakuri III Colliery	U/G	8.8	1190	975	1.0	81.6	582.1	29.8	3.2	NA	12.5	8.2	253.8	2.5
37	Sodepur	Chinakuri III Colliery	D/W	8.7	1105	926	1.2	71.6	543.9	25.2	3.3	15.6	10.1	5.7	247.0	2.0
38	Sodepur	Chinakuri I Colliery	U/G	8.8	1337	1176	0.6	74.9	742.6	0.3	9.7	21.0	16.2	3.9	303.5	3.5
39	Sodepur	Chinakuri I Colliery	D/W	8.7	1228	1055	0.6	70.2	658.5	0.8	8.2	20.1	13.8	4.3	275.0	3.0
40	Sodepur	Dubeshwari Colliery	U/G	8.7	1208	1057	0.8	43.8	643.2	72.5	13.6	15.8	11.0	10.0	244.2	2.5
41	Sodepur	Parbolia Colliery	U/G	8.2	1129	983	1.0	29.0	620.3	61.7	8.8	12.2	12.4	6.1	229.2	2.2
42	Sodepur	Parbolia Colliery	D/W	8.3	1174	1053	0.7	30.7	662.3	68.2	9.3	12.5	13.4	7.1	247.0	2.3
43	Sodepur	Sodepur(R) Colliery	U/G	8.5	1149	960	0.9	61.0	624.1	20.7	4.1	18.4	18.7	21.8	185.8	4.0
44	Sodepur	Bhamoria Colliery (Abnt.)	D/W	9.1	2000	1861	1.1	75.1	1232.0	17.4	7.1	10.9	6.9	2.6	505.0	2.5
45	Sodepur	Sodepur(R) Colliery	D/W	8.3	1033	964	0.9	53.5	627.9	23.0	9.6	19.4	39.2	32.2	154.1	4.1
46	Sripur	Ningha Colliery	U/G	8.5	1175	1134	0.5	67.9	756.8	0.5	10.7	22.6	13.0	18.7	239.5	4.1
47	Sripur	Ningha Colliery	D/W	8.1	1190	1165	0.6	70.0	774.8	0.7	10.3	22.3	27.8	18.6	235.6	4.1
48	Sripur	Kalipahari Colliery	D/W	8.2	927	840	0.5	78.7	402.5	129.0	0.8	14.5	39.8	25.1	146.0	3.6
49	Sripur	Bhanora(W/B) Colliery	D/W	8.5	714	611	0.6	40.9	335.9	41.5	5.6	NA	37.9	29.6	114.8	3.9
50	Sripur	Ghusic Colliery	D/W	8.9	1519	1442	2.3	120.3	918.7	3.9	2.5	NA	8.4	3.5	379.1	2.9
		Minimum		7.5	409	336	0.4	10.6	89.2	0.3	0.3	4.3	6.9	2.6	15.7	2.0
		Maximum		9.1	2000	1861	2.3	205.7	1232.0	585.9	13.6	22.9	134.8	134.7	505.0	15.6
		Average		8.2	991	893	0.9	54.2	373.1	213.1	5.0	12.5	60.7	59.9	109.2	5.8
		Std. Dev.		0.3	289	286	0.4	41.8	238.5	174.3	3.8	4.6	41.7	44.5	110.9	2.8

D\W= Discharge water, U\G= Underground water, I\W =Inlet water, O\W =Outlet water   Units mg /l, Electrical Conductivity (µs cm^{-1}).

Table-2 Range in values of chemical parameters in mine water of study area and WHO and Indian Standards (IS 10500) for drinking water

S.N	Parameters	Range(ppm)	WHO (1993)		BIS (1991) IS 10500	
			Highest permissible	Max. desirable	Highest Permissible	Max. Desirable
1	pH	7.5 - 9.1	6.5 - 8.5	7.0 - 8.5	8.5 - 9.2	6.5 - 8.5
2	EC	409 - 2000	2000	400	-	-
3	HCO_3	89.2 – 1232.0	-	-	600	200
4	Dissolved Silica	4.3 - 22.9	-	-	-	-
5	F	0.4 -2.3	1.5	0.6 - 0.9	1.5	1.0
6	Cl	10.6 – 2 05.7	600	200	1000	250
7	NO_3	0.3 - 13.6	10	-	100	45
8	SO_4	0.3 -585.9	400	200	400	200
9	Na	15.7-505.0	200	-	-	-
10	Ca	6.9-134.8	200	75	200	75
11	Mg	2.6 -134.7	150	50	100	30
12	K	2.0 -15.6	-	-	-	-
13	TSS	14.6 - 248.6	-	-	-	-
14	TDS	336 -1861	1000	500	2000	500
15	TH	28 - 863	500	100	600	300

Units mg /l, EC (μs cm^{-1})

were analyzed for physico-chemical characteristics (Table - 1) by the standard prescribed methods and using sophisticated analytical equipment. The mine water quality data were evaluated in terms of its major ion chemistry (Table - 2), spatial variation in concentration of dissolved ions, sources of the different elements and the possibility and suitability of the mine water of the area for domestic and irrigation uses. For drinking water assessment, the mine water quality data was compared with the prescribed limits of WHO and Indian standard for drinking water (IS-10500). The different parameters like %Na, RSC, SAR, Salinity and Alkalinity Hazard, Magnesium Hazard (MH) were calculated for assessing the suitability of mine water for irrigation uses. Some of the major finding of this study is summarized in following points

The pH of the analyzed mine water samples varies from 7.5 - 9.1 and the average pH was found to be 8.2, indicating alkaline nature of the mine water (Table - 2). The electrical conductivity (EC) values varied from 409 μS cm^{-1} to 2000 μS cm^{-1} for measured samples collected during pre-monsoon season. The spatial differences between the EC values may reflect the wide variation in lithology, mining activities and hydrological processes prevailing in the region.

The anion chemistry of the analyzed water samples shows that HCO_3, SO_4 and Cl are the dominant anions both in underground and open cast mines (Table - 2). Concentration of bicarbonate varies from a minimum of 89.2 mg/l to a maximum value of 1232 mg/l and contributing on an average 50% of the total anionic balance. The concentration of sulphate varies between 0.3 to 585.9 mg/ and contributing on an average 37% to the total anion (Table - 2). High concentration of sulphate is observed in the mine water samples collected from mine of Nirsa, Salanpur and Mugma areas. Relatively low concentration of sulphate is found in the samples of Sodepur and Sripur mining areas. The observed high values of sulphate in the water of the mining affected area may be attributed to the oxidative weathering of pyrites. On an average chloride is contributing 12.4% of the total anionic balance. Fluoride and nitrate have low concentration and have less than 1% contribution to the total anions (Table-2).

The cation chemistry indicate that 34% of the samples are Na>Ca>Mg>K, while the 2% belong to Ca>Na>Mg>K. About 34% of samples have Ca>Mg>Na>K and only 20% samples having Mg>Ca>Na>K. The weathering and cation exchange processes normally control the levels of these cations (Table - 2). In the water samples studied, the concentration of iron (Fe), zinc (Zn) and manganese (Mn) is the most dominant metals. Concentration of Fe ranged from a minimum of 248 ppb to a maximum of 973 ppb and Mn (manganese) from 12 to 356 ppb and Zinc (Zn) concentration varies between 35 - 202 ppb. Thus it emerged that in the mine water of study area, concentration of some heavy metals (Mn, Cr) were present well above the prescribed permissible levels recommended for the drinking water by Indian Standard Institution (ISI, 1991) in some mine water samples (Table - 3).

The water quality assessment for drinking purposes indicate that most of the mine water of the study area is not suitable for direct uses in drinking and domestic purposes and need a proper treatment before its utilization for domestic uses. The values of Total Dissolved Solid (TDS) exceeds

Table-3 Heavy metal contents in mine water of the area and its comparison with WHO and Indian Standards for drinking water

Param.	Range	WHO (1993) Max. Desirable	BIS (1991) IS10500	
			Max. Desirable	Highest Permissible
Cr	45.1- 83.6	50	50	--
Mn	11.8 - 356	100	100	300
Fe	247.5 - 972.9	300	300	1000
Ni	43.1 - 110.2	--	--	--
Cu	11.2 - 51.9	1000	50	1500
Zn	34.7 - 201.7	5000	5000	15000

Units ppb

Table - 4 Calculated Parameters for Assessment Water Quality of Mine Water of Raniganj Coalfield Area

SN	Areas	Sample Site	Water Type	TH	SAR	%Na	RSC	MH
1	Nirsa	Chapapur colliery	U/G	224	1.29	31.4	2.0	53
2	Nirsa	Chapapur colliery	D/W	863	0.88	14.7	-3.4	64
3	Nirsa	Banjana Colliery	U/G	600	0.74	13.7	-1.6	61
4	Nirsa	Banjana Colliery	D/W	653	0.71	13.6	-0.9	69
5	Nirsa	Khudia Colliery	U/G	835	0.64	11.0	-4.3	63
6	Nirsa	Khudia Colliery	U/G	855	0.59	10.3	-4.7	65
7	Nirsa	Khudia Colliery	D/W	801	0.59	10.5	-3.9	67
8	Nirsa	Khudia OCP	Outlet	643	0.37	7.9	-1.8	71
9	Nirsa	Khudia OCP	O/W	614	0.36	7.9	-1.8	73
10	Mugma	Lakhi Mata Colliery	U/G	351	0.44	12.2	-1.0	55
11	Mugma	Lakhi Mata Colliery	D/W	336	0.56	14.7	-0.2	49
12	Mugma	Shyam Pur B Colliery	U/G	375	0.68	17.1	-0.5	51
13	Mugma	Shyam Pur B Colliery	D/W	301	0.57	16.2	-1.1	51
14	Mugma	Mandman Colliery	U/G	476	1.10	21.2	-1.4	45
15	Mugma	Mandman Colliery	D/W	564	0.46	10.2	-1.0	56
16	Mugma	Rajapura OCP	Inlet	472	0.48	11.3	-1.8	76
17	Mugma	Rajapura OCP	O/W	450	0.45	10.6	-1.9	76
18	Salanpur	Mohanpur OCP	I/W	846	0.48	8.5	-3.4	60
19	Salanpur	Mohanpur OCP	O/W	734	0.53	10.2	-2.7	62
20	Salanpur	Mohanpur OCP	S/W	678	0.58	11.3	-2.8	62
21	Salanpur	Sangramgarh OCP	O/W	643	0.64	12.5	-2.6	68
22	Salanpur	Sangramgarh OCP	I/W	707	0.62	11.6	-2.6	60
23	Salanpur	Dabour Colliery	U/G	700	0.69	12.7	-3.5	57
24	Salanpur	Dabour Colliery	D/W	700	0.67	12.4	-3.4	56
25	Salanpur	Benjomeharg OCP	I/W	758	0.32	6.8	-6.2	56
26	Salanpur	Benjomeharg OCP	O/W	760	0.31	6.6	-6.1	57
27	Salanpur	Gorandi OCP	I/W	184	0.54	19.0	0.2	49

28	Salanpur	Gorandi OCP	O/W	182	0.52	18.5	0.1	50
29	Salanpur	Debu (Old) OCP	I/W	182	0.51	17.8	0.0	50
30	Sodepur	Methani Colliery	U/G	484	3.30	43.7	1.6	87
31	Sodepur	Methani Colliery	D/W	499	3.11	41.9	1.4	91
32	Sodepur	Patmohana Colliery	U/G	215	4.06	58.5	4.4	56
33	Sodepur	Patmohana Colliery	D/W	203	4.24	60.2	4.3	57
34	Sodepur	Bejdih Colliery	U/G	383	2.31	37.4	2.4	66
35	Sodepur	Dubeshwari Colliery	D/W	110	3.12	60.3	2.8	41
36	Sodepur	Chinakuri III Colliery	U/G	65	13.70	89.5	8.9	52
37	Sodepur	Chinakuri III Colliery	D/W	49	15.40	91.7	8.4	48
38	Sodepur	Chinakuri I Colliery	U/G	57	17.57	92.2	11.6	28
39	Sodepur	Chinakuri I Colliery	D/W	52	16.57	92.0	10.3	34
40	Sodepur	Dubeshwari Colliery	U/G	69	12.83	88.6	9.9	60
41	Sodepur	Parbolia Colliery	U/G	56	13.32	89.9	9.6	45
42	Sodepur	Parbolia Colliery	D/W	63	13.57	89.6	10.2	47
43	Sodepur	Sodepur(R) Colliery	U/G	136	6.92	75.0	8.9	66
44	Sodepur	Bhamoria Colliery (Abnt.)	D/W	28	30.40	97.5	19.9	38
45	Sodepur	Sodepur(R) Colliery	D/W	230	4.42	59.6	8.0	58
46	Sripur	Ningha Colliery	U/G	109	9.96	82.8	11.3	70
47	Sripur	Ningha Colliery	D/W	146	8.48	78.0	11.2	52
48	Sripur	Kalipahari Colliery	D/W	203	4.46	61.4	4.6	51
49	Sripur	Bhanora(W/B) Colliery	D/W	216	3.40	54.1	3.3	56
50	Sripur	Ghusic Colliery	D/W	35	23.60	95.9	14.7	41
		Minimum		28	0.3	6.6	-6.2	28
		Maximum		863	30.4	97.5	19.9	91
		Average		398	4.6	38.4	2.1	58
		Std.Dev.		275	6.8	32.4	6.0	12

• D\W= Discharge water, U\G=Underground water, I\W=Inlet water, and O\W =Outlet water
Units mg/l, RSC, MH, SAR (meq l-1)

the permissible limit of WHO (1997) and IS-500 in some 28% of the samples indicating higher ionic concentration of dissolved ions.

The total hardness (TH) of the analyzed mine water samples vary between 28 to 863 mg/l (Avg. 398 mg/l) during sampling period, indicating soft to very hard types of water. The data indicate that 32% of the mine water samples have TH values (Table - 4) higher than 600 mg/l, which is the maximum permissible limit (ISI, 1991).

The assessment of water for irrigation use show that the water good to permissible quality and can be used for irrigation uses. However, high values of salinity, %Na, SAR, RSC and Mg-hazard at certain sites restrict its suitability for agriculture uses. The majority of the samples of Sodepur and Sripur area have high value of %Na, salinity and MH values which make water becomes unsuitable for irrigation uses for most of the crops and soils and demand the detail investigation and proper soil management.

Acknowledgement

Authors thankfully acknowledge to The Director, Central Institute of Mining and Fuel Research, Dhanbad for providing laboratory facilities for carried out this work.

References

- APHA, AWWA, WPCF (2005) Standard method for the examination of water and Wastewater; 21st ed. Amer. Pub. Hlth. Assoc. Washington DC; pp-1268.
- ISI (1991) Indian Standards Institution, Indian Standard specification for drinking water. 10500.
- Kumar, Jitendra, Siddharth Singh and Amit Pal (2008) Water quality of Turamdih and Jaduguda uranium mines and adjacent areas, East Singhbhum, Jharkhand, J. Ecophysiol. Occup. Hlth. 8 7-14.
- Maiti, S. K. (2001) Handbook of Method in Environmental studies; Water and Waste Water analysis, ABD Publishers, Jaipur, Vol. 1, pp 86.
- Singh, A. K. (2002) Quality Assessment of Surface and Subsurface Water of Damodar River Basin, Indian J. Environ Hlth, 44, 42- 46.
- Singh, A. K., Jaiswal, R. K. and Mukherjee, S. (1996) Ground Water Quality Assessment of Varanasi City, Hydrol J. 19, 23-31.
- Trivedy, R.K. and Goel, P.K. (1984) Chemical and Biological Methods for water pollution studies, Environmental Publication, Karad (India), pp 8.
- Warhate, S.R., Yenkie, M.K.N., Chaudhari, M.D. and Pokale, W.K. (2006) Impacts of Mining Activities on Water and Soil. J. Environ. Science & Engg. 48, 81-90.
- World Health Organization (1993) Guidelines for drinking water quality, Recommendations, WHO, Geneva.
- Wilcox, L.V. (1965) Effects of Industrial Wastes on Water for Irrigation Use, ASTM Tech Pub. No. 273, 58.

40

EVALUATION OF CERTAIN AQUATIC WORMS AS INDICATORS OF WATER QUALITY IN SHAHPURA LAKE, BHOPAL, MADHYA PRADESH, INDIA

Pranita Verma and Shriparna Saxena

Abstract: The chapter deals with the role of annelida worms a major group of aquatic macroinvertebeates in assessing the water quality of Shahpura Lake at Bhopal. Five species of Oligochaetes and two species Leeches have been collected for seven months in monthly sampling from five selected stations during study period in 2008-2009 out of these five species of oligochaetes and Hirudinea in pollution load. According to organisms, water quality at five sampling stations has been classified into least polluted, moderately polluted and highly polluted categories. This classification in agreement with Physio-chemical parameters.

Keywords: Annelida, Water quality, Tolerant, Pollution indicator, Physio-chemical parameter.

Introduction

Present work reports a study on Shahpura Lake situated in one of the elite localities of Bhopal, the capital city of Madhya Pradesh. The degradation of lake has occurred not only due to effluent inflow but also by siltation, domestic sewage, and immersion of idols and other activities around the lake like washing clothes, cattle, vehicles, and dumping of soil wastes. It receives domestic raw sewage from surrounding habitation. Apart from it a large number of heavy metals is contributed by corrosion of metal pipes in water bodies etc.

Geographical Morphometric features of Shahpura Lake

Latitude 23°12'00"E and Longitude 77°25'30"N

Department of Limnology & Environmental Science, Barkatullah University, Bhopal-462026, M.P. India. E-mail shreepranita@gmail.com

Catchment area	8.29 sq.km
Submergence area	0.96 sq.km
Gross storage	2.29 m cum
Live storage	2.09 m cum
Lowest still level	483.71 m
Full reservoir level	488.30 m
Maximum water level	489.00 m
Water spread at FRL	Approx. 80 ha
Water spread at MWL	Approx. 96 ha

The area under study enjoys a sub-tropical climate. The lake is shallow aquatic ecosystem with a mean depth of 1-2 m. The lake, in most parts is muddy due to the accumulation of silt but towards the NW side, it has hard and clayey substratum. The maximum depth at full tank is 9-10m and the deepest point lies in the centre of the reservoir. The main inlet, the nullah joins the lake at its northern end. The catchment area starches over to 8.29 sq. km. which includes the area of agriculture land no its western side, the barren rocky land mass on its northern side, and hillocks and residential area on its eastern side. The reservoir has a gross storage capacity of 2.29m^3 which sometime submerge the neighboring land lying on the eastern side. Shahpura lake is a man made water impoundment, which was formed in 1974-1975 under the Betwa irrigation scheme, after constructing an earthen dam near Chunabhatti village in the south part of the Bhopal city. After closing down of the oxidation pond in the year 1977-78 the drain brings the storm water in the monsoon and sewage of the settlements came up in the upstream of the drain. The reservoir connects two village, Shahpura at its eastern side and chunabhatti at its western side.

Macro invertebrate organisms occupy specialized habitats and represent the best example of detritus food chain. Many contributors have described Macro invertebrate's organisms in relation to water quality. Some of these organisms have been identified as indicators of pollution. Keeping this in view the present study has been undertaken to investigate the macro invertebrate's organisms of the lake with special reference to their monthly fluctuation and relationship with pollution.

Hence the present study was planned with the following objectives.

To work out the variation of some important physico-chemical properties of water and its correlation with growth of Macro invertebrates.

To work out the monthly variations of qualitative and quantitative studies of Macro invertebrates. Means of management for up gradation of quality of lake water to provide niche to inhabitation.

To Assessment of water quality of the Shahpura lake.

A number of research project have been taken to understand the status of lake. There is a need for continuous evolution of the pollution level in order to promote better living conditions around the lake.

Material and Methods

Five sampling stations were selected based on their different characteristics such as bottom sediment, beach structure, human activities and macro vegetation location near Manisha nullah (S1), Campion nullah (S2), Panchsheel nullah (S3), Chunabatti nullah (S4), Embankment (S5). Each sampling station had two zone, viz. surface and bottom. Samples of sediments of each sampling station were collected in Peterson grab sampler at the entire stations. The sediments sample were sieved (0.2m and 0.5m mesh) and sorted out manually using forceps for animals like, oligochaetes. In each sample the number of individuals of different species and groups was count and their percentage calculated per meter square of bottom area. To the extent possible, oligochaetes worms and Hirudenia were studies in living conditions and organisms preserved 70% alcohol. The samples of the water overlying the surface & bottom were analyzed for pH, Air & Water temperature, TDS, Conductivity, Dissolved oxygen, Free CO_2, Transparency, Total Alkalinity, Total hardness, Chlorides.

To know the faunal diversity, Shannon and Weaver diversity index was calculated using the formula.

$$H = (ni / N) \log (ni / N)$$

Where,

H = The diversity is expressed as number of bits per individual species.

N = Total number of individuals.

ni = Number of individual in 1st species.

Physico-chemical characteristics of surface & bottom water

Maximum and minimum monthly values of the selected physico-chemical parameters at different sites have been shown in fig.1 most of the pollution relevant parameter such as. Water Temperature, pH, Electric conductivity, TDS, D.O., Total alkalinity, Total hardness, Calcium hardness, Magnesium and Chloride.

Results and Discussion

Physico-chemical characteristics of surface & bottom water

Maximum and minimum monthly values of the selected physico-chemical parameters at different sites have been shown in table 1 most of the pollution relevant parameter such as. Water Temperature, pH, Electric conductivity, TDS, D.O., Total alkalinity, Total hardness, Calcium hardness, Magnesium hardness, and Chloride.

Table 1- Range of monthly values of Physico-chemical parameters characteristics of the Water at different stations in Shahpura lake

Parameter	Monthly variation of physico-chemical parameters of the water at different stations				
	S1	S2	S3	S4	S5
Water Temp. (°C)	20-32	21-32	19-33.5	19.5-33	20-33.6
pH	8.0-8.5	8.2-8.6	8.1-8.6	8.1-8.7	8.1-8.7
Electric Cond.(ppm)	630-790	470-730	480-790	490-780	480-780
D.O. (mg/l)	2.1-8.4	2.7-6.4	1.7-8.8	1.3-8.8	1.8-12.0
Total Alkalinity (mg/l)	140-338	144-342	160-370	126-382	152-308
Total hardness (mg/l)	92-338	68-342	84-334	82-340	90-308
Calcium hardness (mg/l)	76-178	62-188	52-186	66-192	66-168
Magnesium hardness (mg/l)	5.83-29.5	18.1-29.1	3.88-39.8	4.86-23.3	5.31-38.8
Chloride (mg/l)	47.9-90.9	47.9-86.9	50.0-89.0	45.5-96.0	61.5-82.8

Table 2- Shannon-Weaver diversity index of different sites of Shahpura lake

Sites	S1	S2	S3	S4	S5
No. of spp.	10	7	8	6	8
Total no. of individuals	476	508	489	444	497
Shannon_H	2.728	2.833	2.703	2.54	2.453

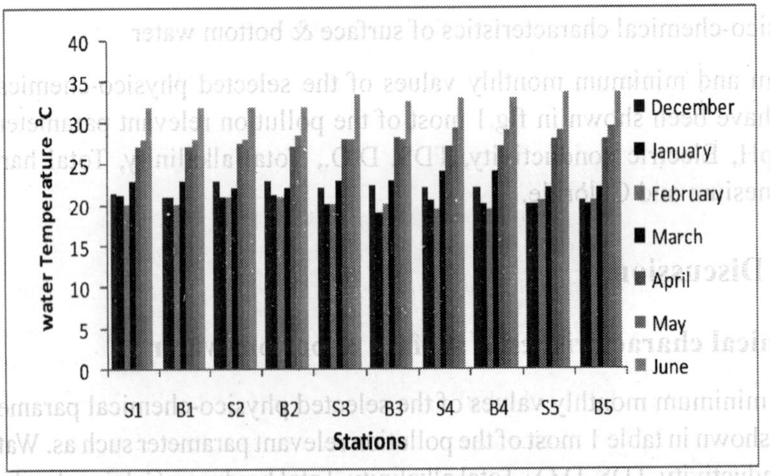

Fig.1. Graph showing monthly variation in Water & Temperature at different stations.

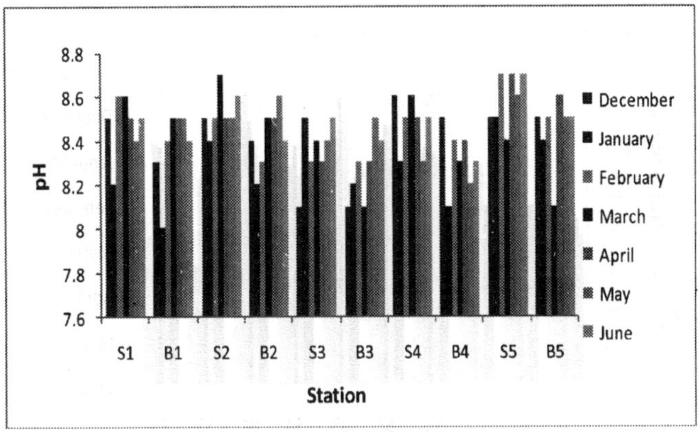

Fig. 2. Graph showing monthly variation in pH at different stations.

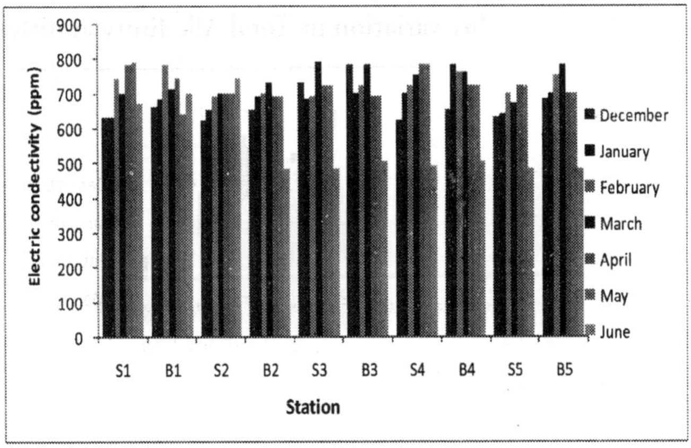

Fig. 3. Graph showing monthly variation in Electric Conductivity, at different stations.

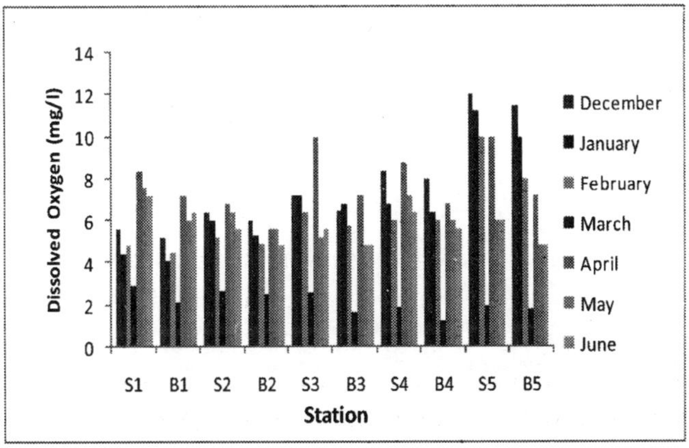

Fig. 4. Graph showing monthly variation dissolved Oxygen at different stations.

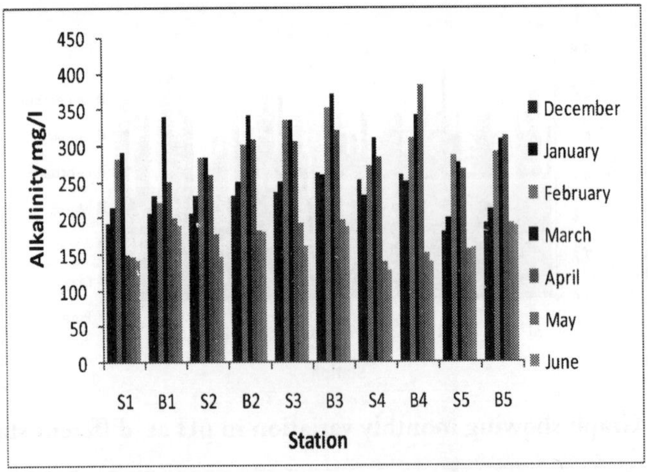

Fig. 5. Graph showing monthly variation in Total Alkalinity at different stations

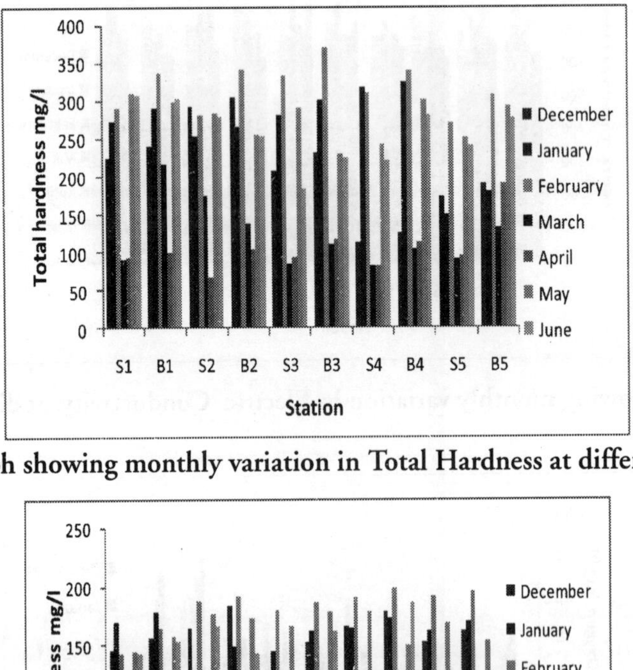

Fig. 6. Graph showing monthly variation in Total Hardness at different stations

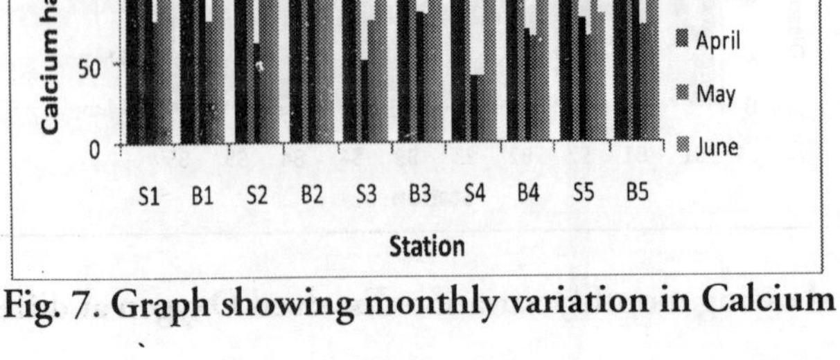

Fig. 7. Graph showing monthly variation in Calcium

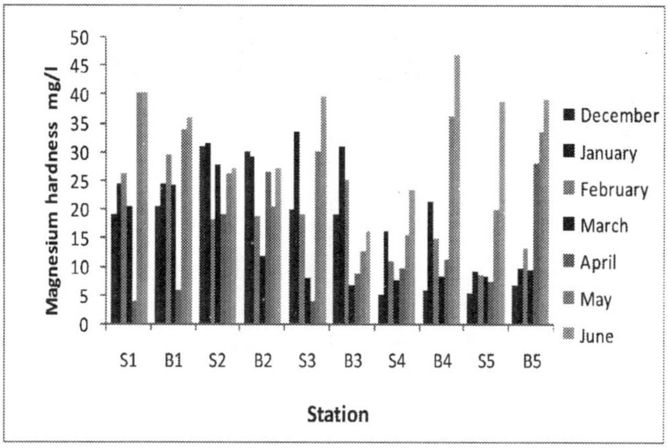

Fig. 8. Graph showing monthly variation in Magnesium hardness at different stations

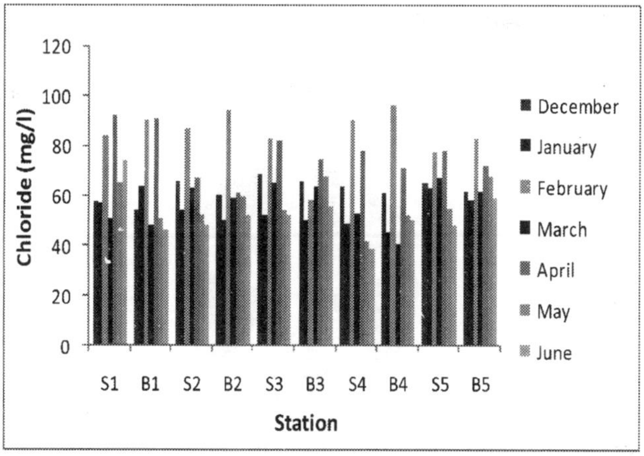

Fig. 9. Graph showing monthly variation in Chloride at different stations

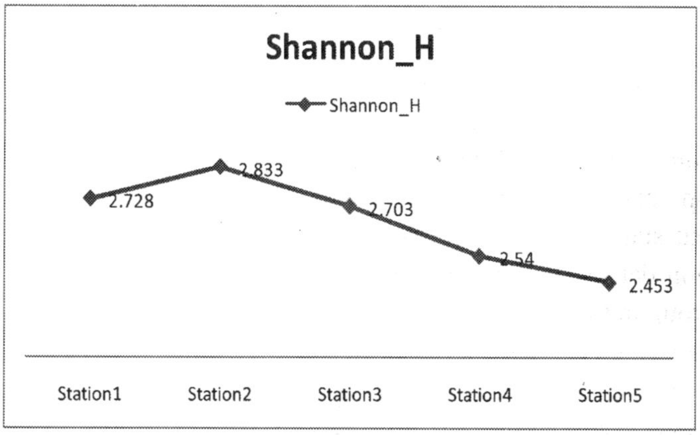

Fig. 10 Shannon-Weaver diversity index of Shahpura lake in during Present study at different stations

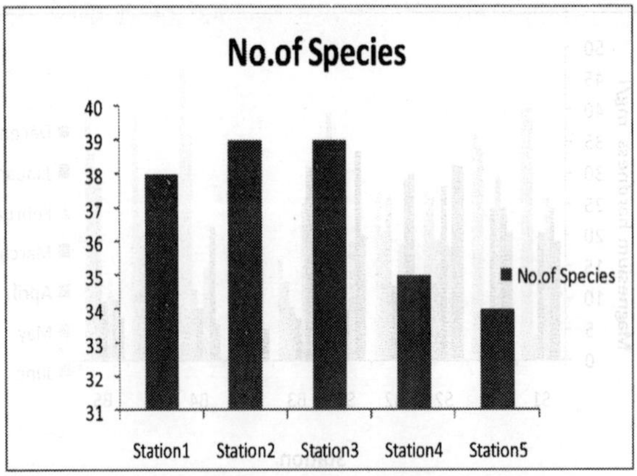

Fig. 11. Relative abundance of the Oligochaetes species in Shahpura lake at different stations

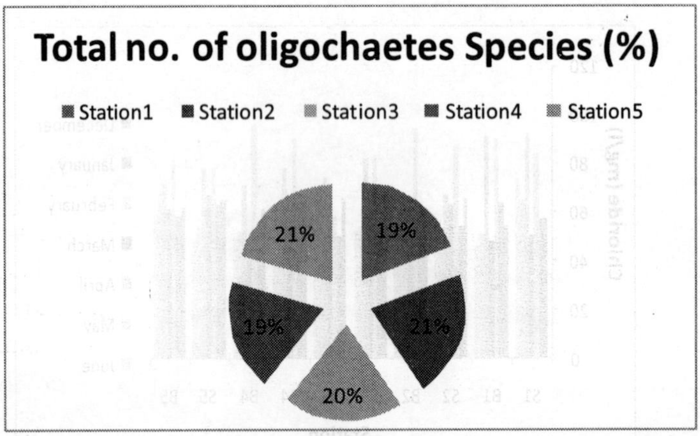

Fig. 12. Total diversity of Oligochaetes species in different station of Shahpura lake during study period

Biological characteristics

During the present study on Shahpura lake Macro invertebrates show a rich diversity in which Oligochaetes were dominant phyla throughout the study period. The sediments of reservoir were found to 10 species of oligochaetes and 2 species of Hirudinea. In class Annelid identified 10 spp. *Hirudinariae* sp., *Lumbriculus* sp., *Autodrillus piqueti* sp., *Branchivora* sp., *Nais* sp., *Tubifex tubifex* sp., *Chectogester* sp., and *Dero dorsalis* sp., *Pristina* sp., *hortensis* sp.The number of organisms recorded at different stations. The Lumbriculus are known as important group of benthos with maximum population density and maximum tolerance to organic pollution. Hirudinea were the second abundant group in the areas of the lake investigated. Their population was almost uniform in density. The oligochaetes are know to exhibit high tolerances to oxygen depletion occurring in deeper depth kaul also reported similar condition in some Lakes of Kashmir valley observed Lumbriculus and *Tubifex tubifex* as the pollution indicator species. Milbrink showed that oligochaetes occur as eutrophic indicator group.

Shannon-Weaver diversity index - For the macro invertebrate of Lake Shahpura at five different study sites. The Shannon's diversity index values were observed in the range of 2.728 at *Manisha nullah*, 2.833 at *Campion nullah*, 2.703 at *Panchsheel nullah*, 2.54 at *Chunabatti nullah* and 2.453 at Embankment. Brian D. Huges (1975), observed the influence of factors other than pollution on the value of Shannon's diversity index for benthic macro invertebrate in lake. During the course of present study the maximum macro invertebrate species diversity was recorded 2.833 on the stone and gravel type ground at Campion nullah (site 2), while as lower range 2.45 at Embankment (site-5), with sand and soil Pliuraite (2001) also reported benthos. Embankment (site-5), with sand and soil. Pliuraite (2001) also reported highest macro-invertebrates.

Conclusion

The present study indicates the current status of water quality. Water temperature, pH values are well with in permissible limit. The D.O. is correlated with growth of oligochaetes.The statistical analysis in the present study exhibited positive correlation between total organisms with water temperature; oligochaetes with pH, temperature and negative correlation were show with oligochaetes total hardness and alkalinity.

References

- Aboo K.M Shastry, C.A and Alex, P.G. (1968) A study of well water on Bhopal city. J. Environ. Hlth. pp189-203.
- APHA (2005) Standard methods for the examination of water and Wastewater 21th addition. American Public Health Association.Washington. DC.
- Aston, R.J.(1971)The effects of temperature and dissolved oxygen concentration on reproduction in Limnodrilus hoffmeisteri and Tubifex tubifex.Vol 5 / 1971 pp122-135.
- Bais, V.S., S. Yatheesh & N.C Agarawal (1992) Benthic macro invertebrates in relation to water and Sediment chemistry. J. Freshwater Biol. 4(3) pp183-191.
- Campbell, I.C., (1978) Biological investigation of an organically polluted urban stream in Victoria, "Aust. J. Mar. Fresh water. Vol. 29/1978 pp275-291.
- Doua Nujattacjerkee (2007) Benthic Invertebrates- A Crucial Tool in Biomonitoring of Lakes. Ministry Of Environment & Forests P 162 Taal 2007.
- Mclntosh, R.P. (1967) An index of diversity and relation of certain concepts to diversity. Ecology 48, pp. 92-404.
- Mount, D.I. (1976) Biotic Monitoring. In Biological monitoring of water and effluent quality. J. CarinsL.Dickson, G.F. Westlake (Eds.) ASTM, STP. Pp 607-620.
- Krishnamurthy, K.P., Sarkar, R. (1979) Macro invertebrates as indicators of water quality. In Proc. Symp. Environ. Biol.; pp133-138.
- Oomachan Licy (1988) Macro invertebrates of lower lake; Number/1988pp-22-31.
- Zutshi, D.P. Subla B.A., Khan M.A. and Wanganeo, A. (1980) Comparative limnology of nine lakes of J& K Himalayas Hydrobiology.Vol.72 pp101-112.

41

BIODIVERSITY AND CONSERVATION OF INDIGENOUS FISH SPECIES OF UPPER LAKE, (RAMSAR SITE) BHOPAL, INDIA

Praveen Tamot and Ashu Awasthi

Abstract: In The 'Lake City, Bhopal', Upper Lake with an catchment area of 361 km^2 was built in 11th Century by Raja Bhoj, is the main source of drinking water for the majority of population of Bhopal. This lake possesses considerable wealth of indigenous species. It is a major source of income for the local fishermen and meeting the nutrition requirement and providing food to people hence, fishermen and people both get benefited. This wetland has poor quality of water due to various anthropogenic activities such as joining of domestic sewage, urbanization, encroachment in the catchment area. This poor water quality, stocking of seed of I.M.C. to increase the production, over-exploitation, introduction of exotic species and indiscriminate fishing puts up a pressure on the indigenous fishes and has lead to the reduction of native fish species in their own natural habitat. Present study focused mainly on water quality monitoring and fish diversity of Upper Lake. The data generated in the present study will help to know the real number of fish species actually present in the wetland (present status) and to find out those species which are endangered, dominant etc. On the basis of this study 'fish atlas' can be prepared. Further this study will play significant role for sustainable management and conservation of natural aquatic resources including fish stocks and to increase the fish production without affecting the indigenous fish diversity of this lake.

Keywords: Indigenous fishes, Wetland, endangered, Anthropogenic activities, Exotic, Nutrition requirement.

Introduction

Fisheries sector has been one of the most important sectors since the time immemorial. Fishes have been the very first source of food for the primitive man. India is one of the 12-mega diversity hot spot contributing to the World's biological resources. India has very rich and diversified fish fauna constituting 2200 species. Out of 630 Indian fresh water species, 327 species are assessed and

Post Graduate Department of Zoology, Govt. Motilal Vigyan Mahavidyalaya, Bhopal (Madhya Pradesh) E-mail: tamot03@yahoo.co.in remaining species are yet to be assessed. The work on fish fauna of Madhya Pradesh was initiated with the first survey of fish fauna of rivers of central India was reported by Hora (1941-1949). There after as per available literature different Ichthyologists has made limited efforts related with the fish biodiversity. Some notable work was done related with the fish fauna of Madhya Pradesh such as Swaroop (1953) worked on fish fauna of Sagar Lakes, Soni (1960,1962) fish fauna of Upper Lake & Lower Lake (Betwa river), Bhopal. Dubey & Verma (1965) conducted a preliminary study of fish fauna of Madhya Pradesh and reported the occurrence of 105 species. This figure has gone to165 in the year 1997 stated by Dubey himself. That it is to be said that there has been an increase of 60 species. Since then many statements regarding number of species were given. The species are increased that may be only due to the introduction of commercial fishes and the species number is reduced then the causes of their depletion are not known. ZSI (2000) reported 172 fish species in Madhya Pradesh. Qureshi (2003) recorded 53 species of fishes from Bhoj Wetland (Upper Lake & Lower Lake), Bhopal, Tamot *et al.* (1989, 1991) worked on limnology of Upper Lake.

In "Lake City Bhopal" large number of lakes and reservoirs are present, which are polluted due to multiple anthropogenic factors in catchment area of wetlands such as unprecedented population growth, consequent urbanization, chemical intensive agriculture, joining of untreated raw sewage, religious activities such as Idol immersion. Anthropogenic interventions results in loss of biodiversity, decline in fish catch and depletion in water quality. There also are drastic alterations in water quality due to inflow of silt, dumping of garbage and reclamation activities. This poor water quality has also lead to the reduction in number of native fishes in wetland. Stocking of seed of fishes, which are productive is also one of the causes for depletion of other indigenous fish species. Today it is well known fact that the environment degradation has caused damage to biodiversity including plants & animals. The damage to biodiversity can range from loss of genetic diversity, population and even consequences leading to extinction of species. The environmental threats could be man-made, natural or in combination with cascading and interlinked impacts. Such threats are wide ranging including over exploitation of resources, habitat alterations, reduction in natural habitat area, construction of dams, diversion or reclamation of river beds for urbanization that reduce water discharge in the lakes, introduction of non-native fishes etc. During last four decades, the Country has witnessed the main thrust on the fast growing Indian Major Carps (I.M.C.), which has singularly contributed to the negligence of native fishes. The local fishes are slowly and steadily getting reduced in their own natural habitat. Beside aquatic pollution, elimination of other species of micro flora and fauna due to dominance of macrophytes, Clandestine introduction and spread of undesirable and predatory fishes into the system, indiscriminate fishing have emerged as the major causes for the dwindling of stocks of many species in our wetlands. Management of water quality and fish diversity is the supreme important aspect to conserve our aquatic heritage for coming generation.

This study is an attempt to know the real number of fish species actually present in the Upper Lake (Present status) and to find out the species, which are endangered, and methods to conserve them. To find out those species which are present in large number and natural ways to control their population and, will also be helpful for sustainable management and conservation of natural aquatic resources including fish stocks and to increase the fish production without affecting the indigenous fish diversity. Therefore, documentation of such information becomes prerequisite, for suitable fish diversity conservation strategies. On the basis of study, Authentic Data Bank on Fish Diversity (Fish Atlas) of Upper Lake, Bhopal can be prepared.

About Upper Lake

Silent Features of the Upper Lake

Bhopal without the Upper Lake reflects the incomplete picture of the city and this lake is said to be the life line. The Upper Lake of Bhopal arguably is the oldest among the large manmade lakes in central part of India. This lake was created in the early 11th century by King Bhoj by construction of an earthen dam across the Kolans River, a rain fed tributary of Betwa River. With passage of time, the administration of Bhopal in year 1956, Bhopal became Capital of Madhya Pradesh. Since then, it has witnessed a tremendous influx of people and consequent urban development, especially on the northern fringe of upper lake. This resulted in an increased demand of portable water and subsequent pressure on upper lake resource. The Upper Lake is a source of portable water and meets 40% of the demand of drinking water for city's growing population. Due to increasing population in the past three decades lake water was insufficient therefore Kolar water supply (35 Kms from city) has been installed to meet growing water demand. In present scenario even both the water bodies are insufficient to meet the growing water needs. Finally Govt. of Madhya Pradesh is going to bring Narmada River water (80 Kms from city) to bridge the gap between demand and the supply of drinking water. Therefore it is necessary to conserve and manage the available aquatic resources in and around the city for future.

Status of Upper Lake

The Upper Lake has both rural as well as urban catchments. With a well protected catchment, its water quality was of portable standard until recently. In last few decades of the 20th century however many sections of lake became surrounded by habitations as city grew. These developments have generated anthropogenic pressure on the Lake, thus accelerating its eutrophication and microbial contamination and making the water unfit for human consumption without proper treatment. The lake is rich in biodiversity however with the principal components being phytoplankton, zooplankton and fish diversity.

Morphometric Features of Upper Lake

- Formation: 11th Century by Raja Bhoj
- Type of Dam: Earthen dam
- Longitude: 77°18'-77°24'

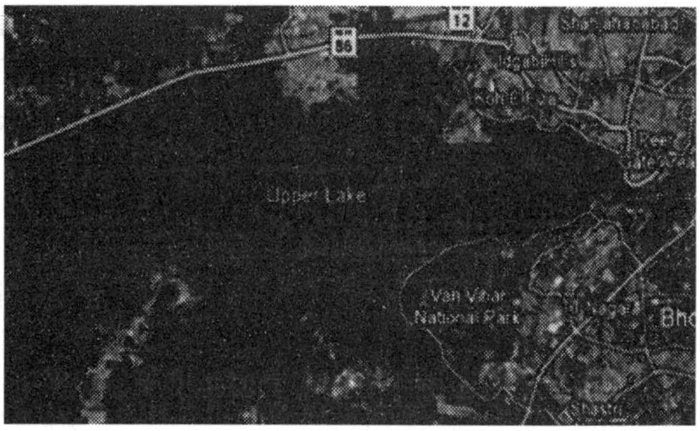

- Latitude: 23°13'-23°16'
- Altitude: 503.5 m
- Catchment Area: 361 Sq. km
- Storage Capacity: 101.5 M. cum
- Surface Area : 36 Sq. km
- Full Tank Level: 1666.80 R.L.(Feet)
- Dead Storage Level: 1656.58 R.L.(Feet)
- Maximum Length: 10.6 Km
- Maximum Width: 3.75 Km Source: google
- Maximum Depth: 13 m

Factors responsible for eutrophication in Upper Lake

- Siltation from agricultural fields.
- Joining of domestic raw sewage.
- Pollution from point & non point sources.
- Encroachment and flourishing growth of aquatic weeds.
- Bathing activity and human intervention in the fringe area of lake.
- Tourism activities.
- Agricultural activities in the catchment area.
- Inflow of residual chemical fertilizers and pesticides and insecticides.
- Trapa cultivation.
- Excessive growth of macrophytes.

Methodology

Scheme of Sample Collection

The monthly surveys were carried out and samples were collected from different sampling stations for the water quality monitoring and study of fish fauna from June, 2009 to March, 2010.

Sampling Stations

S1 - Yatch club, S2 - Behind Hamidia Hospital, S3 - Bairagarh.

Methods

All the methods used for the analysis of the physico-chemical parameters as per "Standard Methods for the Examination of the Water and Waste Water" 19th edition, APHA -APHA-WPCF (1995) and

"Laboratory Manual on Water Analysis, NEERI (1987).The different fish samples will be collected from wetland. These fish species will be identified on the basis of morphometric measurement, and fin formula and the classification given by Jayaram, Jhingaran, Francis Day and Shrivastava.

Observation & Discussion

During the period under study the range of variation in different the physico-chemical parameters are as:

Table 1. Physico-chemical parameters

S.N.	PARAMETER	RANGE
1	pH	7.2 - 8.4
2	Water Temperature	13 – 27°C
3	Transparency	30.0 - 80.0 cm
4	Dissolved Oxygen	4.0 - 12.8 mg/l
5	Free Carbon Dioxide	Nil - 6.0 mg/l
6	Alkalinity	100 - 180 mg/l
7	Total Hardness	48 - 160 mg/l
8	Chloride	14 - 60.4 mg/l
9	B.O.D.	2.2 - 14 mg/l
10	Nitrate	0.2 - 1.2mg/l

On the basis of the observation given in the table has been found that the Upper Lake is Semi eutrophic in nature

Table 2. Fish Species Found in Upper Lake

S	Scientific name	Status	Feeding habit	Economic importance	CAFF 2006
A	F-CYPRINIDAE				
1	*Labeo rohita* (Rohu)	++	HF	MFF	LR-lc
2	*Labeo calbasu* (Kalot)	++	HF	FF	LR-nt
3	*Labeo gonius* (Kursa)	++	HF	FF	LR-nt
4	*Labeo boga* (Rahia)	+	HF	FF	LR-nt
5	*Cirrhinus mrigala* (Naren)	++	OF	MFF	LR-nt
6	*Catla catla* (Catla)	++	NP,WF	MFF	LR-nt
7	*Carrasius auratus* (Rangin)	+++	OF	AF	DD
8	*Amblypharyngodon mola* (Chale)	+++	WF	AF	LR-lc

9	*Ctenopharyngodon idellus* (G. carp)	++	HF	MFF	DD
10	*Hypophthalmichthys molitrix* (S.carp)	+	HF	MFF	DD
11	*Cyprinus carpio* (communis) (C.carp)	++	OF,NP	MFF	DD
12	*Cyprinus carpio* (specularis) (Badhas)	++	OF	FF	DD
13	*Puntius sarana* (Kharpate)	+++	WF	AF	VU
14	*Puntius ticto* (Puthi)	+++	WF	AF	LR-nt
15	*Puntius chola* (Puthi)	+++	WF	AF	VU
16	*Puntius amphibius* (Bhakra)	+++	WF	AF	DD
17	*Garra lamta* (Pattharchatte)	++	OF	FF,LV	VU
B	**F-CULTRINAE**				
18	*Oxygaster* (Salmostoma) *bacaila* (Chale)	++	WF	BF	DD
C	**F-CHANNIDAE**				
19	*Channa marulius* (Sanwal)	++	LV,CF,PD	FF,MV	VU
20	C. *punctatus* (Gaidiya)	++	LV,PD	FF	LR-nt
21	C. *striatus* (Kabra)	++	PD	FF	LR-nt
D	**F-MASTACEMBALIDAE**				
22	*Mastacembelus armatus* (Bam)	++	PD	LV	VU
E	**F- BELONTIDAE**				
23	*Trichogaster faciatus* (Khardi)	++	HF	AF	VU
F	**F-BAGRIDAE**				
24	*Mystus seenghala* (Singhad)	++	PD	MFF	LR-nt
25	*M.aor* (Tengra)	++	PD	FF	LR-nt
26	*M. bleekeri* (Kittu, Kaitiya)	++	PD	FF,AF	VU
G	**F-AMBASSIDAE (CENTROPOMIDAE)**				
27	*Chanda nama* (Khasue)	+++	OF	FF	LR-lc
28	*C.ranga* (Kangsi)	++	WF	AF	VU
H	**F- NOTOPTERIDAE**				
29	*N. Notopterus* (Patola)	++	LV,PD	MV,LV,AF	EN
I	**F- SILURIDAE**				
30	*Ompok bimaculatus* (Gangerwar)	++	PD	GTF	EN
31	*Ompok Pabda* (Pabda)	++	PD	EFF	EN
32	*Wallago attu* (Padin, Lachi)	++	CF,PD	LV,SF	LR-nt

J	F- CLARIDAE				
33	*Clarius batrachus* (Mangoora)	+	OF	FF	VU
K	F-HETEROPNEUSTIDAE				
34	*Heteropneusts fossilis* (Singhan)	++	PD, CF	MV, FF	VU
L	F- BELONIDAE				
35	*Xenentodon cancila* (Suje, Suhia)	++	WF	FF	LR-nt
M	F- RASBORINAE				
36	*Parluciosoma, Rasbora daniconius* (Hinjare)	++	OF	FF	LR-lc
N	F- GOBIIDAE				
37	*Glossogobius giuris* (Gillue)	++	CF	FF	LR-nt

+=Rare, ++=Average, +++=Abundant, HF=Herbivorous Fish, OF=Omnivorous Fish, CF=Carnivorous Fish, WF=Weed Fish, PD=Predatory Fish, PF=Plankton Feeder, LV=Larivorous, MV=Medicinal Value, AF=Aquarium Fish, Bait Fish, GTF= Good Table Fish, EFF=Esteemed Food Fish, SF=Sport Fish, FF=Food Fish, MFF=Major Food Fish, NP=Non Predatory, EN=Endangered, VU=Vulnerable, LR-nt = Low Risk Near Threatened, LR-LC = Lower Risk Least Concern, DD=Data Deficient, CAFF = Conservation Assessment of Freshwater Fish Diversity.

Fish Diversity of Upper Lake

Labeo gonius **(Kursa)**

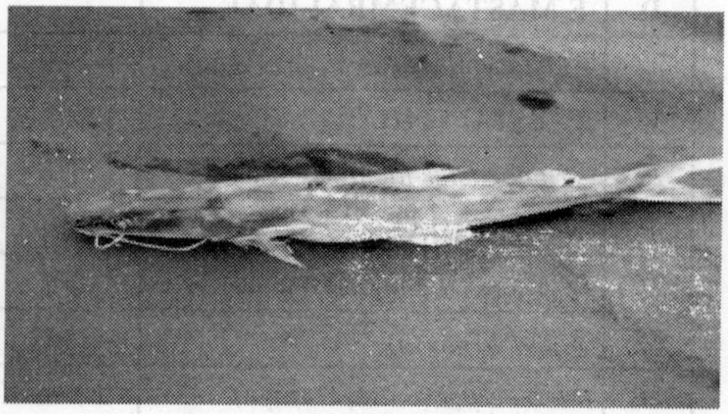

Mystus seenghala **(Singhad)**

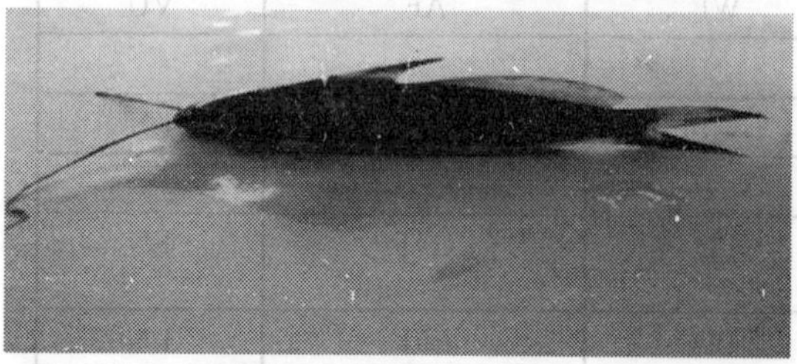

Mystus bleekeri **(Kittu)**

Mastacembelus armatus **(Bam)**

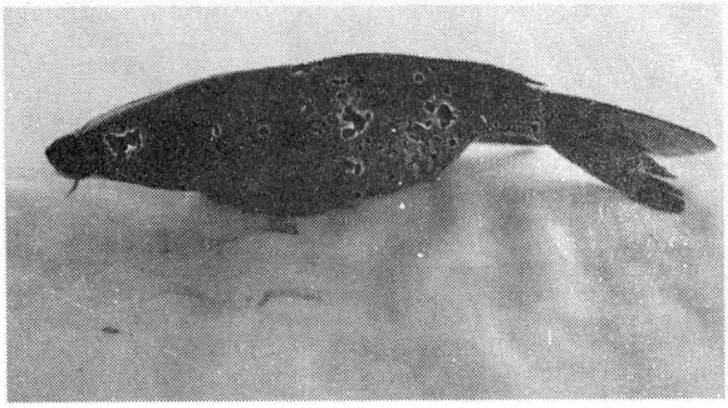

Common carp

Cirrhinus mrigala (Naren)

Catla catla (catla)

Glossogobius giuris (Gillue)

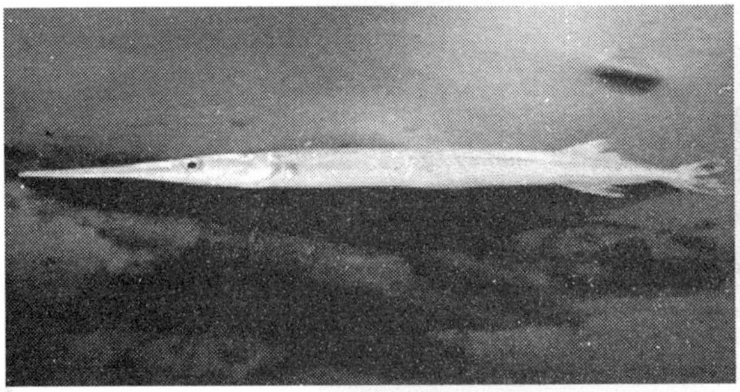

Xenentodon cancila (Suje)

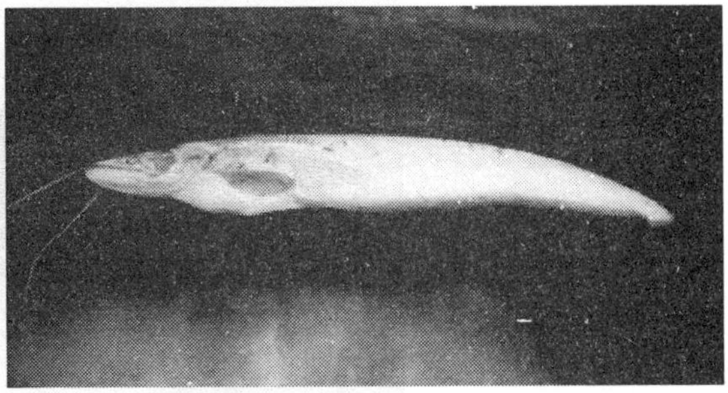

Wallago attu (Padin, Lachi)

Silver carp (Vicket)

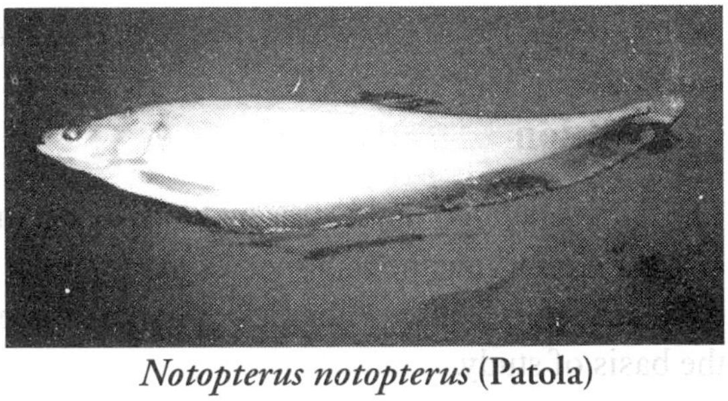

Notopterus notopterus (Patola)

Puntius ticto (Puthi)

Puntius sarana (Kharpate)

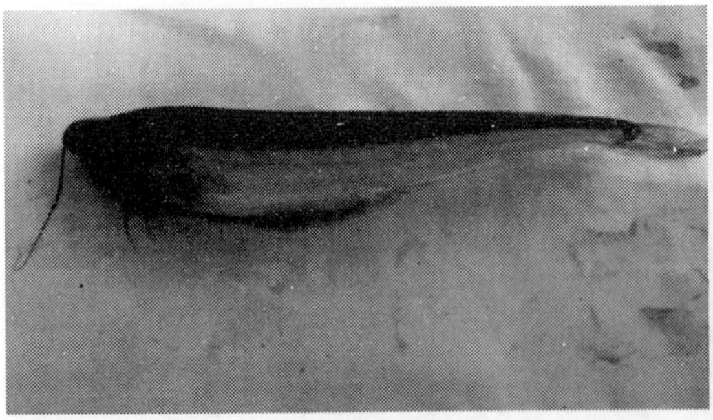

Ompok pabda (Pabda)

Ompok bimaculatus (Gangerwar)

Labeo calbasu (Kalot)

Conclusion

The data generated in the present study is useful for conservation & management of wetlands and also for conserving the fish species in their own natural habitat by sustainable management. Hence, balanced ecosystem will be developed. Authentic Fish Data Bank/Fish Atlas can be prepared on the basis of study.

Acknowledgement

Authors are thankful to M.P. Council of Science & Technology for sanctioning a research project on "Conservation, management & Documentation of Fish Diversity with reference to Water Quality in Tropical Lakes and Reservoirs (Bhopal)". This study was carried out under the project.

References

- APHA (1992): Standard methods for the examination of water and waste water, APHA, AWWA, WPCF, 19th edition, New York.
- Dubey, G.P. and Verma M.N. (1965): A preliminary study of the fish fauna of Madhya Pradesh, Vikram University Journal, Vol.8, No.4, Feb: 52-59.
- Hora, S.L. (1949): The fish fauna of the Rihand River and its geographical significance, Zool. Soc. India, 1-7.
- NEERI (1987): Manual on water and waste water Analysis, National Environmental Research Institute, Nagpur.
- Tamot Praveen, & G.P. Bhatnagar (1989): Studies on raw water quality of Upper lake and its change during various stages of treatment at five M.G.D. water treatment plant (P.H.E.D.), Bhopal (M.P.). In. Hydobiol., Vol. No. I, pp. 35-38.
- Tamot Praveen, S. Khare & P. Shrivastava (1991): Limnological studies of Upper lake, Bhopal during certain religious activities. J. Natcon., Vol.3 No.2, pp.14-146.
- Qureshi, T.A. (2003): Study of fin fish diversity of Madhya Pradesh, Fresh Water Fish Diversity of Central India, NBFGR, pp-35-34.
- Soni, D.D. (1962): Fish fauna of Lower Lake (Betwa River) Proc. First All India Congress (ZSI)-639-64.
- Swaroop, H. (1953): Fish fauna of Sagar Lake, Sagar University Journal, Vol. I, No.3.
- ZSI (2000): List of fresh water fishes of Madhya Pradesh (M.P. & Chhattisgarh)
- CAFF (2006): Conservation Assessment of Freshwater Fish Diversity, Freshwater Fish Diversity of Central India, NBFGR, pp 172-176.

42

COMPETENCY BASED TRAINING AND EDUCATION PROGRAMME FOR INTEGRATED MANAGEMENT WATER RESOURCES

*Dr. Asfa M. Yasin**

Abstract: Natural water bodies are attractive destinations for nature lovers. Fishing is an additional recreation for tourists coming from distant places. Hence, Fishing Tourism, tourism in general and establishing linkages with local and state tourism organization can provide an economically viable venture for resident community. In order to realize the potential of fishing tourism, awareness need to be inculcated amongst community and fisheries agencies to draw economical benefits. From recreational point of view, catch-and-release fisheries has much more potential for fishing tourism, as there is nothing being consumed. In this context, even more importantly, it also seeks to foster ethical values for sustaining fish population amongst anglers or fishing lovers.

Keeping above in view, an attempt has been made in this chapter to carve out a niche for a venture in fishing tourism. Strategies for development of fishing tourism have been discussed. Need to design and developing a competency based modular course has also been emphasized in formal and non-formal mode.

Livelihood and Fishing Tourism

Introduction

Nature is God given gift to mankind. Reaching far off destinations for pleasure visits to enjoy and admire nature may be referred to as nature tourism. Further, ecotourism is a component of a broader category of tourism includes water bodies which are most attractive destinations for enjoying the existence of incredible beauty of nature. "A tourist is one who travels away from home for a distance of at least 50 miles (one way) for any purpose except to commute to work", according to

* Associate Professor and Head, Agriculture Division, PSSCIVE, NCERT, 131, Zone-II, M.P. Nagar, Bhopal – 462 011, M.P.

National Tourism Resources Review Commission, 1973, USA. Water bodies as home of aquatic fauna mainly constituted by fishes, offers much more excitement, when tourism is enjoyed in action such as boating, surfing and fishing.

Fishing tourism or ecotourism refers to fish angling, to catch fish for entertainment, consumption as food or selling in the market as a small scale enterprise. Game fishing or sport fisheries is based on catch-and-release principle where fish is not consumed and released back. Hence, Holland *et al.* (1998, 2000) viewed the fishing on catch-and-release basis as ecotourism, which yield economic benefits for local people as their livelihood by providing facilities for game fishing on charges basis. It is in this backdrop, it is important to develop fishing tourism for generating employment and importation of new revenues. Development of fishing tourism is therefore important from two angles. First, preparing trained people to offer fishing tourism facilities to tourists. The training should aim to protect and sustain fish populations, maintenance of water quality on sustainable basis, conservation goals, besides awareness regarding respect and care for nature and its habitants. Second aspect is how to make use of this natural resource as an enterprise and maximizing benefits. It is therefore important to give educational flavor to fishing tourism for its sustainable development.

In furtherance to this, various dimensions involved in fishing tourism must be listed as competencies to give them space in the educational programme as a course. Besides this, strategies may formulated for the development of 'Fishing Tourism'. Launching an educational programme as a short-term course, we also have to consider, what are the target groups as potential course seekers and their requirements. All these points are discussed in this chapter.

Dimensions of Fishing Tourism and its Management

Fishing as a recreational activity is very popular because of low cost input. Although, considering this as a promising venture, possible items which are required essentially may include transport facilities, accommodation or stay points (hotels or hutments near natural resources), food courts, parks, camp grounds and camping accessories, their supply and management, one-stop stores having fishing equipment. Considering as an enterprise brochures, advertisement, souvenirs', information desk, boat ramps, fishing piers are also to be arranged and managed.

Management of fishing tourism should also include norms, ethics, licensing, associations of anglers, etc. which require legal and administrative support.

Keeping above in place in management of 'Fishing Tourism' venture, one has to under go training before getting in to this venture. It is therefore important to design and offer a course having all these elements, through formal or non-formal system. An overview of requirement in terms of capital, infrastructure, human resources, skills, and their management is necessary for the development of Fishing Tourism. Planning, execution and management are three important components for the development of Fishing Tourism.

Strategies for the development of Fishing Tourism

Development of knowledge system

In the present times there is inflow of modern technologies in all the spheres of life. The same is true for any occupation. Thus, modernization of traditional occupations is an urgent need to enhance and enrich the quality of output in form of services or production of goods by using time and cost effective devices.

Any economically viable project, should also focus on dimensions related to knowledge of legal, administrative and skill development aspects in synchronized and integrated manner. This comprehensive system is referred to as knowledge system. A system which prepare knowledge based society and manpower. The knowledge system equip people with systematic and scientific management to handle the vocation.

Knowledge system (KS) provide know-how of modern gadgets and techniques by which a person will be able to earn more money per unit time and efforts beside ensuring quality and marketing of products/ services. KS therefore is the way of developing skills and competencies in a much shorter time by a large group of people over a period of time.

In context of fishing tourism, which is now-a-days a promising venture for rural areas. It is promising because this is low cost venture but there has to be alertness about rules, ethics and good management practices. Thus, development of competency based module and running the course in formal and non-formal mode will enable development of human resource for this venture.

Development of competency based module on Fishing Tourism

In reference to good management practices to make this venture profitable and be able to provide desired enjoyment to tourists, we need to develop competent persons to run this enterprise. For the purpose, design of a module may include list of competencies required for this venture, then each competency could be detailed out in terms of what knowledge, skills and personality traits/ attitudes/ ethics are required. With a view of testing of competencies, performance criteria may be developed, to know whether or not competency and skills are developed. This competent person, we refer to as a knowledge worker, who in turn can train other persons of the area. Thus, over the time, knowledge system will be developed and these knowledge workers will be able to transform the group of traditional fishermen/ rural people into a knowledge society. A suggestive list of competencies is given below, which may be required to run this venture:

- Competent in legislation, fishing regulations and conservation measures.
- Establishing association for anglers.
- Arranging license for fishing

- Managing tourists stay (lodging, boarding and transportation arrangement by liaisoning with local agencies/ hotels).
- Arranging required resources and accessories for fishing.
- Arranging camps near fishing grounds.
- Handling accounts.
- Registering and maintaining records of tourists.
- Keeping records of catch-and-release of fishes with respect to variety, size and numbers.
- Arranging boat ramps, boats, fish piers and fishing equipment.

Target Groups

Fishing tourism is rural areas based venture for persons who are involved in fishing, agriculture and local unemployed or under employed youth. This enterprise may also become a family venture, because there are many items which even non-technical persons can do after undertaking short term training and some basic know how. Local fishing cooperative may also involve many small groups of students and youth on part-time basis. The decentralized approach can be a good strategy wherein duties may be distributed to every person involved.

Conclusion

Tourism development near aquatic natural sites having rich fisheries can be a promising enterprise. This calls for two pronged strategy. First one is the development of 'Fishing Tourism' as an organized venture. Second thing is that by way of offering vocational educational and training programme, competent workforce and knowledge system should be also be developed. Thus, a competency based modular course for Fishing Tourism should be designed and offered in schools/ polytechnics/ KVKs. Competencies may be identified in consultation with experts and local fisheries agencies/ entrepreneurs involved in the concerned areas.

References

1. Holland, S.M., R.B. Ditton, and A.R. Graefe, 1998. An ecotourism perspective on billfish fisheries. Journal of Sustainable Tourism 6 (2): 1-20.
2. ———— 2000. A response to 'ecotourism on trial – the case of bill fish angling as ecotourism. Journal of Sustainable Tourism 8 (4): 346-351.
3. National tourism Resources Review Commission, 1973. Destination USA, Volume 2, domestic tourism. US Government Printing Office, Washington D.C.

